3-D Seismic Survey Design

Gijs J. O. Vermeer

Craig J. Beasley, Volume Editor
Eugene F. Scherrer, Series Editor

Society of Exploration Geophysicists
Tulsa, Oklahoma

Vermeer, Gijs J. O.
    3-D seismic survey design / Gijs J. O. Vermeer
            p.   cm. -- (Geophysical references; v. 12)
    Includes bibliographical references and index.
    ISBN 1-56080-113-1
    1. Seismic reflection method.   I. Title.   II Series.
TN269.84.C69   2002   98-30018
622' 1592--dc21   CIP

ISBN 0-931830-47-8 (Series)
ISBN 1-56080-113-1 (Volume)

© 2002 by the Society of Exploration Geophysicists
All rights reserved. This book or parts hereof may not be reproduced in any form without permission in writing from the publisher.

Printed in the United States of America.

To Tini

Contents

Foreword *Ian Jack* xi

Acknowledgments xiii

Introduction 1

Chapter 1 2-D symmetric sampling 5

    1.1   Introduction 5
    1.2   The shot/receiver and midpoint/offset coordinate systems in 2-D 5
    1.3   Symmetric sampling 8
    1.4   Symmetric sampling versus asymmetric sampling 11
    1.5   The stack-array approach versus symmetric sampling 13
    1.6   The total stack response 13
    1.7   Concluding remarks 14
           References 16

Chapter 2 3-D symmetric sampling 17

    2.1   Introduction 17
    2.2   Classes of 3-D geometries 18
           2.2.1   Examples of various geometries 18
    2.3   The continuous wavefield 20
           2.3.1   The shot/receiver and midpoint/offset coordinate systems 21
           2.3.2   3-D subsets of 5-D wavefield 21
           2.3.3   The cross-spread 22
           2.3.4   Subsets of zigzag geometry 23
    2.4   3-D symmetric sampling 25
           2.4.1   Areal geometry 27
           2.4.2   Line geometries 28
                 2.4.2.1   Parallel geometry 28
                 2.4.2.2   Orthogonal geometry 29
                 2.4.2.3   Zigzag geometry 33
    2.5   Pseudominimal data sets 33
           2.5.1   Introduction 33
           2.5.2   Building fold with basic subsets 36
           2.5.3   Fold, illumination, and imaging 37
           2.5.4   Construction of pMDSs 38
           2.5.5   A measure of spatial discontinuity 38
           2.5.6   A plethora of OVT gathers 40
    2.6   Application to prestack processing 40
           2.6.1   Introduction 40

		2.6.2	Noise removal 40

        2.6.2   Noise removal 40
        2.6.3   Interpolation and regularization 41
        2.6.4   Muting 41
        2.6.5   First-break picking 42
        2.6.6   Nearest-neighbor correlations 42
        2.6.7   Residual statics 42
        2.6.8   Velocity analysis and DMO 43
        2.6.9   AVO 44
        2.6.10 Amplitude variation with azimuth 45
2.7   Conclusions 45
       References 46

Chapter 3 Noise suppression 49

3.1   Introduction 49
3.2   Properties of low-velocity noise 49
        3.2.1   "Direct" waves 49
        3.2.2   Scattered waves 49
        3.2.3   Discussion 52
3.3   Shot and receiver arrays in 3-D data acquisition 52
        3.3.1   Introduction 52
        3.3.2   "Direct" wave noise suppression 53
        3.3.3   Scattered-wave noise suppression 54
        3.3.4   Analysis of various array combinations 55
        3.3.5   Discussion 57
3.4   Stack responses 61
        3.4.1   Introduction 61
        3.4.2   The 2-D stack response 61
        3.4.3   Multiple suppression by stacking 62
                3.4.3.1   Multiples with small differential moveout 62
                3.4.3.2   Multiples with large differential moveout 62
        3.4.4   3-D stack responses 63
        3.4.5   Discussion 67
       References 67

Chapter 4 Guidelines for design of "land-type" 3-D geometry 69

4.1   Introduction 69
4.2   Preparations 69
        4.2.1   Objective of survey 69
        4.2.2   Know your problem 69
4.3   The choice of geometry 70
        4.3.1   Parallel geometry versus orthogonal geometry 70
        4.3.2   Zigzag geometry versus orthogonal geometry 71

	4.3.3	Slanted geometry versus orthogonal geometry 72
	4.3.4	Comparison of sampled minimal data sets of crossed-array geometries 72
	4.3.5	Areal geometry 74
	4.3.6	Target-oriented geometries 74
4.4	Design criteria and parameter selection 75	
	4.4.1	Spatial continuity 75
	4.4.2	Resolution 75
		4.4.2.1 Resolution requirements and maximum frequency 75
		4.4.2.2 Resolution requirements and spatial sampling 78
		4.4.2.3 Statics and spatial sampling 79
		4.4.2.4 Other processing requirements and sampling 79
		4.4.2.5 Discussion on spatial sampling 79
	4.4.3	Shallowest horizon to be mapped 80
	4.4.4	Deepest horizon to be mapped 81
	4.4.5	Noise suppression 82
		4.4.5.1 Fold as a dependent or independent parameter 82
		4.4.5.2 How to determine desired or required fold 82
		4.4.5.3 Fold as an instrument to suppress multiples 83
		4.4.5.4 The importance of regular fold 83
		4.4.5.5 Shot and receiver arrays 84
	4.4.6	Other survey parameters 85
	4.4.7	The selection of acquisition parameters for areal geometry 86
4.5	The survey grid and the survey area 86	
4.6	Practical considerations and deviations from symmetric sampling 87	
	4.6.1	Logistics and terminology 87
	4.6.2	Harmonizing all requirements 88
	4.6.3	Deviations from symmetric sampling 88
	4.6.4	Different ways of implementing nominal geometry 89
	4.6.5	Multiline roll 90
	4.6.6	Attribute analysis of one-line roll versus multiline roll geometries and orthogonal versus slanted geometries 90
	4.6.7	Conflicting requirements between structural interpretation and AVO 92
	4.6.8	Deviations from nominal due to topography and obstacles 96
4.7	Testing 98	
4.8	Discussion 98	
	4.8.1	Attribute analysis 99
	4.8.2	Model-based survey design 99
4.9	A summary of what to do and not to do in 3-D survey design 100	
	References 100	

Chapter 5 Streamers versus stationary receivers 103

5.1 Introduction 103

	5.2	Geometry imprint 104
	5.3	Streamer acquisition 105
		5.3.1 Shooting direction 105
		5.3.1.1 Dip/strike decision 106
		5.3.2 Multisource, multistreamer acquisition 107
		5.3.2.1 Multisource, multistreamer configurations 108
		5.3.2.2 Multisource effect on fold 108
		5.3.2.3 Crossline-offset variation 108
		5.3.2.4 Irregular illumination 109
		5.3.2.5 Effects of irregular illumination 109
		5.3.2.6 Remedies 113
		5.3.2.7 Operational aspects 113
	5.4	Stationary-receiver techniques 115
		5.4.1 Geometries for stationary-receiver techniques 115
		5.4.2 Vertical hydrophone cable (VHC) 116
		5.4.3 Dual-sensor OBC 117
		5.4.3.1 Ghosting 117
		5.4.3.2 Geometry 117
		5.4.3.3 Logistics 118
		5.4.4 Four-component marine data acquisition 118
		5.4.4.1 Coupling issues 118
		5.4.4.2 SUMIC 119
		5.4.4.3 Other 4-C bottom cable techniques 119
		5.4.4.4 4-C acquisition with buried cables 120
		5.4.4.5 Ocean-bottom seismometers 120
	5.5	Overview and conclusions 121
		References 121

Chapter 6 Converted waves: Properties and 3-D survey design 125

	6.1	Introduction 125
	6.2	Properties of the *PS*-wavefield 125
		6.2.1 Traveltime surfaces and apparent velocity 125
		6.2.2 Illumination 127
		6.2.3 Resolution 127
		6.2.4 Imaging 129
	6.3	3-D survey design for *PS*-waves 130
		6.3.1 Choice of geometry 130
		6.3.1.1 Orthogonal geometry 131
		6.3.1.2 Parallel geometry 133
		6.3.1.3 Areal geometry 134
		6.3.1.4 Parallel versus orthogonal geometry and areal geometry 135
		6.3.2 Sampling 136

		6.3.3 Other considerations 136
6.4	Discussion 136	
6.5	Conclusions and recommendations 137	
	References 138	

Chapter 7 Examples of 3-D symmetric sampling 141

- 7.1 Introduction 141
- 7.2 3-D microspread 141
    - 7.2.1 Introduction 141
    - 7.2.2 Acquisition parameters of 3-D microspread 141
    - 7.2.3 Cross-sections and time slices 142
    - 7.2.4 (f,k)-filtering results 144
    - 7.2.5 Discussion 144
- 7.3 Nigeria 3-D test geometry results 146
    - 7.3.1 Introduction 146
    - 7.3.2 Acquisition geometry 148
    - 7.3.3 Some processing results 149
    - 7.3.4 Interpretation results 150
    - 7.3.5 Discussion 150
- 7.4 Prestack migration of low-fold data 154
    - 7.4.1 Introduction 154
    - 7.4.2 Migration of a single cross-spread 154
    - 7.4.3 Low-fold prestack migration 155
    - 7.4.4 Discussion 156
- References 158

Chapter 8 Factors affecting spatial resolution 159

- 8.1 Introduction 159
- 8.2 Spatial resolution formulas 160
    - 8.2.1 Spatial resolution—The link with migration/inversion 160
    - 8.2.2 Spatial resolution formulas for constant velocity 161
- 8.3 Spatial resolution measurements 163
    - 8.3.1 Procedure for resolution analysis 163
    - 8.3.2 2-D resolution in the zero-offset model 164
    - 8.3.3 2-D resolution in the offset model 165
    - 8.3.4 Asymmetric aperture 165
    - 8.3.5 3-D spatial resolution 166
    - 8.3.6 Sampling and spatial resolution 168
    - 8.3.7 Sampling and migration noise 168
    - 8.3.8 Bin fractionation 170
    - 8.3.9 Fold and spatial resolution 171
- 8.4 Discussion 171

8.5 Conclusions 172
References 173

Chapter 9 DMO 175

9.1 Introduction 175
9.2 DMO in arbitrary 3-D acquisition geometries 175
    9.2.1 Summary 175
    9.2.2 Introduction 176
    9.2.3 The time of a DMO-corrected event 176
    9.2.4 Contributing traces in cross-spread 178
    9.2.5 The DMO-corrected time in the cross-spread 178
    9.2.6 Extension to other geometries 179
    9.2.7 Sampling problems 180
    9.2.8 Conclusions 180
9.3 DMO in cross-spread: The failure of earlier software to correctly handle amplitudes 180
    9.3.1 Introduction 180
    9.3.2 Sampling problem 181
    9.3.3 Geometry effect 181
    9.3.4 Example 181
    9.3.5 The ideal 3-D DMO program 181
    9.3.6 Conclusion 181
9.4 Epilogue 183
    9.4.1 New DMO programs 183
    9.4.2 DMO in pseudominimal data set 183
References 184

Chapter 10 Prestack migration 185

10.1 Introduction 185
10.2 Fresnel zone and zone of influence 186
    10.2.1 Modeling 186
    10.2.2 Migration 186
10.3 Description of model experiments 189
10.4 Prestack migration with minimal data sets 189
10.5 Prestack migration with pseudominimal data sets 191
    10.5.1 Parallel geometry 191
    10.5.2 Orthogonal geometry 191
    10.5.3 Irregular geometries 197
10.6 Velocity-model updating 197
10.7 True-amplitude, prestack migration of regular and irregular data 197
10.8 Discussion 198
References 198

Index 201

Foreword

Any developing technology is heralded by a proliferation of papers of variable quality, ultimately superseded by the definitive work. It is just such a definitive work, in the field of 3-D seismic survey design, that I am proud to introduce here.

In the early days of the seismic method and prior to the implementation of 3-D techniques, the literature is not short of papers and discussions on seismic survey design. Almost every individual operations expert seemed to have his or her own favorite source array or detector array or both, and wished to go into print about their various advantages. Survey design was clearly a popular subject, fueled by the fact that our wavefield sampling was necessarily woefully inadequate–not only was it "2-D," but it was further compromised by constraints such as instrumentation limitations.

Eventually the number of these papers dwindled, because, for most of us, adequate wisdom on 2-D sampling had been expressed by authors such as Nigel Anstey, Leo Ongkiehong, and Henry Askin, in papers with titles such as "Towards the universal seismic acquisition technique," and "Whatever happened to groundroll?"

Then, by the mid-1980s, the use of the 3-D method was growing fast. Again, we experienced a proliferation of experts' favorite designs and opinions, both for marine and for land exploration, especially the latter. For some, the only viable configuration was the "brick," and for others it was the narrow swath. Designs were seen in which the statics solutions for adjacent subsurface lines floated independently of one another. Some designs made unfair demands of field crews, requiring huge overlapping receiver arrays that crossed and recrossed each other in the search for perfection. In some cases, the acquisition staff were so out of touch with the data processors that the latter eventually had to plead with the former not to use their current favorite patented designs! What was sorely needed was the 3-D equivalent of the older 2-D Anstey, Ongkiehong, and Askin papers–a clear and correct explanation of the issues and their solutions.

This is what you now have, as into the breach stepped Gijs Vermeer, who is particularly well qualified to author a book on this subject. His academic background is in physics and mathematics and he has earned his living not just as a theorist, but also with real data as a processor and an interpreter.

Most of his earlier publications have been internal Shell documents, but we are indeed fortunate that from the early 90s we have seen a growing quantity of Gijs's work in the literature. His first major publication was the book entitled "Seismic Wavefield Sampling," published in 1990, which quickly became a necessity on our shelves. Several of his papers since then—with titles such as "3-D symmetric sampling"—showed the directions of his research. Luckily for us, he has recently focused on this subject, and he embarked on a PhD at the University of Delft. This book is the result, and I commend it to you. Within its covers you will find a full insight into the fundamentals of 3-D survey design as well as the common practical issues such as the "footprints" that often plague 3-D data volumes and are especially important as the emphasis on reservoir geophysics increases.

Comprehensively, the book covers not just land and marine surveys, but also ocean-bottom, vertical cable, and converted wave data. The survey design issues and criteria are developed and taken all the way from acquisition through to prestack depth processing. It will become both a textbook and a general reference and will benefit survey designers, data processors, interpreters, and earth scientists both in industry and academia.

Ian Jack, BP.

Acknowledgments

To a great extent this book is the product of many years of research in a stimulating Shell environment. In 1991 I joined the project "Fundamentals of 3D seismic data acquisition" and together with Justus Rozemond I developed the theory of 3-D symmetric sampling, based on earlier work on 2-D symmetric sampling. The main insights were developed during the first year, but thereafter we continued to expand and refine the ideas and we had the opportunity to test the ideas in practice.

It is not really possible to mention all colleagues who have contributed in one way or another to the work described in this book, but I do appreciate their help. I have to make an exception for a few of them. Kees Hornman was my boss during a large part of this period until 1996. He has been a major discussion partner ever since. Jerry Davis, as an adviser to Shell Operating Companies, showed that theory could be put into practice. Whenever there was a technical question, Peter van der Sman found the time to answer it. Since early 1999, Rick Stocker keeps showing me that it is not always easy to put theory into practice.

I am very grateful to Bill Kiel, who made it possible for me to take early retirement from Shell in 1997 and to start 3DSymSam—Geophysical Advice. He allowed me to maintain a good relationship with Shell, while I could devote more of my time to geophysics and to this book than would otherwise have been possible. The good relationship with Shell manifested itself in Shell's permission to show many of their data examples in this publication.

The horizon slices in this publication have been made with a prototype version of Omni Workshop, which was kindly made available by Seismic Image Software Ltd. TNO's permission to include the prototype version of the Acquisition Design Wizard on the CD-ROM of this book is gratefully acknowledged.

I would also like to express my gratitude to Jacob Fokkema and Kees Wapenaar who graciously agreed to be my supervisors for the PhD version of this book. After all the work done for the thesis, I was happy that Craig Beasley, the Editor, came up with some useful suggestions and mild criticism only. It was a great pleasure to work with Jerry Henry, Judy Hastings, and Ted Bakamjian to convert the PhD version into a new tome in SEG's Geophysical Reference Series. My gratitude also extends to Ian Jack who was so kind to write a positively phrased foreword.

Finally, I would like to thank my wife Tini for her kind editing of English grammar in the original thesis, but above all for being there.

Introduction

Three-dimensional (3-D) seismic surveys have become a major tool in the exploration and exploitation of hydrocarbons. The first few 3-D seismic surveys were acquired in the late 1970s, but it took until the early 1990s before they gained general acceptance throughout the industry. Until then, the subsurface was being mapped using two-dimensional (2-D) seismic surveys.

Theories on the best way of sampling 2-D seismic lines were not published until the late 1980s, notably by Anstey, Ongkiehong and Askin, and Vermeer. These theories were all based on the insight that offset forms a third dimension, for which sampling rules must be given.

The design of the first 3-D surveys was severely limited by what technology could offer. Gradually, the number of channels that could be used increased, leading to discussions on what constitutes a good 3-D acquisition geometry. The general philosophy was to expand lessons learned from 2-D acquisition to 3-D. This approach led to much emphasis on the properties of the CMP gather (or bin), because good sampling of offsets in a CMP gather was the main criterion in 2-D design. Three-D design programs were developed that concentrated mainly on analysis of bin attributes and, in particular, on offset sampling (regularity, effective fold, azimuth distribution, etc.).

This conventional approach to 3-D survey design is limited by an incomplete understanding of the differing properties of the many geometries that can be used in 3-D seismic surveys. In particular, the sampling requirements for optimal prestack imaging were not properly taken into account. This book addresses these problems and provides a new methodology for the design of 3-D seismic surveys.

The approach used in this book is the same as employed in my Seismic Wavefield Sampling, a book on 2-D seismic survey design published in 1990: Before the sampling problem can be addressed, it is essential to develop a good understanding of the continuous wavefield to be sampled. In 2-D acquisition, only a 3-D wavefield has to be studied, consisting of temporal coordinate t, and two spatial coordinates: shot coordinate x_s, and receiver coordinate x_r. In 3-D acquisition, the prestack wavefield is 5-D with two extra spatial coordinates, shot coordinate y_s, and receiver coordinate y_r.

In practice, not all four spatial coordinates of the prestack wavefield can be properly sampled (proper sampling is defined as a sampling technique which allows the faithful reconstruction of the underlying continuous wavefield). Instead, it is possible to define three-dimensional subsets of the 5-D prestack wavefield which can be properly sampled. In fact, the 2-D seismic line is but one example of such 3-D subsets.

The 2-D seismic line is a multifold data set with midpoints on a single line only. However, in 3-D acquisition there are many possible 3-D subsets which are single-fold and whose midpoints extend across a certain area. These subsets are called minimal data sets, a term coined by Trilochan Padhi in 1989 in an internal Shell report. A minimal data set represents a volume of data (sometimes called a 3-D cube) that has illuminated part of the subsurface. If there was no noise, a single minimal data set would be sufficient to create an image of the illuminated subsurface volume.

Most acquisition geometries used in practice generate data that can be considered as a collection of sampled minimal data sets. Therefore, the properties of the minimal data sets need to be studied for a better understanding of the acquisition geometries as a whole. This allows an optimal choice of the acquisition geometry (if there is a choice, often the geometry type is dictated by economic or environmental constraints) and of the parameters of the geometry.

The continuous wavefield to be sampled can be reduced to the wavefield of the characteristic minimal data set of the chosen geometry. Proper sampling of that wavefield means that at least two of the four spatial coordinates of the 5-D prestack wavefield will be properly sampled. Next, maximize the useful extent of each minimal data set. Together, these two recommendations ensure a minimum of spatial discontinuities in the total data set. Spatial continuity is maximized and the migrated minimal data sets contain a minimum of artifacts. Other parameters of each acquisition geometry need to be chosen so that requirements of resolution, noise suppression and illumination are satisfied as well.

Based on these basic principles, this book addresses a wide variety of issues. The following provides a summary of each chapter.

Chapter 1. 2-D symmetric sampling

This chapter provides a short summary of 2-D symmetric sampling, which is a recipe for optimal sampling of the 2-D seismic line. Two-D symmetric sampling is based on a corollary of the reciprocity theorem, which affirms that the properties of the common-receiver gather are the same as the properties of the common-shot gather. As a consequence, sampling requirements of shots and receivers are identical.

Chapter 2. 3-D symmetric sampling

Three-D seismic surveys can be acquired using a number of different acquisition geometries. The most important geometries are areal geometry, parallel geometry, and orthogonal geometry. Each geometry has its characteristic 3-D basic subset. If the basic subset is single fold, it is also a minimal data set. In areal geometry, either shots or receivers are acquired in a dense areal grid. If shots are dense, receivers are sparse or vice versa. In the first case, 3-D common-receiver gathers are acquired. These gathers form the basic subset or minimal data set of this particular areal geometry.

Parallel geometry and orthogonal geometry are examples of line geometries in which sources and receivers are arranged along straight acquisition lines, which are more or less widely separated. In parallel geometry the (parallel) shot lines are parallel to the (parallel) receiver lines, whereas in orthogonal geometry shot and receiver lines are orthogonal. The basic subset of parallel geometry is the midpoint line, which runs halfway between the shot line and each active receiver line. The basic subset of orthogonal geometry is the cross-spread, which encompasses all receivers in a single receiver line which are listening to a range of shots in a single shot line. The cross-spread is a minimal data set with limited extent. The difference in properties of the various acquisition geometries is illustrated by the difference in diffraction traveltime surface of the same diffractor for the basic subsets of those geometries.

Two-D symmetric sampling can be readily expanded to 3-D symmetric sampling after recognition of the existence of the basic subsets of each geometry.

For imaging, it would be ideal to have single-fold data sets that extend across the whole survey area, but which possess a minimum of spatial discontinuities so that they would produce a minimum amount of migration artifacts. These data sets are called pseudominimal data sets and can be constructed from so-called offset-vector tiles. In orthogonal geometry, the size of the offset-vector tile is determined by the area between two adjacent shot lines and two adjacent receiver lines. The cross-spread can be split into M disjoint offset-vector tiles (M is fold-of-coverage), in which the x- and the y-components of the offset vector vary over a limited range.

Chapter 3. Noise suppression

Sampling in 3-D acquisition is usually not dense enough to record low-velocity noise without aliasing. To reduce aliasing effects, shot and receiver arrays may be used. The arrays may be linear or areal. For a proper choice of arrays, the properties of the noise need be known. An analysis of the energy distribution of low-velocity scatterers shows that in the cross-spread most energy is concentrated on the flanks of the traveltime surface and there is less energy around the apex. Linear arrays are sufficient to suppress the energy in the flanks. If there is much undesirable energy coming from all directions, circular arrays can be constructed with a circular response.

Often, one of the aims in 3-D survey design is to achieve a regular offset distribution. This is based on Nigel Anstey's stack-array approach for 2-D data, which states that ground roll is best suppressed in the stack by a regular sampling of offsets in each CMP gather. However, this requires high-fold data; if the data are low fold, a random offset distribution tends to have a better stack response. This applies in particular to 3-D data where fold tends to be low, especially if measured in separate azimuth ranges. A wide orthogonal geometry (maximum crossline offset close to maximum inline offset) tends to produce irregular offset sampling in each CMP gather, hence tends to have a better stack response than a narrow geometry.

Chapter 4. Guidelines for design of "land-type" 3-D geometry

The theoretical considerations and observations of the first part of the book are translated into practical guidelines for choice of geometry and selection of parameters for orthogonal geometry. Parallel geometry looks most like 2-D acquisition, the stack response is similar, and processing can use many of the techniques already developed for 2-D processing. It is not suitable for analysis of azimuth-dependent effects. Parallel geometry can be acquired efficiently in the marine environment using streamers, but on land parallel geometry is less efficient than orthogonal geometry. Orthogonal geometry is suitable for analysis of azimuth-dependent effects. It is also used for seabed acquisition using bottom cables. Processing of orthogonal geometry is much more complex than processing of parallel geometry. Zigzag geometry is a geometry devised for efficient acquisition in desert environments. Slanted geometry is similar to orthogonal, but the shot lines cross the receiver lines at an oblique angle. The basic subsets of zigzag and slanted geometry are less suitable for dual-domain processing than the basic subset of orthogonal geometry. Areal geometry is also suitable for analysis of azimuth-dependent effects. It is applied mainly in deep waters when very expensive receiver units are used, such as vertical hydrophone cables and 4-C receiver units (3-component geophone plus hydrophone).

The main parameters of orthogonal geometry are station intervals, line intervals, maximum inline and crossline offsets, and fold. These have to be selected such that requirements of spatial continuity, resolution, mapping of shallowest and deepest horizons of interest, and noise suppression are satisfied. The survey area is always larger than the area to be mapped due to the fold-taper zone and the radius of the migration operator.

In practice, a one-line roll of a nearly square template tends to be quite inefficient. Without compromising the desired acquisition geometry, it is often more efficient to use a full-swath roll. A multiline roll is also more efficient, but it will create strong spatial discontinuities.

Obstacles often prevent acquisition of straight acquisition lines. Spatial continuity then requires the acquisition lines to be smooth. Common practice of moving shots an integer multiple of the receiver interval to the right or to the left produces discontinuities in the receiver gathers leading to migration artifacts. A general requirement in acquisition of parallel and orthogonal geometry is that the receiver gathers should look as good as the shot gathers.

Chapter 5. Streamers versus stationary receivers

In marine seismic data acquisition, the designer has to choose between streamers and stationary-receiver systems. With streamers, multisource multistreamer configurations are used in a parallel geometry. With stationary-receiver systems there is flexibility in the choice of geometry. Streamers are most efficient in deep water without any obstacles. Adjacent boat passes should be acquired antiparallel to minimize illumination irregularities. However, illumination irregularities caused by differential feathering are inevitable. Acquisition with stationary-receiver systems tends to be more expensive than with streamers. Systems in use are vertical hydrophone cable, ocean-bottom cable and node. Nodes are single 4-C units, whereas ocean-bottom cables can be used with a dual-sensor technique as well as a 4-C technique. Repeatability of stationary-receiver systems is better than repeatability of streamer acquisition.

Chapter 6. Converted waves: Properties and 3-D survey design

Survey design for PS-waves is different from P-wave acquisition, owing to the asymmetry of the PS raypath. Differences in PS-illumination by minimal data sets of different geometries are much larger than P-illumination differences. For instance, a cross-spread with a square midpoint area produces an illumination area with rectangular shape even for a horizontal reflector. The raypath asymmetry leads to asymmetric sampling requirements for shots and receivers. Shot sampling interval is determined by P-wave velocity; receiver sampling interval by S-wave velocity. Parallel geometry tends to suffer least from asymmetry effects whereas orthogonal geometry tends to suffer most. For analysis of azimuth-dependent effects, areal geometry might be the best choice.

Chapter 7. Examples of 3-D symmetric sampling

Noise spreads or microspreads are acquired with very dense spatial sampling for an analysis of low-velocity events. A cross-spread with very dense spatial sampling was acquired in The Netherlands. Time slices and cross-sections illustrate the 3-D behavior of the ground-roll cone and of the scatterers inside the cone.

The theory of 3-D symmetric sampling was put to the test in Nigeria, where a cross-spread test geometry was compared with the standard brick-wall geometry. The test geometry produced better results (higher resolution and better continuity) at target level than the standard geometry. The improvement can be attributed to larger width (maximum crossline offset) of the test geometry and (most likely) to its better spatial continuity.

The same Nigerian data set is used to demonstrate that under favorable circumstances very low fold can be sufficient to get acceptable 3-D prestack migration results.

Chapter 8. Factors affecting spatial resolution

The minimal data sets of the various acquisition geometries also have different resolution properties. The main factor influencing the theoretically best resolution is the stretch effect caused by normal moveout. Therefore, zero-offset data potentially have the best resolution. Resolution is not improved by reducing the midpoint sampling intervals while keeping the shot and receiver sampling intervals the same (bin-fractionation technique). Carefully selected "random" coarse sampling may produce less migration artifacts than regular coarse sampling, but in order to eliminate all artifacts, regular dense sampling is best.

Chapter 9. DMO

The theory of DMO correction was developed for 2-D common-offset gathers. Initially, the success of application of DMO correction to 3-D data was not really understood. The theory of DMO application to minimal data sets in general and to cross-spreads in particular dispelled the mystery. The application of existing DMO software to a single-fold data set (cross-spread) revealed serious amplitude and phase artifacts. This prompted improvements in contractor software.

Chapter 10. Prestack migration

The required migration radius is often described in terms of Fresnel zone radius. However, the Fresnel zone radius for broadband data is not large enough for complete imaging. It is better to define the zone of influence for migration (in analogy to what previously has been done for modeling) and to use the radius of that zone in establishing migration-apron requirements. Most minimal data sets have limited extent leading to edge effects in migration. However, using pseudominimal data sets constructed from offset-vector tiles tends to produce better single-fold images than other single-fold subsets of the geometry.

The ideas and results discussed in this book should help to achieve a better understanding of the structure of 3-D acquisition geometries. With this understanding, geophysical requirements can be satisfied with an optimal choice of acquisition geometry and its parameters. Processing techniques can be adapted to honor and exploit the specific requirements of each geometry, especially orthogonal and areal geometry, leading to a more interpretable end product.

This book has been slightly modified from my PhD thesis (defended in February 2001) to correct some mistakes and to make it up-to-date, in particular in the area of marine 3-D/4-C acquisition. The latest developments that could be included date from June 2001.

Together with this book a CD-Rom is provided containing the fully searchable contents of the book. It also contains a prototype version of the Acquisition Design Wizard, a program written in Java allowing the user to design the parameters of a 3-D survey based on the methodology described in this book. The CD-Rom also contains an Excel program allowing survey design optimization. The program is described in my paper "3D seismic survey design optimization," also included on the CD-Rom.

Chapter 1
2-D symmetric sampling

1.1 Introduction

In the 1980s, the theory of seismic data acquisition techniques received renewed interest. In particular, Anstey (1986) and Ongkiehong and Askin (1988) introduced new ideas. These authors argued that ground roll suppression is optimal if the acquisition technique insures a regular distribution of geophones over the common midpoint.

Using field data examples, Morse and Hildebrandt (1989) and Ak (1990) demonstrated the superior performance of the stack-array approach over techniques in which there is no such regular distribution of geophones.

In my book, *Seismic Wavefield Sampling* (Vermeer, 1990), I expand the idea of regularity to the sampling of both receivers and shots. This chapter deals with some highlights of that book, concentrating on the concept of symmetric sampling as the best compromise data acquisition technique.

1.2 The shot/receiver and midpoint/offset coordinate systems in 2-D

Along the 2-D line, each shot with coordinate x_s is recorded by a receiver spread with receiver coordinates x_r. The collection of all common-shot gathers forms the prestack wavefield $W(t, x_s, x_r)$, which is a 3-D data set. The prestack wavefield is smooth and continuous (apart from shot and geophone coupling variations).

The 3-D prestack wavefield (corresponding to a 2-D seismic line) can also be described by traveltime t, midpoint x_m and shot-to-receiver offset x_o. These variables are illustrated in Figure 1.1. The two pairs of spatial coordinates are related by[1]

$$x_m = (x_s + x_r)/2 \qquad x_s = x_m + x_o/2$$

and

$$x_o = x_s - x_r \qquad x_r = x_m - x_o/2. \qquad (1.1)$$

A description of a prestack seismic data set in the two coordinate systems is shown in Figures 1.2a and 2b. Each "×" corresponds to a single trace. In Figure 1.2a, the traces are described in terms of their shot and receiver coordinates. This surface diagram was introduced by Taner et al. (1974), to describe static correction procedures. Figure 1.2b describes the same collection of traces in the midpoint/offset coordinate system. This representation is also called the subsurface diagram.

By keeping one of the spatial coordinates constant, four different subsets can be selected from the seismic data set. These subsets are indicated in Figures 1.2a and 2b. Note that all traces of a common-shot gather with x_s = constant are represented by a horizontal line in the

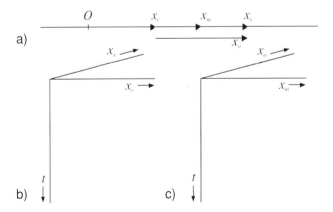

FIG. 1.1. Prestack data coordinate systems. (a) The four spatial coordinates in relation to the seismic line. (b) Shot/receiver coordinate system. (c) Midpoint/offset coordinate system.

This chapter is modified from Vermeer (1991).

[1] With this notation, the offset vector points from receiver to shot. In the next chapter the more logical notation is used in which the offset vector points from shot to receiver.

shot/receiver coordinate system and by an oblique line in the midpoint/offset coordinate system.

By keeping the time coordinate constant, a time slice is generated from the prestack seismic data set. In a time slice, the spatial coordinates vary so that the surface and subsurface diagrams could also be regarded as a description of the data points in a time slice.

We are inclined to think of reflections in prestack data as hyperbolas in the common midpoints. However, it is important to realize that each event represents a surface in the 3-D space of the prestack seismic data set. The three dimensions of the prestack data should not be confused with the three dimensions of the subsurface. In prestack data, offset is the third dimension. Take, for example, the reflection traveltime surface of a dipping plane as shown in Figure 1.3. This figure illustrates the three orthogonal cross-sections: common midpoint, common offset, and common time. The shape of the traveltime surface is a hyperbola in the common midpoint and an ellipse in the time slice.

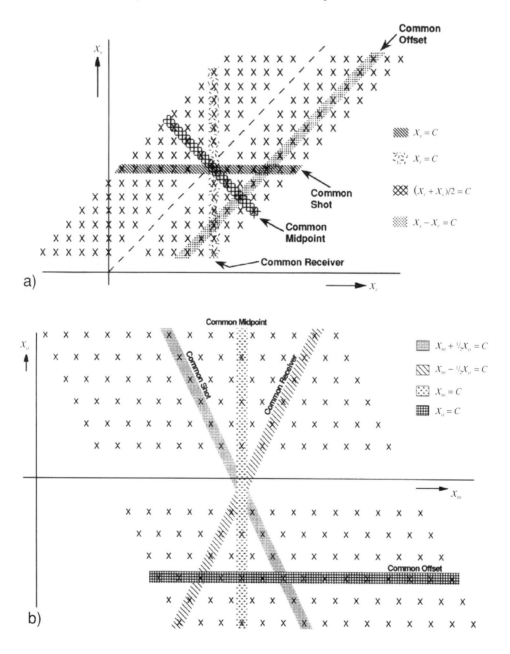

FIG. 1.2. Descriptions of prestack seismic data set in (a) a shot/receiver coordinate system and (b) a midpoint/offset coordinate system. The former is also called a surface diagram or surface stacking chart and the latter a subsurface diagram or subsurface stacking chart.

$$W(t, k_s, k_r) = \iint_{\Omega_S} \times W(t, x_s, x_r) \times \exp[2\pi i(k_s x_s + k_r x_r)]dx_s dx_r, \quad (1.2)$$

A field data example is given in Figure 1.4. Of course, now there is a multitude of events, all having their own spatial and temporal relationships. Actually, the common-offset gather in this example is a stack, which is basically a zero-offset section with a relatively high signal-to-noise ratio. It is possible to follow reflections from dipping layer boundaries through all three cross-sections.

Creating time slices from the prestack data of a 2-D line can be a very rewarding exercise. Time slices increase insight in the characteristics of the data and allow useful diagnostics-at-a-glance of the whole data set. Time slices created after NMO correction allow a quick quality control of the chosen velocities for the level of interest.

For proper sampling of the temporal coordinate, it is important to know the maximum frequency of the data to be sampled. Likewise, for spatial sampling, the maximum wavenumber of the spatial coordinates must be known. A discussion of spatial sampling requires the introduction of four different wavenumbers (k_s, k_r, k_m, k_o) corresponding to the four spatial coordinates (x_s, x_r, x_m, x_o). For instance, k_s and k_r can be defined by the forward Fourier transform

where Ω_s is the range of shots and receivers included in the integration. Equation (1.2) represents the double wavenumber spectrum of a time slice, whereas the triple Fourier transform of $W(t, x_s, x_r)$ can be written as

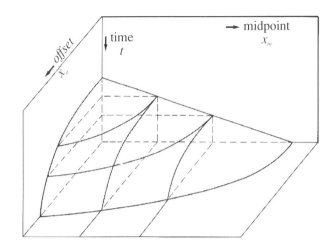

FIG. 1.3. Dipping event in a midpoint/offset coordinate system. The event is a hyperbola in the common-offset panels (a straight line for zero offset), a hyperbola in the CMP, and an ellipse in the time slice.

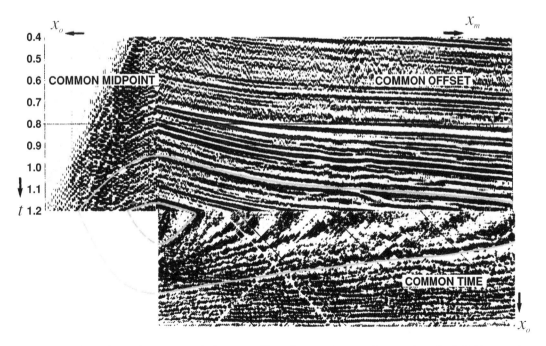

FIG. 1.4. Three cross-sections through a prestack data set. Note that each event is a surface in 3-D (t, x_m, x_o) space.

$$W(f, k_s, k_r) = \int_{\Omega_t} \times W(t, k_s, k_r)\exp[-2\pi ift]dt, \quad (1.3)$$

where Ω_t is the integration time window.

Similar to the pairs of spatial coordinates (x_s, x_r) and (x_m, x_o), there is also a linear relationship between the pairs of wavenumbers

$$k_m = k_s + k_r, \qquad k_s = k_m/2 + k_o,$$

and

$$k_o = (k_s - k_r)/2, \qquad k_r = k_m/2 - k_o. \quad (1.4)$$

These relationships follow directly by substitution of the right-hand equations (1.1) into equation (1.2). It is possible to compute (f,k)-spectra for common-shot gathers, common-receiver gathers, common-midpoint gathers, and common-offset gathers.

Whatever subset is considered, it is important to distinguish the various wavenumber domains from each other, because they represent very different physical effects. In particular, the offset wavenumber k_o describes velocity effects in the common-midpoint gather, and the midpoint wavenumber k_m describes structure effects in the common-offset gather. For instance, for a horizontal earth, there are only horizontal events in the common-offset gather. So the wavenumber spectrum of that gather only shows energy at $k_m = 0$. In more practical cases, there is also energy for positive and negative midpoint wavenumbers.

1.3 Symmetric sampling

The data as described in the surface and subsurface diagrams (Figures 1.2a and 2b) have already been sampled. Figure 1.5 indicates the spatial sampling interval used for the surface diagram. In this case, the shot interval is the same as the receiver (or group) interval. In this section, I want to address the question: What is the best way of sampling the two spatial coordinates x_s and x_r?

To answer this question, we must know the properties of the 3-D prestack wavefield to be sampled. As shot and receiver coordinates are sampled independently, the properties of the wavefield $W(t, x_s, x_r)$ need to be examined both in the common-shot gather and in the common-receiver gather.

The common-shot gather is the result of a physical experiment; therefore, the properties of the wavefield of the common-shot gather are described by elastic-wave theory. On the other hand, the traces of a common-receiver gather are all recorded separately at different times with different shots. So what are the properties of the wave-

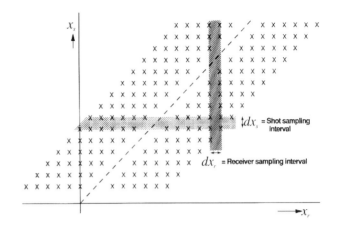

FIG. 1.5. Shot and receiver sampling intervals in the surface diagram.

field in the common-receiver gather? Here we will use the reciprocity theorem. The theorem says that, under certain conditions, two seismic experiments in which the position of shot and receiver are interchanged lead to the same recorded trace (see, e.g., Vermeer, 1990, and Chapter 6 in Fokkema and van den Berg, 1993). A consequence of the reciprocity theorem is that a common-shot gather $W(t, x_P, x_r)$ shot at point $x_s = x_P$ would in its entirety be identical to a common-receiver gather $W(t, x_s, x_P)$ recorded in the same point $x_r = x_P$ (see Figure 1.6). Therefore, the properties of the wavefield in the common-receiver gather, consisting of a large number of different seismic experiments, are the same as the properties of the wavefield of the common-shot gather obtained in a single seismic experiment. As a consequence, the sampling requirements of shots and receivers are the same.

We would like to record the two spatial coordinates without aliasing, just as we do with the temporal coordinate. Figure 1.7 illustrates that the maximum wavenumbers $|k_s|_{\max}$ and $|k_r|_{\max}$ are determined by the maximum frequency $f_{\max}$ of the event with the slowest apparent phase velocity $V_{\min}$. In three-dimensional (f, k_s, k_r) space, the energy of the wavefield is confined to a pyramid-shaped volume with its base at $f = f_{\max}$ (Figure 1.8). Alias-free spatial sampling is achieved if the maximum wavenumbers are properly sampled, which means that

$$\Delta x_s = \Delta x_r \leq \frac{1}{2k_{\max}} = \frac{V_{\min}}{2f_{\max}}. \quad (1.5)$$

In other words, these *basic sampling intervals* Δx_s and Δx_r should not be larger than a half-period of the smallest wavelength. Preferably, the sampling intervals should be somewhat smaller (oversampling) to allow a more accurate reconstruction of the underlying continuous wavefield, especially close to Nyquist (Niland, 1989).

The basic sampling intervals are much smaller than considered practical or affordable (e.g., for f_{max} = 75 Hz, and a ground-roll velocity V_{min} = 300 m/s, the shot and receiver intervals should be ≤ 2 m). As a compromise, seismic field arrays are to be used which act as spatial antialias filters and as resampling operators. As resampling operators they allow the use of more affordable shot and receiver intervals. As spatial antialias filters, they aim to attenuate all energy above the Nyquist wavenumber. Spatial antialias filtering must be applied when sampling both spatial coordinates. In other words, shot arrays are as necessary as receiver arrays and, for optimal results, shot arrays should be identical to receiver arrays. This reasoning leads to the concept of *symmetric sampling* as a prerequisite of consistent data handling:

- Shot interval equal to receiver interval
- Shot arrays equal to receiver arrays

I call this technique "symmetric sampling" because it utilizes the symmetry property of reciprocity and it preserves the inherent symmetry of the prestack wavefield. A consequence of symmetric sampling is that there will be as many traces in the common-receiver gather as in the common-shot gather.

In case of a linear nonweighted array with equal intervals between array elements, the dimension of the array should be such as to achieve a regular uninterrupted sampling by the array elements of the whole length of the receiver line. Basically, this means that the length of the array should be equal to the station interval; if shorter, the wavefield is undersampled, and if longer, the wavefield is oversampled, intra-array statics are larger,

and the signal may lose some of its high-frequency content. Some further remarks on array length are made in Section 1.6.

Figure 1.9 illustrates the concept of symmetric sampling. (In these figures the x_s, x_r coordinate system has been rotated 45° for ease of display.) To keep the pictures simple, I used only three elements per array (when there is an array). Figure 1.9a shows an asymmetric configuration with no shot array and three geophones per geophone array. Each recorded trace is the sum of three elemental traces registered by the three geophones of the geophone array. The groups of three elemental traces are represented by alternating between three open circles and three closed circles. Note that the three ele-

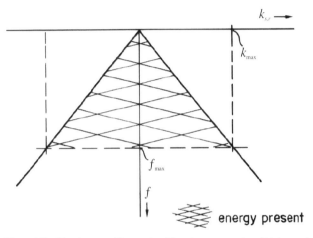

FIG. 1.7. Regions with and without energy in (f, k_s) and (f, k_r). Maximum frequency of event with minimum-phase velocity determines maximum wavenumber and it is the same for shot and receiver coordinates.

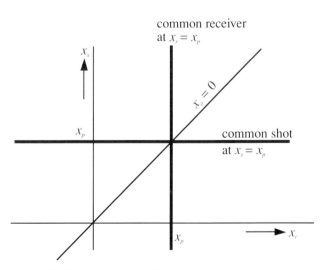

FIG. 1.6. Common-shot gather at $x_s = x_p$ and common-receiver gather at $x_r = x_p$ are identical because of reciprocity.

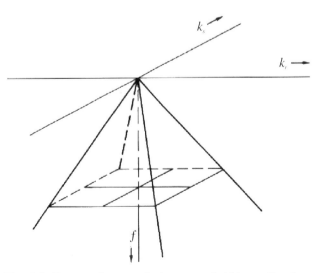

Fig. 1.8. Energy of prestack data wavefield is confined to pyramid shaped volume in (f, k_s, k_r). Base of the pyramid is at $f = f_{max}$.

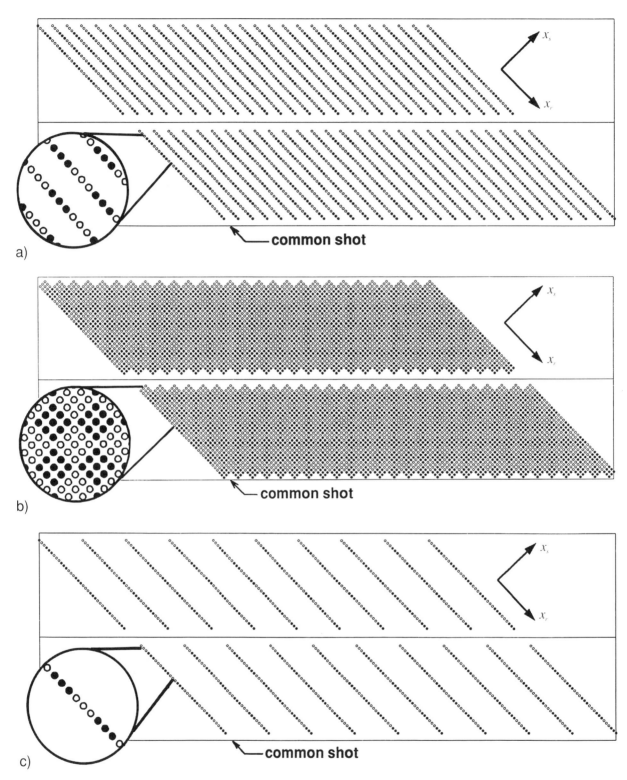

FIG. 1.9. Various center-spread shooting geometries with three elements per array, when there is an array. Each symbol represents an elemental shot/receiver pair; each group of equal symbols (either open circles or closed circles) represents one recorded trace. (a) Asymmetric configuration with shot interval equal to receiver station interval, and a geophone array but no shot array. (b) Symmetric configuration with shot interval equal to receiver-station interval, and both geophone and shot arrays. (c) Asymmetric configuration with a shot interval three times the receiver-station interval. Note the large unsampled area of shot/receiver space.

mental traces for a given recorded trace have different shot-to-geophone offsets and different midpoint positions. Summing these elemental traces causes some damage to the signal, but this is the price to be paid for the antialias effect of the geophone array. In Figure 1.9b, a symmetric sampling configuration is shown. Now there is also a shot array consisting of three elements so that each recorded trace corresponds to nine elemental traces. Note that, again, each of the elemental traces occupies a different position in the shot/receiver coordinate system. Together, all the elemental traces provide a regular two-dimensional sampling of the shot/receiver plane. Compare this with Figure 1.9a, where whole areas of the plane are not sampled. These empty areas may lead to spatial aliasing in the common-receiver domain and also in the common-midpoint and common-offset domains.

Another, perhaps even more common, example of asymmetric sampling is shown in Figure 1.9c. Now the shot interval is three times the group interval. Here even more of the shot/receiver plane is not sampled.

An interesting illustration of the need for arrays with length equal to the station interval is given as Figure 3 in Newman (2000) and reproduced here as Figure 1.10. The shorter length arrays used on the left have a wide passband in the wavenumber domain and do not suppress much of the aliased ground-roll energy. The arrays with length equal to station interval used on the right have suppressed the ground-roll energy better. There is still remaining ground-roll energy which exhibits an odd/even effect (odd traces look more similar to each other than to even traces) in Figure 1.10b, because the first notch of the array response occurs at twice the Nyquist wavenumber k_N corresponding to the station spacing [cf. discussion in Section 1.6 and equations (1.5) and (1.8)]. However, if necessary, the energy passed above k_N may be further suppressed by a two-trace running mix (a convolution with a two-point spatial filter with equal coefficients) in processing.

Whether or not spatial aliasing occurs for a particular shooting geometry (symmetric or asymmetric) and how large the effect is depends on the distance between the array elements and on the shot and receiver intervals. Symmetrically sampled data may still be aliased if the sampling intervals are too large, and asymmetrically sampled data may not show aliasing if the sampling intervals are small enough. A nice compromise to aim for is to use shot and receiver intervals that would record the desired wavefield without aliasing up to the frequency of interest. Then arrays are only necessary to suppress noise and to average out sampling irregularities. This technique is called *full-resolution recording* and the corresponding sampling interval is called *basic signal sampling interval*.

1.4 Symmetric sampling versus asymmetric sampling

Having established that symmetric sampling is necessary to honor the properties of the prestack wavefield, I shall now discuss some effects of asymmetric sampling and enumerate advantages of symmetric sampling.

The effects of asymmetric sampling are different for end-on shooting and center-spread shooting. For the former, asymmetric sampling leads to differences between updip and downdip shooting. Figure 1.11 illustrates there is less difference between the arrival times of the reflections over the length of the receiver array for updip shooting than there is for downdip shooting. Therefore, with asymmetric sampling (for instance, shot array being different from receiver array), the reflection character is less affected by updip shooting than by downdip shooting. This effect would only be visible by careful inspection of two neighboring parallel lines acquired in oppo-

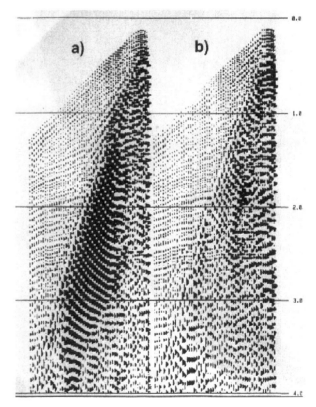

FIG. 1.10. (a) Three-D spread of 60 channels of six phones over 100 ft. Deep hole dynamite source (15 pounds at 60 ft). (b) Parallel line [1320 ft away from line in (a)] of 60 channels with 24 phones laid out over 220 ft from the same shot. The source is between the two lines (from Newman, 2000).

site directions. On the other hand, in a single 2-D line, asymmetric sampling will lead to asymmetries on either side of the apex of diffractions.

In center-spread shooting, the effect of asymmetric sampling is less visible in the stack, as the asymmetry of the sampling is hidden by the symmetry of the spread. However, the differences between updip and downdip shooting, as discussed for end-on shooting, now occur in the recording of one and the same line. Now the effect becomes visible in the common-midpoint gathers. I have simulated the effect in the example shown in Figure 1.12. Figure 1.12a is a CMP with equal shot and receiver arrays, whereas in Figure 1.12b the receiver arrays are three times as long as the shot arrays (75 m versus 25 m). I constructed the right panel using a three-trace running mix in the common-shot gathers, followed by CMP sort. In both CMPs, the traces are sorted according to increasing absolute offset. The right panel now shows jitter in many reflection events. The jitter occurs for events that dip in the common-offset gather. The explanation of the jitter follows from the difference in averaging effects of the arrays on either side of zero offset.

This averaging effect is illustrated in Figure 1.13 in which the two curved lines represent constant time lines of a dipping event in the shot/receiver plane. (As discussed earlier, these lines are ellipses.) The lines are symmetric with respect to the diagonal, which is the zero-offset line. The rectangles represent traces of one common midpoint with each trace formed by a 25 m shot array and a 75 m receiver array. In the top left corner, the rectangle averages across the time lines; in the lower right corner, the rectangles run more or less parallel to the time lines. This difference in averaging leads to a different character between positive and negative offsets. Similar effects can be observed with single-hole dynamite shooting. I am convinced that many seismic processors have noticed those effects. Obviously, it will lead to a suboptimal stack for center-spread shooting.

The severity of asymmetric sampling depends on a number of factors, such as spatial sampling intervals, degree of asymmetry, relative strength of coherent noise, dip (stronger dip, larger effects), and geologic complexity. With some further analysis, some of those effects may be quantified (for instance, the increasing severity with increasing dip).

On the other hand, symmetric sampling has numerous advantages:

- Symmetry of wavefield is preserved

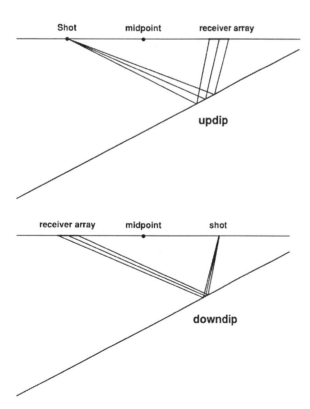

FIG. 1.11. Updip versus downdip shooting for an asymmetric configuration with a receiver array but no shot array.

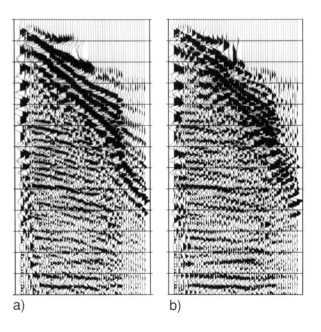

FIG. 1.12. Symmetric versus asymmetric sampling in center-spread geometry. Shot interval equals receiver interval. CMPs are displayed with increasing absolute offset (i.e., adjacent traces originate from opposite sides of the spread). (a) Symmetric data, shot array = receiver array = 25 m. (b) Asymmetric data, shot array = 25 m, receiver array = 75 m. Note the jitter in the asymmetrically sampled CMP.

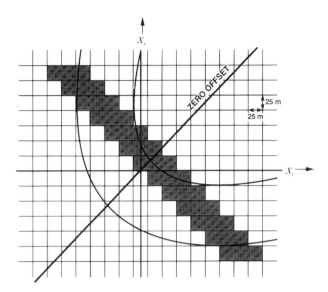

FIG. 1.13. Explanation of jitter in CMP of asymmetrically sampled data. Curved lines represent constant time lines of a dipping event (cf., Figure 1.3). Each rectangle is the convolution of a short 25-m shot array with a long 75-m geophone array. The elemental traces within a rectangle are added to form one recorded trace in the CMP. In the top left corner, the addition takes place across the time lines of the dipping event, whereas in the bottom right corner, the addition is mostly parallel to the time lines.

- Character independent of line direction (end-on shooting)
- Constant character across CMP (center-spread shooting)
- Better coherent noise suppression in the field
- Data better suited for cascaded shot- and receiver-domain processing
- Data better suited for highly sophisticated processes such as amplitude variation with offset (AVO) analysis, migration, inversion

These advantages, in turn, lead to fewer and less severe low quality data zones, better resolution of complex geology, and better reservoir characterization; in short, a more reliable and successful interpretation. Whether or not these advantages materialize depends, to a large extent, on the ability to even out variations in shot strength and geophone coupling with surface-consistent equalization.

1.5 The stack-array approach versus symmetric sampling

Briefly, the difference is that Anstey (1986) emphasizes a regular sampling of geophones (across the CMP), whereas symmetric sampling calls for regular sampling not only of geophones but also of shots. The stack-array approach does not specify the use of shot arrays, leading to an asymmetric sampling technique (as shown in Figure 1.9a). The common-shot gather is properly sampled, but aliasing may occur in the common-receiver gathers. Nevertheless, I would like to stress that Anstey's technique is a tremendous improvement over older techniques using large shot intervals (such as illustrated in Figure 1.9c).

1.6 The total stack response

This section investigates the combined response of field arrays and stacking which can be called the total stack response. This response can be written as the product of three individual responses: the shot-array response $p(k_s)$, the receiver-array response $p(k_r)$, and the so-called CMP-array or stack response $p(k_0)$, or

$$S(k_m, k_o) = p(k_s) \times p(k_r) \times p(k_0). \quad (1.6)$$

The three responses involve three different wavenumbers, because each array operates in its own spatial domain. Equation (1.4) has been used implicitly to describe the total response $S(k_m, k_o)$ as a function of two wavenumbers only. Each response can be described by a discrete spatial Fourier transform

$$p(k_i) = \sum_{j=1}^{N} w_j \exp(2\pi i k_i x_{ij}) / \sum_{j=1}^{N} w_j, \quad (1.7)$$

where N is the number of elements in the array, w_j is the weight factor for element j, k_i represents one of the wavenumbers k_s, k_r, or k_o and x_{ij} is the corresponding spatial variable (i.e., x_{sj}, x_{rj}, or x_{oj}).

For linear arrays with equal weight factors and constant element spacing d, equation (1.7) turns into a geometric series and can be simplified to (neglecting a phase factor)

$$p(k_i) = \frac{\sin(N\pi k_i d)}{N \sin \pi k_i d}, \quad (1.8)$$

so that the first notch of this array occurs at $k_i = (Nd)^{-1}$. Nd is called the length of the array. [The length of the array is not equal to $(N-1)*d$, which is the distance between the first and last elements of the array. Doubling the number of elements should lead to an array length which is twice as large, but this is not the case when using $(N-1)*d$ as definition of array length.] In the forthcoming examples, I will use 50-m field arrays and 1200-

m offset range, hence the first notches of the three arrays occur at $k_s = 1/50$, $k_r = 1/50$, and $k_0 = 1/1200$ m^{-1}. [This discussion does not take NMO into account; see Ongkiehong and Askin (1988) for the effect of NMO on the array responses.]

In displays of array and stack responses usually the absolute value of equation (1.8) is taken, thus neglecting the phase of the response. For a 2-D seismic line, the offset distribution tends to be regular with an equal number of traces and a constant offset interval between traces in all midpoints, but with a slightly shifted range of offsets between neighboring midpoints. For such data the absolute value of the stack response is the same for all midpoints, whereas the range shift leads to a phase variation in $p(k_0)$ causing the checkerboard or odd/even effect (Vermeer, 1990, Section 5.11.4).

Although the stack response is formulated mathematically in the same way as the array response, the desired response is quite different. The ideal array response looks like an antialias filter response with a passband and a cut-off wavenumber. The ideal stack response passes all energy at $k_0 = 0$, and rejects all energy with $k_0 \neq 0$.

Figures 1.14, 15, and 16 represent displays of equation (1.6) in the midpoint/offset wavenumber domain for different situations. Figure 1.14 shows the response of a 50-m shot array combined with a 50-m receiver array and no CMP array. Lines of constant shot and receiver wavenumber run obliquely in the midpoint/offset wavenumber domain. Note the diamond-shaped central passband of the two arrays in this domain.

A common simplification is to compute the product of the two array responses as a function of only one wavenumber. This product describes the effect of the arrays on a horizontal earth with no midpoint dependence. The horizontal earth response is found for $k_m = 0$, i.e., along the vertical axis of Figure 1.14. However, any dipping events will contain energy away from the vertical axis and will be affected differently by the field arrays. So, the correct representation uses the double wavenumber domain.

Figure 1.15 shows the total stack response for a symmetric sampling technique with center-spread shooting and an offset range -1200 to 1200 m. Taking reciprocity into account, this configuration effectively produces a 50-m offset interval in each CMP leading to a first alias in the stack response at $k_0 = \pm 1/50$ m^{-1}. Note that the stack produces notches parallel to the horizontal k_m-axis. The diamond-shaped passband of the field arrays has now been reduced to a narrow passband centered on the midpoint wavenumber axis. Everywhere else the combination of field arrays and stack is supposed to suppress all energy. As is clear from the picture, the suppression is certainly not uniform although it is symmetric. There are areas of very good suppression where all three arrays are effective, and there are also areas of less good suppression. (The parameters of this example should not be taken as recommended symmetric sampling field parameters; usually, smaller intervals are necessary for good results.)

How much unwanted energy is left after application of the three arrays depends on

- energy distribution of the prestack wavefield,
- choice of field parameters (shot and receiver interval, and fold), and
- choice of pre- and poststack processing parameters.

Leaving out the shot array has a dramatic effect on the total stack response (as shown in Figure 1.16). The severity of not using a shot array or any other form of asymmetric sampling depends on the energy distribution of the original continuous wavefield in the (k_m, k_0)-wavenumber domain. If there are many rapid variations as a function of midpoint, asymmetric sampling will do more harm than if the geology varied more slowly.

Finer sampling (shorter shot and receiver intervals with array lengths equal to those intervals) pushes the filter notches out toward larger wavenumbers. As a consequence, a larger part of the original wavefield will fall in the passband of the combined field arrays. In the passband, more of the suppression of the unwanted events is then left to the stack and to other digital processes. Digital processes such as (f,k)-filtering are usually required to compensate for the reduced effect of the two field arrays.

1.7 Concluding remarks

This chapter serves as a summary of 2-D symmetric sampling as described in more detail in Vermeer (1990). Symmetric sampling is the preferred recording technique for 2-D seismic surveys, so it should be high on the "wish list" of every interpreter. In case an asymmetric technique has been used in the past, for instance with a shot interval that is larger than the receiver interval, there is always scope for improvement by repeating the survey with symmetric sampling parameters.

Yet, it should be realized that the parameters of symmetric sampling are still a compromise and need to be established after an evaluation of the geologic and geophysical problems at hand. In particular, time and again improvements have been achieved by using smaller station spacings leading to noisier field records, but allowing better noise removal in processing.

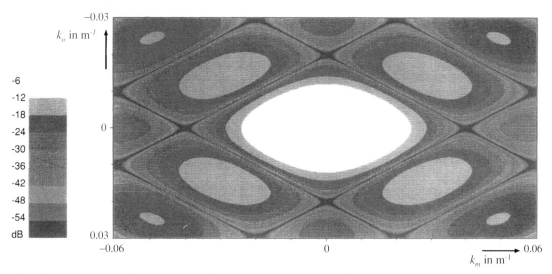

FIG. 1.14. Combined response of shot and receiver arrays in midpoint/offset wavenumber domain. Oblique dark blue lines represent notches in the shot-array and receiver-array responses.

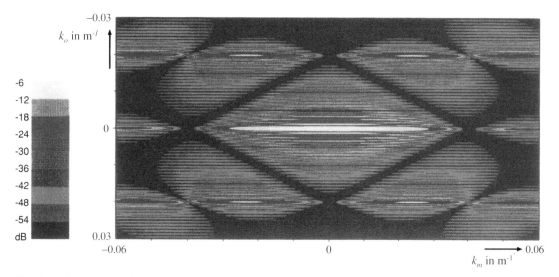

FIG. 1.15. Total stack response for a symmetric sampling technique. The notches of the CMP array run parallel to the k_m axis.

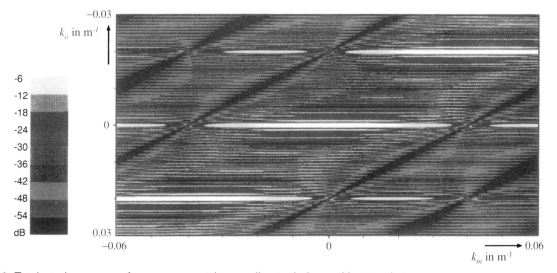

FIG. 1.16. Total stack response for an asymmetric sampling technique without a shot array.

A better understanding and knowledge of the energy distribution in (f, k_m, k_0) would help predict the effect of any choice of the acquisition parameters. Noise spreads and very densely sampled multiple-coverage data can be used to help gain such information.

References

Ak, M. A., 1990, How effective is the stack-array?: Presented at the 52nd Eur. Assn. Expl. Geophys. Conference.

Anstey, N., 1986, Whatever happened to ground roll?: The Leading Edge, **5**, No. 3, 40–45.

Fokkema, J. T., and van den Berg, P. M., 1993, Seismic applications of acoustic reciprocity: Elsevier Science Publ. Co., Inc.

Morse, P. F., and Hildebrandt, G. F., 1989, Ground-roll suppression by the stackarray: Geophysics, **54**, 290–301.

Newman, B. J., 2000, Spatial aliasing and 3-D bin size: The quest for cleaner, cheaper data: The Leading Edge, **19**, 158–161.

Niland, R. A., 1989, Optimum oversampling: J. Acoust. Soc. Am., **86**, 1805–1811.

Ongkiehong, L., and Askin, H., 1988, Towards the universal seismic acquisition technique: First Break, **5**, 435–439.

Taner, M. T., Koehler, F., and Alhilali, K. A., 1974, Estimation and correction of near-surface time anomalies: Geophysics, **39**, 441–463.

Vermeer, G. J. O., 1990, Seismic wavefield sampling: Soc. Expl. Geophys.

———1991, Symmetric sampling: The Leading Edge, **10**, No. 11, 21–27.

Chapter 2
3-D symmetric sampling

2.1 Introduction

Since the early 1980s, there has been a steady increase in the number of acquired 3-D surveys. Continuing improvements in technology have made it possible to make 3-D seismic data acquisition more and more efficient and cost-effective. Yet, a clear theory as to what constitutes a good 3-D acquisition geometry has not been available, and much of the design of 3-D acquisition geometries has been based on earlier experience—what seemed to have worked in the past was adopted for the future—and on the possibilities and limitations of the available equipment (Stone, 1994). Cordsen et al. (2000) rely on some rules of thumb and guidelines to help them "through the maze of different parameters that need to be considered." In this chapter, I will provide a theoretical framework for the design of 3-D acquisition geometries suitable for both marine and land data acquisition.

Quite rightly, many current design techniques for 3-D geometries attempt to extend to 3-D what had been learned from the design of 2-D geometries. A breakthrough in thinking about 2-D geometries was provided in Anstey's paper (1986) "Whatever happened to ground roll?" Anstey argued that the combination of field arrays and stacking takes care of adequate suppression of ground roll, provided the offset distribution in the common midpoint (CMP) is regular and dense—the so-called stack-array concept. Ongkiehong and Askin (1988) proposed the hands-off seismic data acquisition concept. They argued that the distance between elements in an array and the length of the contiguous arrays is fully determined by signal velocity and required bandwidth. These ideas are encompassed and reexplained by the symmetric sampling theory introduced in Vermeer (1990, 1991). In symmetric sampling, both shots and receivers have to be sampled in the same way, including the shot and receiver arrays. In this theory, a regular offset distribution in the CMP gather is a consequence of the requirement of symmetric sampling.

Anstey's (1986) considerations on offset distributions, valid for 2-D, could be applied also to 3-D marine streamer acquisition because it is basically 3-D by repeating 2-D. However, these considerations are not generally applicable to land-type acquisition geometries such as the orthogonal arrangement of shot and receiver lines, unless very high fold is used. On the other hand, 2-D symmetric sampling theory can be extended to 3-D for all types of common 3-D geometries. As we shall see, symmetric sampling of the 2-D seismic line is in fact a special case of *3-D symmetric sampling*. At the 1994 SEG annual meeting, I first proposed the 3-D symmetric sampling technique (Vermeer, 1994). This chapter provides a more comprehensive description.

In 2-D, the sampling problem is one of sampling the 3-D wavefield $W(t, x_s, x_r)$ with temporal coordinate t, shot coordinate x_s, and receiver coordinate x_r. In 2-D symmetric sampling, the two spatial (shot and receiver) coordinates are sampled in the same way. Using sufficiently small sampling intervals allows the faithful reconstruction of the underlying continuous wavefield, i.e., it maintains the spatial continuity of the wavefield $W(t, x_s, x_r)$ (see also Section 1.3 and Vermeer, 1990).

In 3-D, we are faced with the sampling of a 5-D wavefield $W(t, x_s, y_s, x_r, y_r)$, now with shot y_s and receiver y_r as additional spatial coordinates. It would be prohibitively expensive to completely sample this 5-D wavefield, as this would mean filling the whole survey area with a dense coverage of both shots and receivers. As a compromise, 3-D symmetric sampling settles for the more affordable aim of correct sampling of overlapping single-fold 3-D subsets of the 5-D wavefield $W(t, x_s, y_s, x_r, y_r)$. Such correctly sampled subsets are suitable for imaging of the subsurface with the right resolution (provided the source wavelet has a suitably wide bandwidth) using prestack migration (Beylkin, 1985; Beylkin et al., 1985; Cohen et al., 1986; Bleistein, 1987;

The first part of this chapter is an adaptation of Vermeer (1998a), whereas Sections 2.5 and 2.6 stem from Vermeer (2000).

Schleicher et al., 1993). The subsets have to be spatially overlapping (multifold acquisition) to gain redundancy for an adequate signal-to-noise ratio and to allow velocity analysis.

To set the scene, I will first show that the geometries most commonly used are either members of the class of areal geometries or members of the class of line geometries. The line geometries can be subdivided further into parallel and crossed-array geometries. In Section 2.3 I extend some of the results of Vermeer (1990) to a description of some properties of the continuous wavefield and various 3-D subsets of that wavefield. These properties are used in Section 2.4 to describe the requirements of symmetric sampling of the two spatial coordinates of each subset.

Having established what is intuitively a good way of approaching the sampling problem in 3-D data acquisition, it should be made plausible that the subsequent processing would also benefit from symmetric sampling. For parallel geometry this is readily acceptable since acquisition with parallel geometry more or less mimicks 2-D acquisition. For the crossed-array geometries, new processing approaches need be considered to fully exploit the better sampling of the input data. For those geometries, Section 2.5 introduces new subsets in addition to the subsets which are the core of symmetric sampling. Section 2.6 describes how each prestack processing step can benefit from the most suitable choice of subset to be input to that processing step. Different processing steps require different gathers of input data.

2.2 Classes of 3-D geometries

Alias-free sampling of all four spatial (surface) coordinates of the 5-D prestack wavefield $W(t, x_s, y_s, x_r, y_r)$ would mean that each shot should be recorded by a dense areal grid of receivers and that the shotpoints should also occupy a dense areal grid. Virtually nobody can afford this full sampling of $W(t, x_s, y_s, x_r, y_r)$. Instead, a wide variety of geometries has been devised based on a sparser sampling of shots and/or receivers.

Most solutions to the seismic sampling problem fall into one of two main classes: (1) the receivers "listening" to each shot still occupy a dense areal grid, but the shots are sampled in only a coarse grid (or the other way around), and (2) the receivers listening to each shot are densely sampled along parallel receiver lines, whereas the shots are densely sampled along parallel shot lines. The geometries in the first class are called *areal* geometries, whereas those in the second class are called *line* geometries. Depending on the orientation of the shot lines with respect to the receiver lines, the line geometries can be subdivided into parallel and crossed-array geometries. Figure 2.1 provides a pictorial description of areal and line geometries. Note that the shot lines in the main types of line geometries are parallel to each other, whereas the receiver lines are also parallel to each other. In crossed-array geometry, shot lines and receiver lines cross each other; in orthogonal geometry the lines make an angle of 90° with each other, whereas in slanted geometry they cross at an angle α, either with tan α = 2 or tan α = 1 in most cases. Zigzag geometry is a special case of crossed-array geometry; in this geometry there are two sets of parallel shot lines making an angle of ±45° with the parallel receiver lines. In this chapter only orthogonal geometry is discussed in more detail; in Sections 4.3.2, 4.3.3, and 4.3.4 comparisons are made between orthogonal geometry and other crossed-array geometries.

In Vermeer (1994), I used the term "patch" for the areal geometry. The term "patch" was adopted from the geometry described in Crews et al. (1989). They use areal patches of geophones listening to a sparse grid of shots. Unfortunately, patch is used in the geophysical industry also for particular implementations of line geometries. Therefore, I have now opted for the name *areal* to emphasize the difference with line geometries. Of course, all geometries want to achieve an areal coverage.

Virtually all commonly used geometries can be classified as areal or line geometries. *Random* geometries are characterized by the absence of regularity in the shot and receiver positions. Random geometries are only used when the surface conditions (obstacles) preclude a regular layout of shots and receivers.

2.2.1 Examples of various geometries

Areal geometry provides either 3-D common-shot gathers (as defined by an areal grid of receivers listening to a shot in the center of the grid) or 3-D common-receiver gathers. The idea of acquiring 3-D common-shot gathers with a 2-D array of receivers was patented as early as 1960 (Becker, 1960). Walton (1971) called the 3-D common-shot gather "The dream." Part of his dream was to hover with a helicopter over the area on a dark night and watch the geophones light up when a sound wave hit them. [A modern version of this idea using laser interferometry is described in Berni (1994).] It turned out to be more practical to invoke reciprocity and to use an areal grid of thumper positions being recorded in a geophone patch in the center. Esso (now Exxon) used this technique in several surveys (Walton, 1971), but abandoned it in favor of the more cost-effective "X" spread technique.

The technology in the 1970s was not yet advanced enough to allow multiple-coverage areal geometries. This changed in the 1980s, and Crews et al. (1989) contains an acquisition technique that is reminiscent of multiple-coverage areal geometry. However, instead of a full areal grid, each shot is recorded by a checkerboard pattern of geophone stations. It would require double the effort to acquire a true areal geometry. On land, this is kind of a tall order, as obstructions usually abound.

With the advent of stationary recording systems in marine data acquisition, it is becoming feasible (though still quite time-consuming) to record 3-D common-receiver gathers with receiver stations located on the sea bottom (or anchored to the sea bottom) and with shots fired in a dense areal grid. A geometry closely approaching the ideal areal geometry is described in Moldoveanu et al. (1994). They used a dual-hydrophone Digiseis system for undershooting of platforms. An interesting aspect of this geometry is that z_r, the depth coordinate of the receiver, is sampled twice.

In the introduction, I omitted the depth coordinate from the prestack wavefield $W(t, x_s, y_s, x_r, y_r)$ because this coordinate is not a variable being sampled in surface seismic data acquisition. (Of course, in VSP acquisition, depth is a major spatial coordinate.) An areal geometry in which z_r is sampled up to 16 times is described in Stubblefield (1990) and in Krail (1991, 1993).

Depending on the conditions in the survey area, one of the various line geometries is usually the most efficient in terms of progress per square kilometer, and this might be the decisive factor in choosing the type of line geometry. It goes without saying that the most efficient geometry does not necessarily produce the best quality.

Parallel geometry is basically an extension of 2-D geometry where the shot lines and receiver lines are collinear. It is used mainly for marine data acquisition, using multisource and multistreamer configurations (e.g., quad/quad geometry; Naylor, 1990), but it has also been used on land (e.g., Dickinson et al., 1990). In quad/quad geometry, four sources are alternately fired into four streamers. Each source records its own four midpoint lines, leading to 16 parallel midpoint lines. As the seismic vessel has to maintain speed during the firing cycle, the distance between shots in a midpoint line must be large, leading to relatively low fold. This shortcoming has been solved by recent developments in marine acquisition technology, allowing towing of 8 to 16 streamers by one vessel. With two sources, modern seismic vessels can also produce 16 or more midpoint lines in one boat pass, while doubling the fold as compared to using four sources.

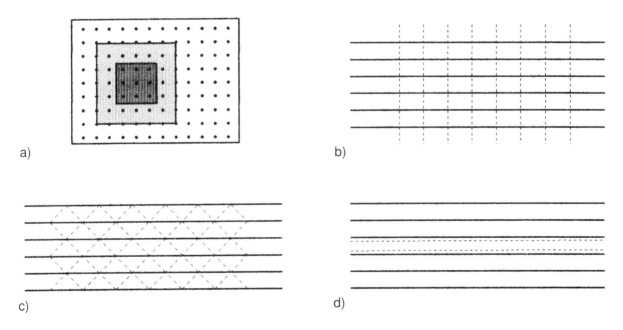

FIG. 2.1. Classes of 3-D acquisition geometries: (a) areal, (b) orthogonal, (c) zigzag, and (d) parallel. Areal geometry is based on widely spaced shot stations covered areally by receiver stations (or the other way around). For a shot in the center of the squares in (a), the small square and the large square indicate the midpoint area and the receiver area, respectively. Orthogonal geometry is characterized by widely spaced parallel shot lines perpendicular to widely spaced parallel receiver lines. In zigzag geometry, two families of widely spaced parallel shot lines make angles of ±45° with widely spaced parallel receiver lines. In parallel geometry, both shot and reciver lines are parallel to each other; the lines may or may not be widely spaced.

An interesting example of marine data acquisition using parallel acquisition lines is (concentric) circle shoot geometry (Durrani et al., 1987; Reilly, 1995). In this geometry, shot and receiver lines are (nearly) concentric circles. It is a typical example of a target-oriented geometry, the center of all circular lines being the known position of a salt dome. A similar geometry on land is the spider-web geometry (see Section 4.3.6). This is a geometry with radial receiver lines and circular shot lines. Constance et al. (1999) describe a real implementation of this geometry.

In parallel geometry, the survey area is still covered rather densely with shots and receivers. In the 1960s, it was already discovered that an orthogonal arrangement of a shot line and a receiver line could produce areal midpoint coverage without requiring an areal coverage of shots and receivers (Ball and Mounce, 1967). In the late 1960s, Esso acquired single-fold 3-D surveys consisting of single "X" spreads or cross-spreads (Walton, 1971, 1972). Properties of the cross-spread and interpretation techniques based on time slices through cross-spreads are discussed in Dunkin and Levin (1971). In those days, the single-fold cross-spreads were still a big burden to the interpreter, but, with the advent of digital processing, the data from partially overlapping cross-spreads could be stacked and migrated for easier interpretation (Dürschner, 1984). More recently, Lee et al. (1994) discussed migration results of partially overlapping cross-spreads.

The idea of areal midpoint coverage by orthogonal shot and receiver lines is fully exploited in orthogonal geometry. In this geometry, widely spaced parallel shot lines are perpendicular to widely spaced parallel receiver lines (see Figure 2.1b). This is typical land geometry, allowing 3-D coverage with a minimum of field effort. There are numerous variations on this theme, with brick-wall geometry (Wright and Young, 1996) and cross-spread geometry (Dickinson et al., 1990) as the two main implementations. In brick-wall geometry, staggered shot lines are used; in cross-spread geometry, the shot lines are sampled more or less regularly. Orthogonal geometry may also be used for marine data acquisition using ocean-bottom cables.

Zigzag geometry is another land geometry, but now the shots are fired along zigzag lines between the receiver lines. Zigzags between adjacent pairs of receiver lines are arranged such that two sets of parallel shot lines are obtained eventually (see Figure 2.1c). The zigzag geometry is very efficient for data acquisition in deserts (Onderwaater et al., 1996; Wams and Rozemond, 1997).

A special case of the zigzag geometry is the double zigzag (Onderwaater et al., 1996; Wams and Rozemond, 1997). In this geometry, two zigzags (both with the same zigzag period) are traversed instead of one. The second zigzag is separated from the first zigzag by one quarter of the zigzag period to produce an optimal offset distribution (in a four-line geometry with line spacing four times station interval). The advantage of this geometry is the much better stack response as compared to the single zigzag geometry. More recently, the triple zigzag has been introduced (Al-Mahrooqi et al., 2000). It is still a four-line geometry but now with line spacing six times station interval.

Another recent addition to the family of zigzag geometries is the inverted zigzag in which the role of sources and receivers is interchanged (Lansley et al., 2000). This geometry was used to reduce the number of shots relative to the number of receiver stations.

Apart from the three main types of line geometries, the seisloop method (Ritchie, 1991)—an early attempt at cost-effective 3-D land acquisition—may also be mentioned. In this geometry areal midpoint coverage is reached by distributing shots and receivers along a closed loop of (curved) lines as, for instance, provided by a road system.

An example of random geometry is described in Bertelli et al. (1993), where it is applied in an area surrounding the city of Milan, Italy.

2.3 The continuous wavefield

In the literature dealing with migration and inversion (e.g., Beylkin et al., 1985; Cohen et al., 1986; Schleicher et al., 1993), it is often tacitly assumed that the seismic wavefield is a continuous function of its temporal and spatial variables. The assumption of continuity, of course, is justified for the wavefield generated by a single source (apart from near-field discontinuities in case of dynamite as a source). The justification of the assumption of continuity as a function of source coordinates is based on an idealized world in which there are no source wavelet variations. In the following, I also assume that $W(t, x_s, y_s, x_r, y_r)$ can be considered as a continuous function of its variables.

This section deals with the properties of this continuous wavefield to establish requirements for proper sampling. In the acquisition of seismic data, the 5-D wavefield $W(t, x_s, y_s, x_r, y_r)$ is sampled at individual source and receiver locations. The assumption of continuity means that small shifts in source or receiver position would lead to only small changes in the wavefield. Proper sampling of the continuous wavefield allows full reconstruction of that wavefield.

2.3.1 The shot/receiver and midpoint/offset coordinate systems

As in the 2-D case discussed in Vermeer (1990) and in Chapter 1, we can express the wavefield not only in the shot and receiver coordinates, but also in the midpoint and offset coordinates. It is often convenient to use half-offset rather than offset. The midpoint and half-offset coordinates ($\mathbf{x}_m$, $\mathbf{h}$) can be expressed in the shot/receiver coordinates ($\mathbf{x}_s$, $\mathbf{x}_r$)

$$\mathbf{x}_m = (\mathbf{x}_r + \mathbf{x}_s)/2$$
$$\mathbf{h} = (\mathbf{x}_r - \mathbf{x}_s)/2 \qquad (2.1)$$

in which vector notation is used for each coordinate pair. The orientation of $\mathbf{h}$ (h_x, h_y) describes the shot-to-receiver azimuth, whereas h_x and h_y describe what are also called inline half-offset and crossline half-offset, respectively. The offset vector $\mathbf{h}$ plays an important role in the selection of subsets discussed in Section 2.5.

2.3.2 3-D subsets of 5-D wavefield

It is interesting to consider various 3-D subsets (cross-sections) of the 5-D prestack wavefield. In these subsets, we keep the temporal coordinate together with two spatial coordinates. For instance, in case the two varying spatial coordinates are x_r and y_r, then the subset corresponds to a single shot. It turns out that (except for random geometry) each of the acquisition geometries introduced in Section 2.2 has its own subsets. I call these subsets *basic subsets* of the geometry. Table 2.1 lists the most important basic subsets. In a common-offset-vector (COV)-gather, the offset vector (X, Y) is the same for each trace (inline offset X, crossline offset Y); it is also called common-offset gather with constant azimuth. Note that with the description of shot and receiver coordinates, it is assumed implicitly that each subset is a continuous function of its variables. Figure 2.2 illustrates how the various subsets can be constructed, keeping two coordinates fixed, while allowing two other coordinates to vary.

An areal geometry is either a collection of single-fold 3-D common-shot gathers or a collection of single-fold 3-D common-receiver gathers. For the time being, we assume a continuous areal coverage of receivers for the areal geometry with widely spaced shots and, similarly, a continuous areal coverage of shots for the areal geometry with widely spaced receivers.

Considering each shot line and each receiver line in the line geometries as a continuous coverage of shots and receivers along those lines leads naturally to the basic subsets of the line geometries. A basic subset is formed by all traces that have a shot line and a receiver line in common. For orthogonal geometry, the basic subset is called the cross-spread (also for brick-wall geometry, see next section). In zigzag geometry, we have zig- and zag-spreads (because of the two orthogonal families of shot lines), and in parallel geometry the combination of a shot line and a receiver line is just the midpoint line. In the ideal parallel geometry, the COV gather is another 3-D subset. Figure 2.3 schematically illustrates the subsets of these line geometries.

All basic subsets are also single-fold, except the midpoint line. The midpoint line does not provide areal coverage, whereas the other subsets do. The number of overlapping single-fold subsets at any point determines the fold-of-coverage in that point (see also Section 2.5.2).

Because each subset is generated in its own specific way, each subset will see the same subsurface structure in a different way. This is illustrated in Figure 2.4 for a diffractor and for a dipping plane in a constant-velocity medium. The traveltime contours are shown for a 3-D common-shot gather, a cross-spread, a zig-spread, and for a COV gather. The contours are displayed as a function of the (x, y)-coordinates of the midpoints. The traveltime surfaces in Figure 2.4a are all versions of Cheops pyramid (Claerbout, 1985), but each one is computed in a different 3-D subspace of the 5-D space containing the prestack wavefield. Figure 2.4 illustrates that each subset represents a spatially continuous domain in the 5-D prestack wavefield.

The COV gather (also called COA gather, for common-offset with constant azimuth, Ferber, 1998) covers the whole survey area, whereas the other subsets have a limited extent. The COV gather is therefore better suited for prestack migration than any of the other subsets. As we shall see, however, it is virtually impossible to acquire COV gathers at a reasonable cost. A disadvantage of COV gathers is the single shooting direction. Some subsurface structures can best be illuminated using a wide range of azimuths (cf. O'Connell et al., 1993; Reilly, 1995).

All single-fold subsets mentioned in Table 2.1 lend themselves to true-amplitude 3-D prestack migration. In fact, various authors dealing with prestack migration implicitly or explicitly assume a 3-D single-fold subset and derive formulas for the migration of such data sets (Beylkin et al., 1985; Cohen et al., 1986; Schleicher et al., 1993; Vermeer, 1995). The subsets are also suitable for imaging with dip moveout (DMO) (Vermeer et al., 1995; Pleshkevitch, 1996; Collins, 1997; Padhi and Holley, 1997). Padhi and Holley (1997) named those single-fold subsets "minimal data sets" (MDSs, i.e., data sets minimally required for imaging). This general suitability for imaging of the various basic subsets suggests that their sampling must get due attention.

Before discussing sampling, however, it will be helpful to first discuss the subsets of orthogonal geometry and zigzag geometry in some more detail.

2.3.3 The cross-spread

Orthogonal geometry consists of more or less straight acquisition lines, which may be widely spaced. In the field, the data are acquired according to templates, which may consist of a series of shots (sometimes called a shot salvo) shooting center-spread into the active receivers of an even number of receiver lines (see left part of Figure 2.5). Other template implementations are discussed in Section 4.6. Cross-spreads can be extracted from the orthogonal geometry by collecting all traces that have a shot line and a receiver line in common. Hence, there are as many cross-spreads as there are intersections between shot lines and receiver lines.

The right part of Figure 2.5 highlights the shots and receivers corresponding to one cross-spread in an orthogonal 3-D survey. The gray square indicates the midpoint area of the cross-spread. The maximum inline offset of this geometry is equal to the length of the receiver spread divided by two, and the maximum crossline offset is given by the length of the shot spread divided by two. The ratio of these two lengths (crossline/inline) determines the aspect ratio of the cross-spread, which is the same as the aspect ratio of the template. (Note that for this to be true, the crossline dimension of the template has to be taken as $N * d$, where N is number of receiver lines, and d is the interval between receiver lines; compare the discussion of array length in Section 1.6.) Figure 2.5 represents a wide acquisition geometry with an aspect ratio 1. In a narrow geometry the aspect ratio may be as low as 0.2 or even lower. It is interesting to note that this gathering of cross-spreads from orthogonal geometry is the subject of a patent (Thomas, 2000).

Figure 2.6 illustrates some of the properties of the cross-spread. The trace at midpoint M is a member of a common-shot gather, a common-receiver gather, a common-offset gather, and a common-azimuth gather. The midpoints of the common-offset gather form a circle; therefore, horizontal layers show up as circles in the time slices of a cross-spread (Figure 2.7). The midpoints of a common-azimuth gather lie along a straight line through the origin of the cross-spread.

Each trace in the 3-D survey is an element of a unique cross-spread. The neighbors of the trace in the cross-spread have been shot by the same or by adjacent shots, and have been recorded by the same or by adjacent receivers. In other words, the spatial attributes of the traces around M vary slowly, making the cross-spread a spatially continuous data set. On the other hand, the maximum useful offset limits the extent of each cross-spread (in a time-variant way), and the edges of the cross-spreads form spatial discontinuities in orthogonal geometry.

It is interesting to compare the spatial continuity of the cross-spread with that of the template used in the field (see Figure 2.5). The midpoint area acquired by the template consists of small strips, each strip corresponding to the shot salvo shooting into one of the receiver lines. Inside these strips, there is also spatial continuity, but the

Table 2.1. Basic subsets of various 3-D geometries in 5-D prestack wavefield.

Basic subset	Source coordinates*	Receiver coordinates*	Acquisition geometry
Midpoint line	(x_s, Y_1)	(x_r, Y_2)	Parallel
COV gather	(x_s, y_s)	$(x_s + X, y_s + Y)$	Parallel
3-D shot	(X, Y)	(x_r, y_r)	Areal
3-D receiver	(x_s, y_s)	(X, Y)	Areal
Cross-spread	(X, y_s)	(x_r, Y)	Orthogonal
Slanted spread	$(x_s, Y + (x_s - X)\tan\alpha)$	(x_r, Y)	Slanted
Zig-spread	$(x_s, Y + x_s - X)$	(x_r, Y)	Zigzag

*X and Y are fixed, lower case coordinates vary.

edges of each strip form spatial discontinuities because of the jump from one receiver line to the next. The midpoint area of the cross-spread has the same size as that of the template, but it has no internal discontinuities.

If staggered shot lines are used (as in brick-wall geometry), the shot lines are only partially sampled, leading to cross-spreads that are split into a number of strips (the same strips as in the template). The number of edges in this geometry is much larger than in the continuous shot-line geometry; spatial continuity in this geometry is therefore degraded (see also Section 7.3).

2.3.4 Subsets of zigzag geometry

In the field, zigzag geometry is acquired by zigzagging (at 45° angles with the receiver lines) with the sources between two adjacent receiver lines. A swath may consist of four or more receiver lines listening to each shot. The maximum crossline offset is equal to $N/2 + 1$ receiver-line intervals, where N is the number of active receiver lines. Figure 2.8 shows how the pattern of source trajectories can be arranged such that all zig parts form continuous lines across the receiver lines, whereas the zag parts form another set of straight shot lines. In this arrangement, zig-spreads as well as zag-spreads can be gathered from the recorded data. The maximum crossline offset in the zig- and zag-spreads is equal to the maximum crossline offset in the swath.

Normally, each shot is recorded center-spread, which means that the active receivers move with each shot. As a consequence, the number of traces in a common-receiver gather (in a zig- or zag-spread) is not constant,

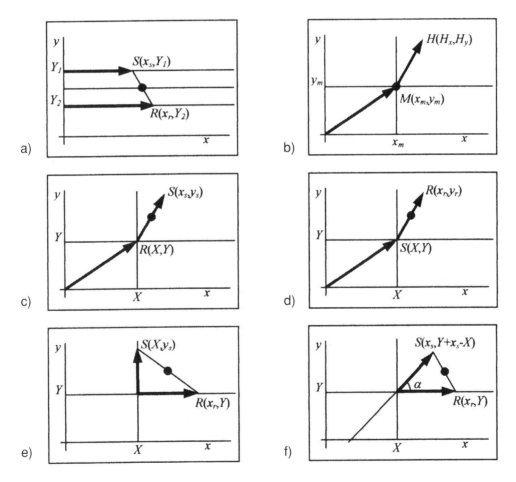

FIG. 2.2. Generating 3-D subsets of the 5-D prestack wavefield: (a) midpoint line, (b) common-offset-vector gather, (c) 3-D receiver, (d) 3-D shot, (e) cross-spread, (f) zig-spread (tan α = 1), or slanted spread (usually, tan α = 1 or 2). X, Y are fixed, whereas lower case coordinates vary in the subset; S is shot, R is receiver. Midpoint position is indicated by a black circle. The midpoint line is a multifold subset, because many shot/receiver combinations may share the same midpoint. With the exception of the midpoint line, the midpoints of a subset can occupy any position in (x,y), and each midpoint corresponds to a unique combination of shot and receiver, i.e., all other subsets are single-fold data sets.

24   Chapter 2   3-D symmetric sampling

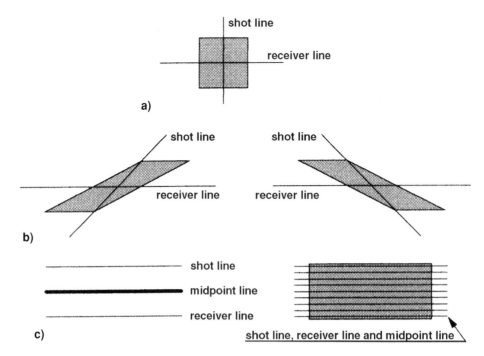

FIG. 2.3. Basic subsets of line geometries: (a) orthogonal, (b) zigzag, and (c) parallel. Shaded areas indicate the midpoint areas of the subsets. The basic subset of orthogonal geometry is the cross-spread. Zigzag geometry can be decomposed into subsets consisting of zig- and zag-spreads. Parallel geometry has two possible basic subsets: the midpoint line (left) or the common-offset-vector gather (right). The latter may be acquired using repeated 2-D surveys.

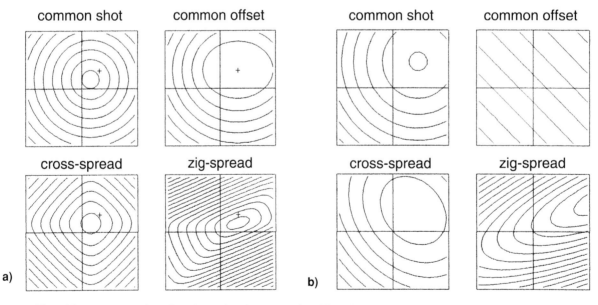

FIG. 2.4. Traveltime contours in 3-D subsets for the case of a diffraction (a) and a dipping plane (b). The contours are displayed as a function of the (x,y)-coordinates of the midpoints. The position of the diffractor at (500,500,500) is indicated by the symbol +.

but the number of traces in a common-inline-offset gather is. Current practice is to move the shots in the inline direction over a distance equal to the receiver station interval. This leads to a shot interval that is the square root of two times the receiver interval. In this geometry, the acquired offsets are the same as in an orthogonal geometry with the same spread length and the same number of receiver lines, but the offset distribution is different.

Figure 2.9 illustrates some properties of the zig-spread. Any trace in this spread is a member of a common-shot gather, a common-receiver gather (parallel to the shot line), a common-inline-offset gather (parallel to the edges of the zig-spread), a common-offset gather, and a common-azimuth gather (see also Figure 4.1b). Note that the midpoints of the common-offset gather now form an ellipse. If the maximum crossline offset equals the maximum inline offset, the corresponding offset ellipse will touch all four edges of the midpoint area of the zig-spread.

2.4 3-D symmetric sampling

Symmetric sampling was first introduced for 2-D lines in my book *Seismic Wavefield Sampling* (Vermeer, 1990). Some main points of the book are discussed in Chapter 1. In this section, 3-D symmetric sampling is introduced. It turns out that 2-D symmetric sampling is just a special case of the more general case of 3-D symmetric sampling.

One approach to 3-D survey design (e.g., mega-bin survey technique, Goodway and Ragan, 1997) attempts to sample all four spatial coordinates of the 5-D prestack wavefield as well as possible. Because of the high cost of dense sampling, this objective leads to coarse sampling of the four spatial coordinates with ensuing difficulties in the application of spatial filters and prestack migration. Alias-free sampling of the whole 5-D prestack wavefield is clearly too expensive. Often, it is also impractical, since it requires free access to the whole survey area. Instead, in the 3-D symmetric sampling approach, we attempt to properly sample the single-fold subsets of the chosen areal or line geometry. If we succeed in that more modest objective, the continuous wavefield of the subset underlying the samples can be reconstructed fully. This more modest aim is achieved by dense enough sampling of the varying coordinates in each subset (cf. Figure 2.2 and Table 2.1). Usually, sampling of a subset will provide a single-fold data set (except in the case of sampling the midpoint line). To achieve M-fold data, M subsets need be overlapping in the full-fold area of the survey.

Of all basic subsets listed in Table 2.1, only the COV gather may extend across the whole survey area. All other single-fold basic subsets have limited areal extent in practice, because offset increases toward the edges in those subsets (cf. Figure 2.6 for cross-spread) and the

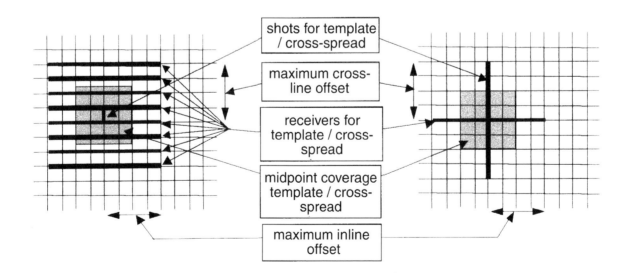

FIG. 2.5. The same orthogonal geometry with template (left) and with cross-spread (right). Horizontal lines are reveiver lines; vertical lines are source lines. The template represents the way in which the data are acquired in the field; in this case there are eight receiver lines with a number of shots in the center of the template. The cross-spread gathers all data for receivers along the same receiver line that have listened to a range of shots along the same source line.

26   Chapter 2   3-D symmetric sampling

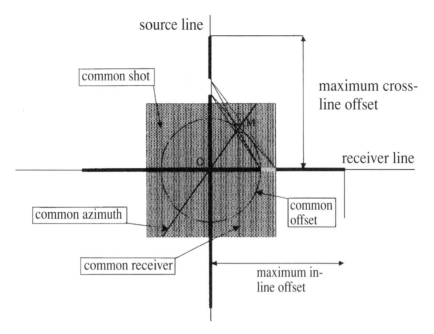

FIG. 2.6. Properties of cross-spread. The half-offset of a trace at M equals the distance to the center O of the cross-spread (i.e., traces with same offset lie on a circle). The trace is both part of a common-shot gather (horizontal through M) and part of a common-receiver gather (vertical through M). All traces close to M correspond to neighboring shots on the source line and to neighboring receivers on the receiver line.

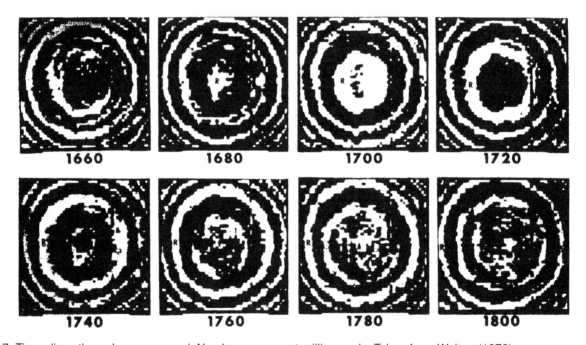

FIG. 2.7. Time slices through cross-spread. Numbers represent milliseconds. Taken from Walton (1972).

target depth has a maximum useful shot-to-receiver offset. Often the extent of those subsets is maximized in only one spatial direction. A large extent in both spatial directions would fully exploit the potential of each geometry. Therefore, besides alias-free sampling of the basic subsets, we should maximize the (useful) areal extent of the subsets with limited extent. This prescription maximizes the spatial continuity in the 3-D survey and, for a given fold, minimizes the number of edges in the survey.

Together, alias-free sampling of the basic subsets and maximizing the extent of each subset form a generic prescription of 3-D symmetric sampling.

The requirements of 2-D symmetric sampling—equal shot and receiver sampling intervals, and equal shot and receiver arrays—apply without change to the sampling of the subsets of the various 3-D line geometries. However, apart from the 2-D symmetric sampling criteria, each 3-D line geometry needs some additional criteria to fully satisfy 3-D symmetric sampling requirements. Areal geometry has its own requirements to satisfy the prescription of 3-D symmetric sampling. This extension to 3-D is discussed in the following sections.

2.4.1 Areal geometry

In areal geometry, the basic subsets are either 3-D common-shot gathers acquired with widely spaced shots or 3-D common-receiver gathers recorded with widely spaced receivers. Alias-free sampling of 3-D common-shot gathers requires that receivers be sampled at the basic sampling interval in x as well as in y (see Section 1.3 for definition of basic sampling interval and basic signal sampling interval).

On land, the basic sampling interval is usually so small that sampling at that interval becomes prohibitively expensive. An alternative to this very fine sampling is to use coarser receiver-station intervals, where alias protection is provided by areal geophone arrays. But this would mean that the whole survey area still has to be covered with geophones. A practical alternative to plastering the area with areal geophone arrays might be the use of an areal shot array (with the same dimensions as would be required for the areal geophone arrays). Even though the effect of a single areal shot array is not identical to that of areal geophone arrays, it might come close. Another alternative is to use deep shot holes so that hardly any ground roll is generated, leading to a larger basic sampling interval. But even then, the areal geometry is very labor-intensive, making it much more expensive than an equally satisfactory orthogonal geometry.

For deep-water acquisition, the basic sampling interval is equal to the basic signal sampling interval. In that environment, areal arrays are not needed to suppress unwanted coherent energy. Moreover, covering the survey area with closely spaced shots need not be prohibitively expensive, so that the recording of 3-D common-receiver gathers using a grid of widely spaced stationary receivers might be affordable.

The areal geometry can be implemented most efficiently using a hexagonal distribution of sources and receivers. In this sampling the sample points are chosen at the vertexes of equilateral triangles (Figure 2.10). Hexagonal sampling of a 2-D function of which the wavenumber spectrum is limited by a circle requires fewer samples than rectangular sampling (Petersen and Middleton, 1962). It leads to a reduction of 13.4% in the number of required source points in the areal geometry

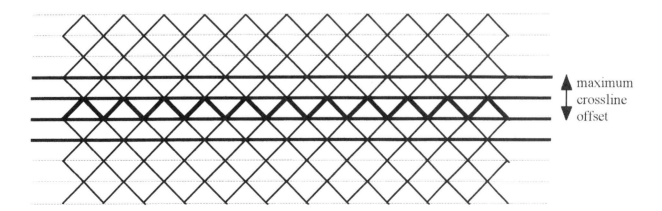

FIG. 2.8. Zigzag geometry. The sources (usually vibrators) follow zigzag line (heavy line in the figure) between two adjacent receiver lines, while (in this case) four spreads of receivers record each shot. A zig-spread can be gathered by taking data acquired with adjacent swaths. Four zig segments make up the shot line in the zig-spread.

(Bardan, 1996). Similarly, a hexagonal arrangement of the receivers allows a lower density of receivers for the same "largest minimum offset." Another advantage of this geometry is that the shape of the subsets can be arranged to be hexagonal, allowing a better distribution of the long offsets over azimuth. More efficient signal processing operators can be designed on basis of a hexagonal grid (Mersereau, 1979).

2.4.2 Line geometries

Alias-free sampling of the 3-D subsets of line geometries requires sampling of shots and receivers along their respective acquisition lines using the basic sampling interval. Again, arrays can be used as antialias filters and resampling operators to allow sampling at the basic signal sampling interval. Linear arrays along the acquisition lines are sufficient to take care of the problem of aliasing noise with low apparent velocities. However, if needed, noise suppression can be improved by using areal shot and/or receiver arrays (see Section 3.3).

2.4.2.1 Parallel geometry

In parallel geometry, it is not sufficient for the midpoint line to be sampled without aliasing; the distance between the midpoint lines also has to be considered. If that distance is small enough, the COV gathers can also be sampled alias free in both spatial dimensions. Repeated acquisition, at small intervals, of 2-D lines produces the ideal parallel geometry (for each midpoint line, its shot line and its receiver line are collinear; i.e., $Y_1 = Y_2$ in Figure 2.2a). In case of center-spread acquisition, the COV gather is properly sampled straightaway. In marine acquisition with end-on shooting and equal shot and receiver intervals, the odd/even signal pattern (checkerboarding) in the midpoints can be remedied by de-aliasing of the common-offset gathers (by interpolation in the common-shot and the common-receiver gathers; Vermeer, 1990).

Another—quite hypothetical—way of acquiring properly sampled COV gathers is to have a constant (nonzero) crossline offset between the source track and the receiver line. Moving this arrangement for the next midpoint line over a small distance (half the basic sampling interval) also leads to well-sampled COV gathers. In this setup, each COV gather would have its own shot-to-receiver azimuth.

In marine streamer acquisition, parallel geometry is more or less the rule. Unfortunately, with this, 3-D symmetric sampling is far from the rule. The first marine 3-D surveys were often shot as a series of 2-D lines, which often satisfied the 2-D symmetric sampling criteria; but these surveys used too large line spacings, consequently requiring later reshoots. Modern streamer acquisition uses multisource multistreamer configurations. Though common–inline-offset gathers can be extracted from such surveys, the crossline offset varies between midpoint lines (see also Section 5.3.2). These geometries lead to irregular subsurface illumination, even if the surface sampling is regular (Beasley and Mobley, 1995; Beasley, 1996).

The potential irregularity of subsurface sampling is illustrated in Figure 2.11, which shows illumination pat-

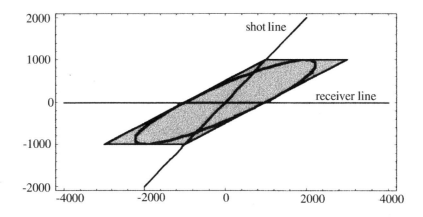

FIG. 2.9. Zig-spread with equal maximum inline and maximum crossline offsets. Note that the receiver spread moves with the shot, ensuring center-spread acquisition for all shots. The rhomboid gray area is the midpoint area of the zig-spread. Horizontal lines represent common shots; oblique lines parallel to the edges of the zig-spread represent common-inline offsets. The ellipse inscribed within the rhomboid represents midpoints with offset equal to the maximum inline offset.

terns of various multisource, multistreamer configurations for a plane dipping layer in a constant-velocity medium. Each graphic consists of the (x, y)-coordinates of the reflection points for 24 adjacent midpoints in a crossline of the geometry. Each vertical or near-vertical line in Figure 2.11 connects the coordinates of the reflection points corresponding to one midpoint. A horizontal or near-horizontal line connects the reflection points corresponding to the same long offset. The shape of the reflection point trajectories can be understood if one realizes that the reflection point moves updip, that is, toward the source when shooting downdip (sailing updip) and away from the source when shooting updip. Note that the crossline shift of the reflection points is largest for the long offsets, even though the azimuth variation is smallest for the long offsets.

In these multisource multistreamer configurations, the shortest offsets sample the subsurface in a regular way, but the longer offsets sample the subsurface irregularly, the irregularity increasing with the range of crossline offsets. The irregularity also depends on the inline dip: the larger this dip, the more irregularly will the reflector be illuminated. Only the single-source single-streamer geometry samples the subsurface in a regular way. Another reason why properly sampled subsets are not obtained in streamer acquisition is differential feathering between successive midpoint lines or boat passes. This causes even more variation of shot-to-receiver azimuth in the 3-D common-offset gathers.

Figure 2.12 shows illumination patterns of the same geometries as Figure 2.11, but now including random feathering between boat passes and assuming constant feather within a boat pass. In this case, even the single/single geometry fails to illuminate the subsurface in a regular way.

It may be noted that feathering turns the midpoint line into a midpoint area that has basically single-fold coverage. Owing to differential feathering, however, these midpoint areas (single-fold subsets of the "feather" geometry) do not overlap in a regular way. Normally, the feathering is not large enough to permit prestack migration of individual midpoint areas (i.e., these midpoint areas do not qualify as MDSs; Padhi and Holley, 1997). Only if the feathering could be made constant across the whole survey would the single-source single-streamer geometry again be ideally suited for prestack migration. In that case, each 3-D common-offset gather would have constant azimuth.

In the inline direction, the variation in illumination caused by both multisource, multistreamer acquisition and differential feathering is far less than in the crossline direction, leading to striping of the amplitudes seen in horizon slices. Various techniques have been proposed to correct for these irregularities (e.g., Beasley and Klotz, 1992; Gardner and Canning, 1994; Huard and Spitz, 1997; Albertin et al., 1999), but a fully satisfactory solution seems to be impossible [an improvement to the technique proposed in Albertin et al. (1999) is suggested in Section 10.7]. A better solution would be to prevent differential feathering by application of steerable streamers (Bittleston et al., 2000) or by using stationary-receiver systems (see Chapter 5).

2.4.2.2 Orthogonal geometry

Besides equal shot and receiver intervals and equal shot and receiver arrays, 3-D symmetric sampling of orthogonal geometry also requires as many receivers in the common shot as shots in the common receiver, and the center-spread acquisition of both shots and receivers. This recipe ensures the acquisition of square cross-spreads (the aspect ratio of the geometry equals one, as in Figure 2.5). The shot-line interval and the receiver-line interval should preferably also be the same for symmetric sampling. However, allowing some difference in shot- and receiver-line density in the case where shots and receivers differ in cost, would be quite acceptable generally.

Figure 2.13a illustrates that a linear geophone array mixes (mixing is filtering with positive filter coefficients only, often with equal coefficients) midpoints in a common-shot gather, thereby reducing the aliasing in that gather. Figure 2.13b illustrates what happens when a linear shot array (along the shot line) is introduced as well: it reduces aliasing in the common-receiver gather. Together, the linear shot and receiver arrays ensure sampling of the whole cross-spread with minimal aliasing. It

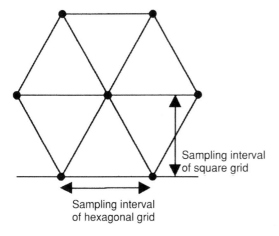

FIG. 2.10. The hexagonal sampling interval is $2/\sqrt{3}$ times the corresponding square sampling interval.

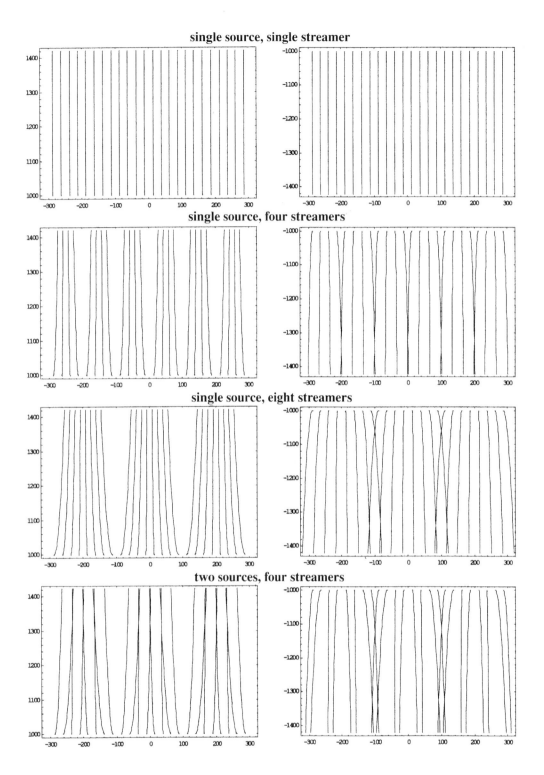

FIG. 2.11. Nominal illumination patterns of multisource, multistreamer configurations in the case of downdip shooting (left column) and updip shooting (right column). Each vertical or near-vertical line connects the (x,y)-coordinates of the reflection points as seen by one of 24 midpoints. The 24 midpoints are adjacent in the crossline direction. In every case, a reflector with 30° dip is illuminated in a constant-velocity medium. Depth of reflector is 2309 m in $y = 0$. Maximum inline offset is 3000 m. The short offsets sample the reflector regularly (at $y = 1000$ m and $y = -1000$ m in the left column and right column, respectively). Note the irregular sampling of the long offsets in the crossline direction in all cases except for the single-source, single-streamer configuration.

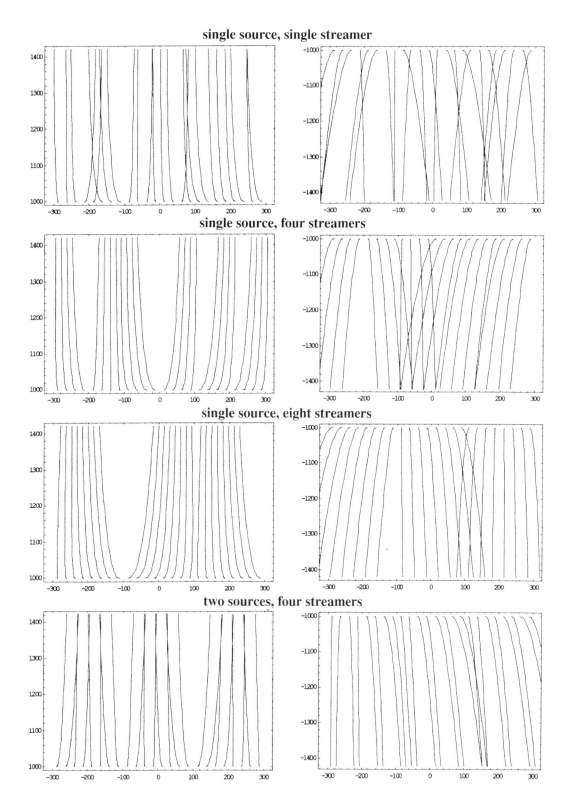

FIG. 2.12. Illumination patterns of multisource, multistreamer configurations with feathering. Each vertical or near-vertical line connects the (x,y)-coordinates of the reflection points as seen by one of 24 midpoints. For each boat pass, a constant feathering angle was randomly selected from a uniform distribution between $-2.5°$ and $2.5°$. Otherwise, the acquisition geometries and subsurface are the same as in Figure 2.11. Note the dramatic departure from regularity for the single-source, single-streamer configuration.

should be realized that shot arrays are as important as geophone arrays; since geophone arrays will not prevent aliasing in the common-receiver gather, they are fully complementary (see also Smith, 1997, and Section 3.3).

In addition to serving as antialias filters and resampling operators, arrays also serve to suppress noise, such as ground roll. The first arrival of the ground roll has the shape of a cone centered on the center of the cross-spread. A common-shot cross-section through this cone has the shape of a hyperbola. The ground roll near the apex of the hyperbola will not be suppressed by the receiver arrays. This flat part of the hyperbola is centered on the shot line. The common-receiver gathers, however, cut the same part of the ground-roll cone at much larger angles. Hence, in that area, the shot arrays will suppress the ground roll. The same reasoning can be applied with shots and receivers interchanged. In other words, in the cross-spread, shot and receiver arrays are fully complementary with respect to ground-roll suppression. In those places where the shot array is less effective in suppressing ground-roll energy, the receiver array is at its best, and vice versa. If the noise is very strong, noise suppression may be improved by using areal rather than linear arrays.

In areas where shots are particularly expensive, areal receiver arrays may be considered in combination with single shots. At least for first-arrival ground roll in a homogeneous medium, the action of an N-element shot array convolved with an M-element receiver array is identical to the action of an $N \times M$-element receiver array convolved with a single shot (apart from shot strength effects). For noise traveling in other directions—backscattered noise and side-scattered noise—the response would be different. Theoretically, an areal receiver array would not protect as much against aliasing in the common-receiver gather as would the combination of a linear shot array and a linear receiver array. As far as noise suppression is concerned, however, the areal array has a small advantage: it will always suppress energy with slow apparent velocity, irrespective of the traveling direction of the energy. Therefore, if an areal geophone array is cheaper than the combination of linear shot and receiver arrays, such a departure from symmetric sampling might be the best option (see further Section 3.3).

The case for center-spread acquisition and equal maximum crossline offset and maximum inline offset is supported strongly by the time slices of a square cross-spread shown in Figure 2.14. In the top time slices in Figure 2.14, the traveltime contours are circular, corresponding to reflections from horizontal layering, whereas in the bottom time slices in Figure 2.14, the traveltime contours are elliptical, corresponding to plane-dipping reflectors. Note the similarity of these latter contours to those for the cross-spread in Figure 2.4b. If the maximum crossline offset were much smaller than the maximum inline offset, the spatial continuity of the cross-spread would not be fully exploited. Figure 2.14 also illustrates the need for equal shot and receiver intervals; the wavefield clearly behaves in the same way in both spatial directions. Doubling the shot interval would cause aliasing in the common-receiver gathers and, hence, largely hamper the usefulness of k- or (f, k)- filters in that domain.

Even though cross-spreads have limited extent, it is possible to create single-fold coverage across the whole survey by a tiling of adjacent cross-spreads. In such a single-fold gather, the data would be piecewise continuous with discontinuities between the adjacent cross-spreads. Figure 2.15 shows the illumination by four adjacent cross-spreads of a reflector with 15° dip and a reflector with 45° dip. Each cross-spread covers the reflector with its own "blanket." Around the edges of these blankets gaps and overlaps exist. Within each blanket, illumination can be considered as continuous (provided the cross-spread is sampled alias free), but illumination is discontinuous across the edge of each blanket (see also Section 10.5).

A choice for 3-D symmetric sampling has significant consequences for the distribution of offsets over the offset range. This is illustrated in Figure 2.16, which compares the offset distributions for narrow and wide (approximately symmetric) geometries. The top two graphs show a comparison where the area of the cross-spread of the wide geometry was limited by the number of available channels. In this comparison the narrow geometry builds up fold fastest for shallow levels (short offsets), whereas at deeper levels the fold of the wide geometry is larger. The middle two graphs compare geometries where the length of the receiver lines is the same. The wide geometry has 12 active receiver lines compared to six for the narrow geometry, and the line intervals are twice as large in the wide geometry as in the narrow geometry. The bottom two graphs show perhaps the most realistic comparison: fold and line intervals are kept the same. It is interesting to note that in this case fold build-up is fastest for the wide geometry, whereas in the comparison with equal receiver spread lengths, the fold builds up fastest for the narrow geometry.

Note that in all situations, the offset density starts building up linearly as a function of offset. This linearity stops when the minimum of the maximum inline offset and maximum crossline offset is reached. Hence long offsets dominate in wide geometry. The preponderance of long offsets in wide geometry gives greater weight to the long offsets than to the short offsets, leading to better suppression of multiples with a small differential moveout (cf. Section 3.4.3.1). On the other hand, resolution suffers from the NMO stretch effect associated with long offsets. This is an important dilemma to be solved in 3-D survey design (see Section 4.4).

2.4.2.3 Zigzag geometry

Alias-free sampling of the zig- and zag-spreads would require that the spacing of the traces in the common-receiver gather be the same as the trace spacing in the common-shot gather. Similarly to other geometries, this requirement would mean equal shot and receiver intervals, the shot interval being measured along the shot line.

As mentioned before, in actual practice the shot interval is the receiver interval times the square root of two. This means that alias-free sampling of the common-receiver gathers would require oversampling of the common-shot gathers. The zig- and zag-spreads (cf. Figures 2.9 and 4.1b) have a constant number of traces N in the common-inline–offset gather, whereas the number of traces in the common-receiver gathers varies from one to N.

The maximum useful extent of the zig-spread is reached if the offset ellipse of the maximum useful offset touches the edges of the zig-spread as shown in Figure 2.9. In that case, the maximum crossline offset equals the maximum inline offset.

Zigzag geometry is particularly efficient in a desert environment surveyed with vibrators. The distance to be traveled by the vibrators is a factor square root of two shorter than in an equivalent orthogonal geometry, and it is much easier to avoid driving over geophones, because no sharp turns have to be made. These considerations only apply in case the vibrators have to stay between two adjacent receiver lines. A full-swath roll approach to acquisition (see Section 4.6.4) can make orthogonal geometry acquisition very efficient.

2.5 Pseudominimal data sets
2.5.1 Introduction

Minimal data sets of crossed-array geometry and of areal geometry have limited extent. Yet, for quite a few processing steps, it would be helpful to avail of MDSs that extend across the whole survey area. Since these do not exist in those geometries, we have to look for *pseudo-minimal data sets* (pMDSs), which can be constructed from the available data, extend across the whole survey area, and which are as close as possible to an MDS.

The only type of MDS, which extends across the whole survey area is the COV gather. This is the MDS of the ideal parallel geometry and is never acquired in practice. Yet, also for parallel geometry it is useful to avail of single-fold data sets which extend across the whole survey area and are as close as possible to a true MDS.

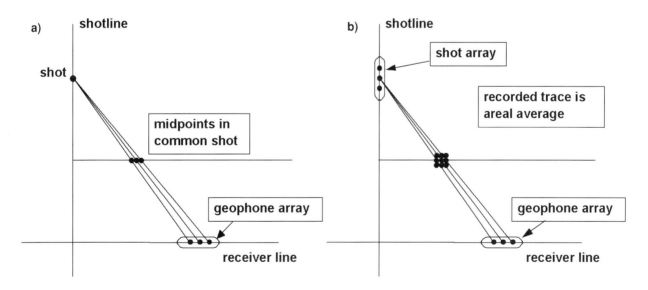

FIG. 2.13. Antialiasing by geophone array, alone (a) and in combination with shot array (b). A geophone array reduces aliasing in a common-shot gather, whereas a shot array reduces aliasing in a common-receiver gather. Together, they take care of reduced aliasing in the cross-spread. To avoid clutter, only three of the nine contributing shot/receiver segments have been drawn in (b).

34   Chapter 2   3-D symmetric sampling

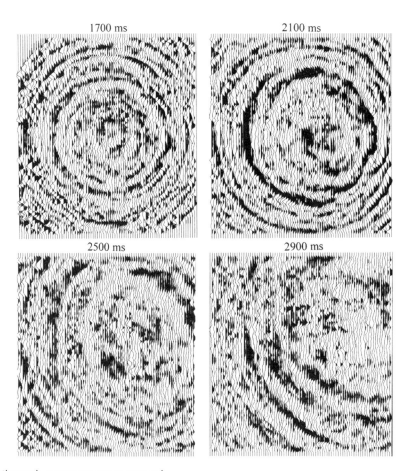

FIG. 2.14. Time slices through a square cross-spread.

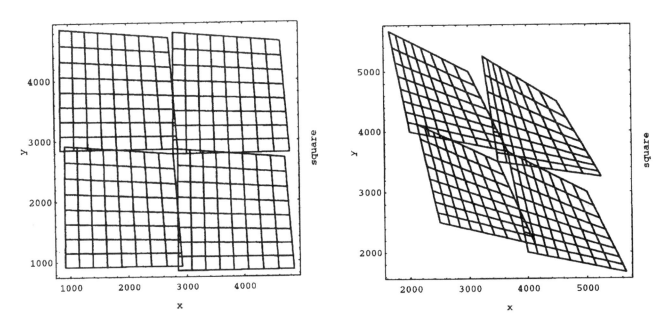

FIG. 2.15. Illumination of 15° (a) and 45° (b) dipping reflectors by four adjacent cross-spreads. Note that illumination fold can be considered a continuous function inside the cross-spreads, whereas it is discontinuous across the edges of the cross-spreads.

A common feature of all MDSs is that two of the four spatial coordinates in $W(t, x_s, y_s, x_r, y_r)$ or in $W(t, x_m, y_m, h_x, h_y)$ are well-sampled in such data sets. In the MDS that extends across the whole survey area, the COV gather, these spatial coordinates are x_m and y_m. The other two coordinates h_x and h_y are fixed. Similarly, for a single-fold data set to extend across the whole survey area, x_m and y_m must be well-sampled as well. Hence, to establish data sets that are suitable as pMDS, h_x and h_y should vary as little as possible.

A logical way of constructing pMDSs from marine multisource multistreamer data is to collect all data recorded with the same channel, i.e., with the same h_x. The crossline offset h_y would vary depending on the width of the geometry. However, this data set would not yet have a trace in each midpoint. Depending on the

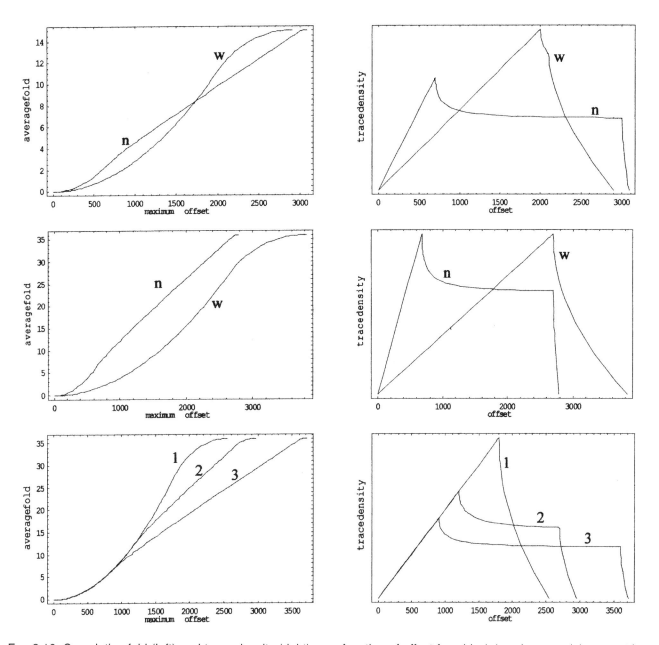

FIG. 2.16. Cumulative fold (left) and trace density (right) as a function of offset for wide (w) and narrow (n) geometries. Maximum fold is 15 in the top graphs and 36 in the middle and bottom graphs. Trace density can be viewed as the derivative of the cumulative fold function shown on the left. Top: Equal number of channels; this is a comparison for the narrow production geometry and wide test geometry discussed in Section 7.3. Middle: Equal maximum inline offset. Bottom: Equal line intervals. Parameters of the geometries are given in Table 2.2.

source interval and the number of sources, a number of consecutive channels has to be combined for complete single-fold coverage. The construction of pMDSs can become considerably more complicated in case of differential feathering.

In the remainder of Section 2.5, I will deal with the construction of pMDSs in orthogonal geometry. It is shown that a plethora of pMDSs may be constructed from regularly sampled orthogonal geometry. With some modification, the construction is also applicable to other crossed-array geometries and to areal geometry. In Section 2.6, processing with pMDSs is discussed.

2.5.2 Building fold with basic subsets

For a better understanding of the various forms of pMDS, it may be helpful to describe first how fold-of-coverage is built in an orthogonal geometry.

Consider the cross-spread in Figure 2.5. The width of the midpoint coverage in the inline (receiver line) direction W_x is

$$W_x = L_R / 2, \qquad (2.2)$$

Coverage M_x equals the number of times the shot-line interval S fits on the width of the inline coverage

$$M_x = W_x / S. \qquad (2.3)$$

Similarly, the width of the midpoint coverage in the crossline (shot-line) direction W_y is

$$W_y = L_S / 2, \qquad (2.4)$$

where L_S is shot spread length, which is the part of the shot line being recorded by the receivers in the receiver spread. The crossline fold M_y equals the number of times the receiver-line interval R fits on the width of the crossline coverage:

$$M_y = W_y / R. \qquad (2.5)$$

Total fold-of-coverage M is

$$M = M_x M_y. \qquad (2.6)$$

The total fold equals the number of overlapping midpoint areas (the gray areas in Figure 2.5) in any point. This is further illustrated in Figure 2.17, where overlapping cross-spreads are shown for a geometry with $M_x = 4$ and $M_y = 2$.

If M_x or M_y are not integer, then the number of traces in the CMPs of the geometry is not the same everywhere. Therefore, for regular fold, it is necessary that $W_x = n\,S$ and $W_y = n\,R$.

In Figure 2.17, coverage is shown for a single unit cell (the dark area in the lower part of the figure). The size of the unit cell equals the area between two adjacent receiver lines and two adjacent shot lines. Figure 2.17 illustrates that for fold M, the area of the cross-spread can be subdivided into M areas with the size of a unit cell.

Section 2.3.1 discussed that, in 3-D, offset can be described by x- and y-components: the inline offset and the crossline offset. Half offset as $\mathbf{h} = (h_x, h_y)$. Therefore, an appropriate name for the unit-cell–sized subareas in the cross-spread is offset-vector tile (OVT). Each OVT is built from a limited range of shots along the shot line and a limited range of receivers along the receiver line (Figure 2.18). These two ranges restrict the range of offset vectors to a small area. Figure 2.19 illustrates the variation of offset and azimuth of the center of each OVT in

Table 2.2. Parameters used in geometry comparisons of Figure 2.16.

Figure 2.16	Geometry	Max inline offset	Max crossline offset	Shot-line interval	Receiver-line interval
Top	Narrow	3000 m	700 m	400 m	350 m
	Wide	2000 m	2100 m	400 m	700 m
Middle	Narrow	2700 m	675 m	225 m	225 m
	Wide	2700 m	2700 m	450 m	450 m
Bottom	1	1800 m	1800 m	300 m	300 m
	2	2700 m	1200 m	300 m	300 m
	3	3600 m	900 m	300 m	300 m

a cross-spread. OVTs are important building blocks for pMDSs.

An OVT can be characterized by four parameters, OVT = OVT (h_x, h_y, Δh_x, Δh_y), where h_x and h_y are the half-offset coordinates of the center of gravity of the OVT, and Δh_x and Δh_y describe the range of half-offsets in x- and y-direction. (In a cross-spread centered coordinate system, h_x and h_y equal the midpoint coordinates: $x_m = h_x$, $y_m = h_y$.) In a cross-spread which is symmetric with respect to both axes (center-spread acquisition for both receiver spread and source spread), each OVT has counterparts in the other three quadrants with the same absolute values of its four parameters. Of these four OVTs, the pairs in opposite quadrants have also opposite, i.e., similar shot-to-receiver azimuths (cf. Figure 2.19).

2.5.3 Fold, illumination, and imaging

In the previous section fold-of-coverage was introduced on the basis of coverage by a receiver spread, not by a series of receiver points. Similarly, total fold-of-coverage is counted by the number of overlapping midpoint areas of the MDSs of the geometry, not by counting the number of traces in a bin. Defined in this way, fold is a piece-wise continuous function of the midpoint coordinates. Discontinuities may exist at the edges of the midpoint areas of individual MDSs. Therefore, to achieve constant fold throughout a survey area, the midpoint area of one MDS must take over where another one stops.

Next to the midpoint area of an MDS, we can define an illumination area (the area on the reflector illuminated by all shot-receiver pairs of the MDS), and an image area (the area on the reflector for which correct imaging is possible). Similar to the definition of fold-of-coverage we may define

"illumination fold": number of overlapping illumination areas, and

"image fold": number of overlapping image areas.

For *P*-wave acquisition, illumination fold will in general not be very different from fold-of-coverage, though it may be locally higher or lower. Image fold is the same as illumination fold, if we neglect edge effects. Fold-of-coverage (in the case of stacking) or image fold (in the case of prestack migration) provides a statistical means of suppressing noise. If the data are properly sampled and do not show spatial discontinuities, fold is not necessary to improve the migration result itself, because single-fold data are sufficient for imaging.

For *PS*-wave acquisition, illumination fold can be

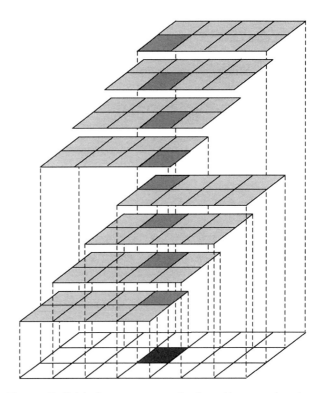

FIG. 2.17. Fold-of-coverage can be found by counting the number of overlapping cross-spreads. In this case inline fold is four and crossline fold is two: there are eight overlapping cross-spreads in each point.

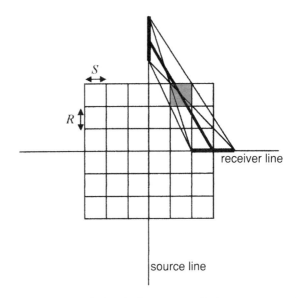

FIG. 2.18. Unit-cell-sized offset-vector tile in cross-spread of 36-fold geometry. Heavy lines along source line and receiver line indicate range of shots and receivers contributing to OVT. Heavy line through middle of OVT indicates average offset and average shot-receiver azimuth.

considerably smaller or larger than fold-of-coverage due to the asymmetric raypaths (see Section 6.2.2).

In the previous section, fold-of-coverage was defined on the basis of the underlying continuous wavefield. Fold-of-coverage is equal to the number of overlapping MDSs. This means that fold-of-coverage does not depend on the sampling density of the MDSs, hence, is independent of bin size. Similarly, illumination fold and image fold are independent of bin size. Unfortunately, it is common practice to state that fold-of-coverage or illumination fold does depend on bin size (e.g., Cordsen et al., 2000, Sections 2.9 and 12.5). However, only stacking fold might depend on bin size in case neighboring traces of the same MDS turn up in the same bin. Of course, these neighboring traces would have sampled different parts of the subsurface. Stacking of such data would lead to loss of resolution.

2.5.4 Construction of pMDSs

Even though cross-spreads have limited extent, it is possible to create single-fold coverage across the whole survey area by a tiling of adjacent cross-spreads. In such a single-fold gather, the data are piecewise continuous, with discontinuities between the adjacent cross-spreads (see Figure 2.20). Figure 2.15 shows the illumination by four adjacent cross-spreads of a reflector with 15° dip and a reflector with 45° dip. Each cross-spread covers the reflector with its own illumination area. Around the edges of these areas, gaps and overlaps exist. Within each illumination area, illumination can be considered as continuous (provided the cross-spread is sampled alias free), but illumination is discontinuous across the edge of each area.

A tiling of adjacent cross-spreads as in Figure 2.20 is the first example of a pMDS (Vermeer, 1998b). The number of different such tilings equals the fold-of-coverage. It is clear from Figure 2.15 that these tilings cannot produce good images of the subsurface everywhere. Locally, the images will show considerable artifacts, depending on the dip of the reflectors being imaged. Therefore, it would be desirable to find a single-fold gather (or 100% cube as it is sometimes referred to) using data with smaller spatial discontinuities. As the discontinuities of the cross-spreads are a given, the only way to reduce their effect is by spreading the discontinuities thinly over the survey area. This can be done by selecting tilings of OVTs as illustrated in Figure 2.21. In such a tiling or OVT gather, the frequency of spatial discontinuities is much higher than in adjacent cross-spread tilings. Their magnitude, however, is much smaller. Note that there is perfectly continuous single-fold midpoint

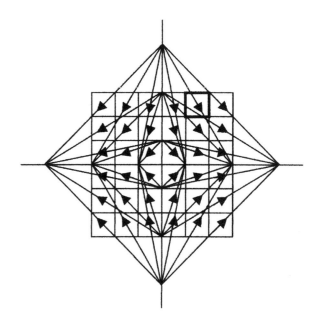

FIG. 2.19. Cross-spread with its OVTs. For each OVT, offset and azimuth of the central trace are indicated. The OVT with heavy lines is used in Figure 2.21 to generate a pMDS.

coverage in Figure 2.21; the spatial discontinuities referred to only pertain to the offset-vector.

Cary (1999) also introduced the OVT gather as a basic building block of wide-azimuth surveys. He called them common-offset vector (COV) gathers, which would be a bit too optimistic as offset still does vary across each tile of the gather. Yet, I like the expression "offset vector," and therefore, I introduced here the expression offset-vector tile, which was called offset/azimuth slot in Vermeer (1998b). COV gather is a more appropriate name for the subset of the ideal parallel geometry.

2.5.5 A measure of spatial discontinuity

Let us consider a subdivision of a cross-spread into OVTs as in Figure 2.19. Then the horizontal width of the OVT Δh_x [cf. equation (2.3)]

$$\Delta h_x = W_x / M_x = S, \tag{2.7}$$

and the vertical width Δh_y

$$\Delta h_y = W_y / M_y = R. \tag{2.8}$$

The offset discontinuity across the vertical edges of an OVT equals Δh_x. This discontinuity occurs along a length Δh_y. So, a representative measure of the total discontinuity across the length of a vertical edge of an OVT might be $\Delta h_x \Delta h_y$. The same expression is found for the discontinuity across each horizontal edge, for a total discontinuity of 4 $\Delta h_x \Delta h_y$. The OVT shares this discontinuity with four other OVTs, so the average discontinuity per OVT D_{OVT} may be characterized by

$$D_{OVT} = \Delta h_x \Delta h_y = S R, \qquad (2.9)$$

which is the area of the OVT. Hence, the spatial discontinuity in an OVT gather per unit area equals 1.

In a tiling of adjacent cross-spreads, the spatial discontinuity across a cross-spread D_X could be derived in a similar way as for an OVT, leading to

$$D_X = W_x W_y, \qquad (2.10)$$

which equals the area of the cross-spread. Therefore, the spatial discontinuity in a tiling of adjacent cross-spreads also equals 1.

My definition of spatial discontinuity implies that the amount of spatial discontinuity for a given geometry is invariable, but that its local density can be varied. The smaller the unit cell of a geometry, the smaller the discontinuities inside OVT gathers can be.

It should be noted that the measure of spatial discontinuity introduced here is not sufficient to predict the effect of the discontinuity. The effect also depends on the average absolute offset of the OVT gather; the larger that offset, the stronger the effect in general. Moreover, it depends on the dip of the events, the larger the dip the larger the discontinuities.

The measure of spatial discontinuity applied to a multiline roll geometry (see Section 4.6.5) would lead to a discontinuity per unit area that is larger than 1, suggesting that these configurations are not optimal for spatial continuity. On the other hand, the measure of spatial discontinuity does not discriminate against narrow geome-

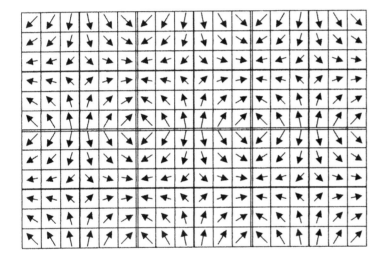

FIG. 2.20. Tiling with (six) adjacent cross-spreads. Spatial continuity exists inside the cross-spreads, but large discontinuities occur across the edges of each cross-spread, in particular in the corners. From the corners to the axes of the cross-spreads, the discontinuities decrease.

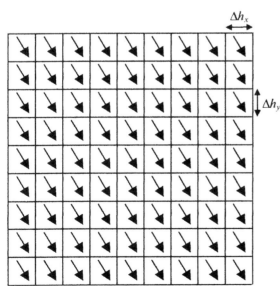

FIG. 2.21. Pseudo-minimal data set constructed from offset-vector tiles. In this case, the generating OVT is the upper central OVT in the first quadrant of all cross-spreads (cf. Figure 2.19). In this OVT gather the spatial discontinuities are spread thinly across the whole survey area.

tries; a narrow geometry acquired with single-line roll would also have a spatial discontinuity per unit area equal to 1. This suggests that my (dis)continuity criterion is not sufficient as a quality measure. It would be valuable to avail of a measure of spatial continuity, which could be used as a discriminator between acquisition geometries. Unfortunately, as yet, I have not been able to find one.

2.5.6 A plethora of OVT gathers

Up till now, the cross-spread has been subdivided into OVTs, which taken together fill the whole cross-spread. However, a single-fold OVT gather can also be constructed using a generating OVT (h_x, h_y, Δh_x, Δh_y), which still has the size of a unit cell, but which can be located anywhere inside the cross-spread, i.e., OVT (h_x, h_y, S, R), with $|h_x| < (W_x - S)/2$ and $|h_y| < (W_y - R)/2$. This will increase the flexibility of selecting suitable OVT gathers considerably.

A generating OVT may also consist of $n \times m$ unit-cell sized areas together. Taking the same area of each cross-spread in this way will lead to $n \times m$ fold OVT gathers. Higher fold in an OVT gather may be useful for high-fold data, or for noisy data.

For any single-fold tiling of the survey area it is necessary that the tiles have dimensions $S \times R$ or multiples thereof. However, in some cases it may be desirable to construct the tiles from smaller OVTs. For instance, along the x-axis, OVT (h_x, 0, $S/2$, R) may be combined with OVT ($-h_x$, 0, $S/2$, R) to form a complete tiling (Figure 2.22). This implies the use of an OVT with the area of half a unit cell and its mirror image. Similarly, along the y-axis we may combine OVT (0, h_y, S, $R/2$) and OVT (0, $-h_y$, S, $R/2$). It is of interest to investigate the spatial discontinuity of these OVTs.

In the juxtaposed bottom corners of the OVT along the x-axis, the offset vectors are ($H_x + S/2$, $-R/2$) and ($-H_x - S/2$, $-R/2$) (Figure 2.22). Using reciprocity, the second offset vector may also be written as ($H_x + S/2$, $R/2$). Hence, the discontinuity in offset vector at that point equals $-R$. Along the x-axis the juxtaposed offset vectors are ($H_x + S/2$, 0) and ($-H_x - S/2$, 0). With reciprocity these two are the same, i.e., there is no discontinuity along the x-axis. Using the same reasoning for the juxtaposed top corners of the OVT, there the discontinuity equals R. Hence, the discontinuity along the vertical varies between 0 and R along a distance R. So, the measure of spatial discontinuity across the vertical equals $R\,R\,/2$.

Across the horizontal boundaries, the same OVTs are found, with a constant jump of R in the y-coordinate and no discontinuity in the x-coordinate. Hence, along the horizontal the measure of spatial discontinuity equals $R\,S\,/2$. For $S = R$, the spatial discontinuity associated with each OVT again equals its size, i.e., $D_{\text{OVT}} = R\,S\,/2$. If $R < S$, the spatial discontinuity of OVTs along the x-axis is smaller than the OVT size, whereas for OVTs along the y-axis it would be larger than the OVT size, and vice versa for $R > S$.

For situations where azimuth does not play a role, unit-cell sized tiles may be constructed from four small OVTs (Figure 2.22).

Figure 2.23 illustrates the tilings that can be constructed from the smallest offset-vectors of the geometry and the tiling that can be constructed from the largest offset-vectors. The largest offset in the tiling of smallest offset-vectors is sometimes called LMOS, the largest minimum offset. The smallest offset in the tiling of largest offset-vectors is also called minimum maximum offset and is referred to as X_{minmax}. The reason for this nomenclature is that in any full-fold bin of the geometry the smallest offset is not larger than LMOS and the largest offset is not smaller than X_{minmax}.

2.6 Application to prestack processing

2.6.1 Introduction

In the following sections, ideas are put forward for the most suitable input gathers for noise removal, interpolation and regularization, muting, first-break picking, residual statics picking, velocity analysis, AVO, and AVAzimuth. Velocity-model updating and prestack migration are discussed in Chapter 10.

As different tasks need different data gathers, either much sorting has to be done to feed the different gathers to the various processing steps or random access should be available. Sorting is very time-consuming, whereas random access is fast, but it requires a database with pointers to the correct trace positions. Eventually, random access is likely to take over (Jack, 1999).

2.6.2 Noise removal

Ground roll tends to be partially aliased, because of its slow velocity. The nonaliased part of the ground roll (and even a bit more) can be removed by prestack velocity filtering. The obvious input gather for this process is the cross-spread, so that noise can be removed either by cascaded application of shot- and receiver-domain (f, k)-filtering, or by a 3-D velocity filter.

In an OVT gather with $S \times R$ sized tiles (or smaller) the spatial discontinuities of the nonaliased part of the ground roll tend to be even larger than across cross-spread boundaries. Across cross-spread boundaries there is usually no ground roll, except perhaps at larger traveltimes. Each OVT that cuts through the ground roll shows discontinuities in the noise at its edges. Therefore, it is important to remove the ground roll as much as possible prior to any spatial processes applied to OVT gathers. A particularly powerful technique, which also removes much of the aliased ground roll, was discussed in Miao and Cheadle (1998).

2.6.3 Interpolation and regularization

Interpolation and regularization can be applied easiest in a domain in which only one spatial variable varies (Zwartjes and Duijndam, 2000). Therefore, the cross-spread lends itself best as input gather to these processes. If there is a missing shot, interpolation can best be carried out in the common-receiver gathers of the cross-spread. Interpolation may also be carried out to generate better-sampled data for other processes; for instance, Cooper et al. (1997) interpolated cross-spreads for better DMO results.

2.6.4 Muting

At first sight, it might seem strange to require a specific input sorting for an optimal mute application. Indeed, the idea here is not to use a different sorting, but to learn from the insights gained in Section 2.5 about the many different OVTs into which a cross-spread may be subdivided.

The unit cell of a regular orthogonal geometry represents the 2-D periodicity of the acquisition geometry. Usually, the acquisition imprint shows this same periodicity. The visibility of the acquisition imprint may be caused by two main factors: (1) variability of fold inside the unit cell for times where traces with larger offsets are muted, and (2) the unit-cell periodicity in the offset distribution. There is little one can do about the periodicity, but the variability of fold can be easily removed.

Consider Figure 2.22. Taking eight quarter-unit-cell sized OVTs as indicated with the checkered squares and the striped squares, then each of those OVTs has the same absolute-offset distribution. The same mute time can be assigned to all traces inside these squares. Tapering in time of the step function should reduce any jumps in fold caused by identical mutes for tiles with the same offsets (but different azimuths). If this procedure is carried out for all OVTs with the same absolute-offset distribution, the effective fold-of-coverage will be constant for constant time. This should reduce the acquisition imprint of the geometry. It would be interesting to check this using real data acquired with a regular orthogonal acquisition geometry.

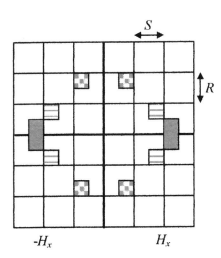

FIG. 2.22. Special case OVTs. Together, the two rectangular OVTs (dimension 1/2 S x R) can be used to construct an OVT gather with small spatial discontinuity between the OVTs. Together, the four checkered squares (dimension 1/2 S x 1/2 R) may be used to construct an OVT gather in case azimuth does not play a significant role. The locations of these OVTs may be selected anywhere inside the cross-spread, provided the pairs or quartets occupy mirrored positions. The eight small squares may be assigned the same mute time to achieve constant fold.

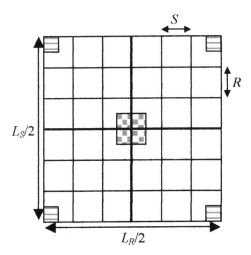

FIG. 2.23. Smallest and largest offset OVTs. The four small checkered squares form a single-fold tiling of the shortest offsets of the geometry. The largest offset in this tiling is the maximum minimum offset or largest minimum offset LMOS. The four small squares in the corners can be used to form a single-fold tiling of the largest offsets of the geometry. The smallest offset of this tiling is the minimum maximum offset X_{minmax} of the geometry.

Hill et al. (1999) show a clear correlation between time slice amplitude and the fold of data contributing to the time slice. They used synthetic data acquired with zigzag geometry. The muting proposed here for orthogonal geometry could also be adapted to other regular acquisition geometries. If applied to the data discussed in Hill et al. (1999), the acquisition footprint would be removed almost entirely.

2.6.5 First-break picking

In first-break picking, only the near-surface variation plays a role. The first-break times depend on shot static, receiver static, and on offset. While picking, it would be helpful to use data gathers in which the trace-to-trace variations would depend on one variable only. To some extent this can be achieved by picking in OVT gathers. In these gathers the offset variation across the gather is kept to a minimum, whereas within each tile the shot static is constant in the inline direction and the receiver static is constant in the crossline direction. All M OVT gathers are potential candidates for picking, but some of them may drop out due to quality problems.

In case there are serious picking problems, it may be beneficial to combine OVT gathers with mirror OVTs in the opposite quadrant, as these have about the same azimuths. It may be more difficult to combine mirror OVTs in adjacent quadrants, as these have different azimuths and may have different traveltimes.

An alternative to picking in gathers of (S, R)-sized OVTs is picking on a per cross-spread basis. The advantage of this alternative is that the area with spatial continuity in a cross-spread is much larger than in an OVT gather. The disadvantage is that the large spatial discontinuity between cross-spreads might necessitate to start picking afresh for each cross-spread.

The more flexible approach is to combine picking in the OVT gathers with picking in the cross-spreads. Especially in combination with the nearest-neighbor approach to picking (see next section), this should give the best results.

2.6.6 Nearest-neighbor correlations

Conventional first-break picking and reflection-time picking techniques are based on a sequential approach in which traces in a midpoint as a function of absolute offset or in a shot as a function of receiver station are picked one after the other (Cox, 1999). Often, mispicks are only identified in the statics computation. Here I would propose to carry out the picking in an areal approach, using nearest neighbors. This approach allows verification of the time picks as part of the picking procedure, followed by feeding the verified picks to the statics computation procedure (Marcoux, 1981; Vermeer, 1990, Section 5.7).

In the nearest-neighbor approach, each trace is cross-correlated with its eight nearest neighbors. This has the advantage of comparing traces with a minimum of difference in character between them. Another advantage is that it leads to redundant picking, which allows correction of mispicks before these are used in the statics computation procedure. Redundancy exists for every closed loop between traces: the sum of the corresponding time shifts should equal zero. Once all mispicks have been solved, all time shifts can be integrated into a single time surface across the area of the picked times.

This procedure was proposed in Vermeer (1990, Section 5.7) for 2-D data, but it applies just as well or even better to 3-D data. All mispicks might first be solved for a number of single-fold OVT gathers, and by making links between the gathers (via cross-spread continuity), the picks might even be made consistent in a 3-D sense (x, y, and fold).

It should be realized that the spatially nearest neighbors in an OVT gather are not always nearest neighbors in 5-D space, because of the spatial discontinuity which still exists across the edges of neighboring OVTs. Again, the picking redundancy should help to solve any problems in linking time shifts across these boundaries.

2.6.7 Residual statics

Picking of time shifts for residual statics analysis in 3-D data usually takes place in bins or in a small group of bins. Each trace in a bin corresponds to a different cross-spread; therefore, consecutive traces sorted according to absolute offset, may have entirely different shot-to-receiver azimuth and originate from widely spaced cross-spreads. This is illustrated in Figure 2.24, where trace positions are displayed according to their (h_x, h_y)-coordinates inside each bin. Traces with mirrored positions inside these bins have about the same absolute offset.

Determining time shifts between traces using nearest neighbors (as proposed in the previous section) ensures that the difference in character between traces that are to be compared is as little as possible. Moreover, it allows removal of mispicks even before the statics computation procedure is entered.

The time differences established in nearest-neighbor communities are not only composed of static differences, but also of structure and velocity differences. Moreover, there is picking noise. To compute the statics from the time-shift surfaces across the survey area, new algorithms are required. These algorithms should make use of the special properties of static differences, which are very different from differences due to structure variations or velocity variations. Note that velocity determination prior to residual-statics determination is no longer necessary. A very rough NMO correction may be applied, or no NMO at all, prior to the time-shift measurements. This is an advantage, especially for wide orthogonal geometries, because velocity determination is best carried out after DMO, whereas statics should be determined prior to DMO.

An alternative to picking of nearest neighbors in OVT gathers might be picking of nearest neighbors in bins as displayed in Figure 2.24. Time shifts would be measured only between traces with offset vectors that differ as little as possible.

2.6.8 Velocity analysis and DMO

Conventional velocity determination after DMO splits the input data into small offset ranges; each offset range has DMO applied separately; this is followed by gathering of the results per bin and semblance analysis. In a parallel geometry or in a narrow orthogonal geometry, this procedure should work satisfactorily. However,

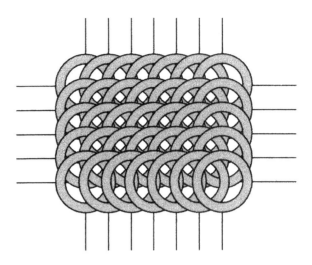

FIG. 2.25. Offset-range gather in orthogonal geometry. Each ring represents traces in midpoint domain with a narrow range of absolute offsets.

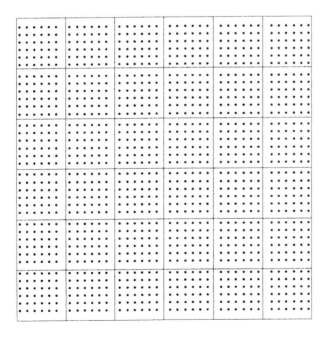

FIG. 2.24. Unit cell with offset distribution in each bin for a 36-fold geometry. Each square represents a bin. The 36 dots in each bin represent traces, which correspond to 36 different cross-spreads. Each bin has its own (h_x, h_y)-coordinate system centered in the bin. Nearest neighbors inside the bin have at least one different acquisition line.

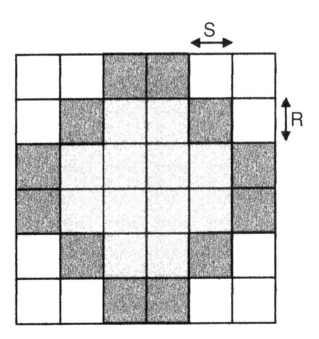

FIG. 2.26. Nears, mids, and fars in 36-fold geometry. The data set is split over three 12-fold subsets, the nears (light gray area), mids (dark gray area), and fars (the white areas in the corners of the cross-spread). Stacking these subsets provides regular 12-fold substacks across the entire full-fold area of the survey.

in a wide geometry, common-offset-range gathers have a very irregular fold, and are not likely to produce well-resolved DMO images. A common-offset-range gather is shown in Figure 2.25. It illustrates the irregular fold, and shows the many edges in such a gather. An alternative technique is to sort the data in each bin according to absolute offset, subdivide the traces in groups with an equal number of traces, and use those groups as input to the DMO and velocity determination procedure. This technique as well has the disadvantage of grouping traces with widely different positions in 5-D prestack space.

Several authors showed that cross-spreads are suitable for DMO (Vermeer et al., 1995; Collins, 1997; Padhi and Holley, 1997). It should be possible to obtain good quality DMO images for the interior part of each cross-spread. However, offset varies inside a cross-spread, and as a consequence, each image is made up of different offsets and the offset attached to each DMO image is not known anymore. To determine velocity, it is still necessary to split the data over offset ranges. However, rather than splitting the data over absolute-offset ranges, splitting the data over offset-vector ranges should be considered.

The smallest offset-vector ranges, which still give complete single-fold coverage, can be found along the acquisition lines as indicated in Figure 2.22 with the gray rectangles. For an inline fold of six, there are six different OVT gathers with disjoint offset-vector ranges. If the geometry would also be six-fold in the crossline direction, another five OVT gathers can be made from OVTs along the shot line. For a maximum inline offset and a maximum crossline offset of 3000 m, the range of offsets in any OVT gather would still be at least 500 m. Hence, the uncertainty about the offset at the image point is still quite large.

In a low-relief geology, the DMO shift is small, and it would be sufficient to select points in the center of the tiles of each OVT gather as locations for velocity determination. The offsets in these points can be used to estimate the velocity in those points.

In a steeper dip situation, the (unknown) offset of the image trace and the offset of the input location will differ considerably, and this would lead to systematic errors in the velocity estimates. In these situations, it may be better to try a velocity scanning procedure (i.e., apply DMO after many different NMO corrections) rather than a semblance measurement. Usually, the velocity determination is restricted to some discrete points across the survey area. Using only a restricted subset of the input data—the offset-vector tiles around the acquisition lines—a scanning procedure would still be cost-effective.

Of course, there are many variations possible on this theme. The main point is to select good input data gathers to ensure the best possible images with the least amount of edge effects.

Should the total fold along the two orthogonal directions not be sufficient for accurate measurements, OVT gathers using different OVTs may be used, in particular those in the far corners of the cross-spread having the largest absolute offsets. The measurement of velocity in OVT gathers taken along two orthogonal directions, also allows recognition of velocity anisotropy under suitable circumstances.

2.6.9 AVO

The determination of amplitude variation with offset (AVO) parameters from data acquired with orthogonal geometry is one of the most challenging tasks. The main problem is that proper common-offset gathers are not available for analysis; moreover the trace density per offset increases with increasing offset. It is also difficult to give a general recipe for AVO analysis, because there are so many different types of problems. In some cases, one would like to scan a large time window for possible AVO anomalies; in other cases specific horizons are to be investigated, and then these horizons may or may not need prestack migration.

A technique that is often used in AVO analysis is to generate substacks of near and far offsets, or substacks of nears, mids, and fars (e.g., Purnell et al., 2000). If absolute-offset ranges are used for those substacks, fold variation at target levels may cause undesirable amplitude effects. Therefore, one should try to achieve regular fold in each substack. For the deeper levels, this can be achieved quite simply as indicated in Figure 2.26. Here a 36-fold geometry has been split into three regular 12-fold subsets, which can serve as input to near, mid and far substacks. In more complex geometries, different subdivisions will have to be found, which may or may not overlap partially. If the basic building blocks of the subdivisions are either unit-cell sized OVTs, or pairs of rectangular half a unit-cell sized OVTs, or quadruplets of quarter unit-cell sized OVTs, regular fold is ensured over the full-fold part of the survey area (cf. Figure 2.22).

The type of subdivision indicated in Figure 2.26 does not provide for regular fold of substacks at shallower levels. These levels, if important for AVO analysis, need subdivisions based on a smaller range of offsets.

In the remainder of this section, I will give some suggestions to be tried for measuring horizon amplitudes as a function of offset. I will propose two different approaches, both taking into account offset as a vector.

The first approach is bin-oriented and the second approach is unit-cell oriented.

In the bin-oriented approach (Starr, 2000), the starting point is a sorting of all traces inside a bin according to their inline and crossline offset as depicted in Figure 2.24. After picking the horizon of interest, an amplitude map is obtained for each bin and a 2-D surface can be fitted to this map. In this way, a best fit can be obtained for intercept time and a 2-D amplitude gradient. The technique is different from conventional AVO analysis in that it takes amplitude variations caused by shot-to-receiver azimuth variation into account. For this approach to work, there should be a good signal-to-noise ratio in each trace and fold should be high to increase the redundancy of the fit. DMO and migration move traces around and would disturb the relationships between the traces in each bin. Therefore, this approach would work best for (sub)horizontal geology.

The unit-cell oriented approach to AVO analysis would be less sensitive to fold because in this approach all traces in a unit cell take part in one analysis. The basic input would be OVT gathers of unit-cell sized disjoint OVTs, i.e., M gathers of OVTs as indicated in Figure 2.19. Depending on the problem, these gathers would be either NMO-DMOed or prestack migrated, followed by stacking.

The next step would be to pick the horizon on the stacked data volume, followed by making horizon slices according to these picked times in the contributing OVT gathers. Accepting that the spatial resolution of the AVO analysis will be restricted to approximately the size of a unit cell, the horizon amplitudes can now be analyzed by averaging in a ring-shaped area corresponding to some range of offsets as indicated in Figure 2.27.

The procedure described here will break down if the migration distance becomes significant. Then there will no longer be a direct relationship between position inside a tile and the offset of the migrated image. A solution of this problem is discussed in Section 10.6. Tura et al. (1998) show the importance of prestack migration for AVO analysis for data acquired with parallel geometry.

If the tiles are small, offset does not vary much across each tile and the average amplitude in the tile may be considered representative for the average offset of the tile. If shot-to-receiver azimuth variation does not affect AVO, another acceptable way of reducing the size of the tiles is to use M disjoint ($S/2$, $R/2$) sized tiles as indicated by the checkered tiles in Figure 2.22. These M quarter unit-cell sized tiles may be mapped such onto one quadrant of a cross-spread that absolute offset is continuous across the mapping (see Figure 2.28).

2.6.10 Amplitude variation with azimuth

For analysis of azimuth-dependent effects, the same unit-cell oriented procedure can be applied as proposed for AVO in the previous section. Now unit-cell sized areas of the survey have to be split over the M different OVTs. Pie slices taken from the collection of data represent data with the same azimuth range (Figure 2.27). In this case amplitude behavior has to be analyzed on a per pie slice basis. Note that the arrows indicating the average azimuth in each tile do not have the same direction as the orientation of the pie slice.

2.7 Conclusions

For all intents and purposes, it is impossible to properly sample the whole 5-D prestack wavefield. Three-dimensional symmetric sampling prescribes the next best alternative: the proper sampling of single-fold basic subsets (minimal data sets) of 3-D geometries. Such sampling allows optimal prestack processing, and it takes care of a design criterion that is often overlooked: spatial continuity.

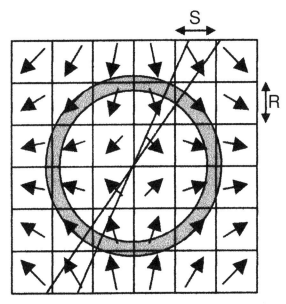

FIG. 2.27. Basic unit for AVO and amplitude versus azimuth analysis. All OVTs corresponding to the same unit-cell sized part of the survey area are displayed next to each other for further analysis. Amplitudes for the same offset can be averaged along rings with a constant absolute-offset range. Repeating this for all relevant positions in the survey area allows analysis of the spatial variation of the AVO effect. Azimuth-dependent effects can be analyzed using pie-slice shaped areas, which contain data with the same azimuth range.

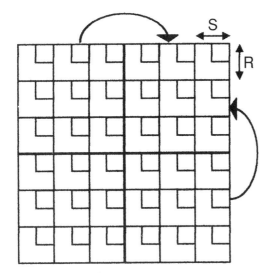

FIG. 2.28. Creating a gather, which is continuous in absolute offset by using quarter unit-cell sized OVTs. In the example, the upper right corner of each unit-cell sized OVT is taken. First, all OVTs in the lower half of the figure are mirrored around the horizontal axis; next, all OVTs in the upper left quadrant are mirrored around the vertical axis. This procedure fills the whole upper right quadrant with OVTs. Azimuth is not continuous in that gather, but absolute offset is.

The basic subsets of all common acquisition geometries, except parallel geometry have limited extent. This constitutes a limiting factor for the spatial continuity that can be obtained with those geometries. The selection of appropriate gathers of offset-vector tiles for all prestack processing steps mitigates problems associated with those geometries.

References

Albertin, U. et al., 1999, Aspects of true amplitude migration: 69th Ann. Internat. Mtg., Soc. Expl. Geophys., Expanded Abstracts, SPRO11.2, 1358–1361.

Al-Mahrooqi, S., Duijndam, B., and van der Schans, C., 2000, Slip-sweep seismic acquisition in Petroleum Development Oman: GeoArabia, **5**, 28.

Anstey, N., 1986, Whatever happened to ground roll?: The Leading Edge, **5**, No. 3, 40–45.

Ball, J. D. and Mounce, W. D., 1967, Apparatus for converting lineal seismogram sections into an areally presented seismogram: U.S. Patent 3 327 287.

Bardan, V., 1996, A hexagonal sampling grid for 3-D recording and processing of 3-D seismic data: 58th Conf., Eur. Assoc. Geosc. and Eng., Extended Abstracts, P061.

Beasley, C. J., 1996, Statistical measures of subsurface illumination: 58th Conf., Eur. Assoc. Geosc. and Eng., Extended Abstracts, B043.

Beasley, C. J., and Klotz, R., 1992, Equalization of DMO for irregular spatial sampling: 62nd Ann. Internat. Mtg., Soc. Expl. Geophys., Expanded Abstracts, 970–973.

Beasley, C. J., and Mobley, E., 1995, Spatial sampling characteristics of wide-tow marine acquisition: 57th Conf., Eur. Assoc. Geosc. and Eng., Extended Abstracts, B031.

Becker, C. H., 1960, Method of geophysical exploration: U.S. Patent 2 925 138.

Berni, A. J., 1994, Remote sensing of seismic vibrations by laser Doppler interferometry: Geophysics, **59**, 1856–1867.

Bertelli, L., Mascarin, B., and Salvador, L., 1993, Planning and field techniques for 3-D land acquisition in highly tilled and populated areas—Today's results and future trends: First Break, **11**, No.1, 23–32.

Beylkin, G., 1985, Imaging of discontinuities in the inverse scattering problem by inversion of a causal generalized Radon transform: J. Math. Phys., **26**, 99–108.

Beylkin, G., Oristaglio, M., and Miller, D., 1985, Spatial resolution of migration algorithms: in Berkhout, A. J., Ridder, J., and van der Wal, L. F., Eds., Proc. 14th Internat. Symp. on Acoust. Imag., 155–167.

Bittleston, S., Canter, P., Hillesund, O., and Welker, K., 2000, Marine seismic cable steering and control: 62nd Conf., Eur. Assoc. Geosc. and Eng., Extended Abstracts, L-16.

Bleistein, N., 1987, On the imaging of reflectors in the earth: Geophysics, **52**, 931–942.

Cary, P. W., 1999, Common-offset-vector gathers: an alternative to cross-spreads for wide-azimuth 3-D surveys: 69th Ann. Internat. Mtg., Soc. Expl. Geophys., Expanded Abstracts, SPROP1.6, 1496–1499.

Claerbout, J. F., 1985, Imaging the earth's interior: Blackwell Scientific Publishing, Inc.

Cohen, J. K., Hagin, F. G., and Bleistein, N., 1986, Three-dimensional Born inversion with an arbitrary reference: Geophysics, **51**, 1552–1558.

Collins, C. L., 1997, Imaging in 3-D DMO, Part I: Geometrical optics model: Geophysics, **62**, 211–224.

Constance, P. E. et al., 1999, Simultaneous acquisition of 3-D surface seismic data and 3-C, 3-D VSP data: 69th Ann. Internat. Mtg., Soc. Expl. Geophys., Expanded Abstracts, BH/RP 4.5, 104–107.

Cooper, N. J., Williams, R. G., Wombell, R., and Notfors, C. D., 1997, An improved 3-D DMO implementation for orthogonal cross-spread acquisition geometries: 59th Conf., Eur. Assoc. Geosc. and Eng., Extended Abstracts, A051.

Cordsen, A., Galbraith, M., and Peirce, J., 2000, Planning land 3-D seismic surveys: Soc. Expl. Geophys.

Cox, M., 1999, Static corrections for seismic reflection surveys: Soc. Expl. Geophys.

Crews, G. A., Henderson, G. J., Musser, J. A., and Bremner, D. L., 1989, Applications of new recording systems to 3-D survey designs: 59th Ann. Internat. Mtg., Soc. Expl. Geophys., Expanded Abstracts, 624–631.

Dickinson, J. A., Fagin, S. W., and Weisser, G. H., 1990, Comparison of 3-D seismic acquisition techniques on land: 60th Ann. Internat. Mtg., Soc. Expl. Geophys., Expanded Abstracts, 913–916.

Dunkin, J. W., and Levin, F. K., 1971, Isochrons for a three-dimensional seismic system: Geophysics, **36**, 1099–1137.

Durrani, J. A., French, W. S., and Comeaux, L. B., 1987, New directions for marine 3-D surveys: 57th Ann. Internat. Mtg., Soc. Expl. Geophys., Expanded Abstracts, 177–180.

Dürschner, H., 1984, Dreidimensionale Seismik in der Exploration auf Kohlenwasserstoff-Lagerstätten: J. Geophys., **55**, 54–67. English translation *in* Graebner, R. J., Hardage, B. A., and Schneider, W. A., Eds., 2001, 3-D seismic exploration: Soc. Expl. Geophys.

Ferber, R., 1998, Is common-offset common-azimuth DMO really that bad?: 60th Conf., Eur. Assoc. Geosc. and Eng., Extended Abstracts, 1-13.

Gardner, G. H. F., and Canning, A., 1994, Effects of irregular sampling on 3-D prestack migration: 64th Ann. Internat. Mtg., Soc. Expl. Geophys., Expanded Abstracts, 1553–1556.

Goodway, W. N., and Ragan, B., 1997, "Mega-Bin" land 3-D seismic: Toward a cost effective "symmetric patch geometry" via regular spatial sampling in acquisition design with co-operative processing, for significantly improved S/N & resolution: Proc. Soc. Expl. Geophys. Summer Research Workshop.

Hill, S., Shultz, M., and Brewer, J., 1999, Acquisition footprint and fold-of-stack plots: The Leading Edge, **18**, 686–695.

Huard, I., and Spitz, S., 1997, An application of 3-D filtering to the restoration of missing traces: 59th Conf., Eur. Assoc. Geosc. and Eng., Extended Abstracts, AO 38.

Jack, E., 1999, The future is random: Archiving seismic data to random access media: 69th Ann. Internat. Mtg., Soc. Expl. Geophys., Expanded Abstracts, SACQ 3.8, 683–686.

Krail, P. M., 1991, Case history vertical cable 3-D acquisition: 53rd Conf., Eur. Assoc. Expl. Geophys., Extended Abstracts, 206.

——— 1993, Sub-salt acquisition with a marine vertical cable: 63rd Ann. Internat. Mtg., Soc. Expl. Geophys., Expanded Abstracts, 1376.

Lansley, R. M., Elkington, G. J., and Battaglino, N. J., III, 2000, Configuration of source and receiver lines for 3-dimensional seismic acquisition: U.S. Patent 6 028 822.

Lee, S. Y., et al., 1994, Pseudo-wavefield study using low fold 3-D geometry: 64th Ann. Internat. Mtg., Soc. Expl. Geophys., Expanded Abstracts, 926–929.

Marcoux, M. O., 1981, On the resolution of statics, structure, and residual moveout: Geophysics, **46**, 984–993.

Mersereau, R. M., 1979, The processing of hexagonally sampled two-dimensional signals: Proceedings of IEEE, **67**, No. 6, 930.

Miao, X., and Cheadle, S., 1998, Noise attenuation with wavelet transform: 68th Ann. Internat. Mtg., Soc. Expl. Geophys., Expanded Abstracts, SP 1.1, 1072–1075.

Moldoveanu, N., Adessi, D., Lang, J. T., Stiver, K., and Chang, M., 1994, Digiseis-enhanced streamer surveys (DESS) in obstructed area: A case study of the Gulf of Mexico: 64th Ann. Internat. Mtg., Soc. Expl. Geophys., Expanded Abstracts, 872–875.

Naylor, R., 1990, Positioning requirements for complex multi-vessel seismic acquisition: Hydrographic J., **58**, 25–32.

O'Connell, J. K., Kohli, M., and Amos, S., 1993, Bullwinkle: A unique 3-D experiment: Geophysics, **58**, 167–176.

Onderwaater, J., Wams, J., and Potters, H., 1996, Geophysics in Oman: GeoArabia, **1**, 299–324.

Ongkiehong, L., and Askin, H., 1988, Towards the universal seismic acquisition technique: First Break, **5**, 435–439.

Padhi, T., and Holley, T. K., 1997, Wide azimuths—Why not?: The Leading Edge, **16**, 175–177.

Petersen, D. P., and Middleton, D., 1962, Sampling and reconstruction of wave-number limited functions in *N*-dimensional Euclidean spaces: Information and Control, **5**, 279.

Pleshkevitch, A., 1996, Cross gather data—A new subject for 3-D prestack wave-equation processing: 58th

Conf., Eur. Assoc. Geosc. and Eng., Extended Abstracts, P137.

Purnell, G., Sukup, D., Higginbotham, J., and Ebrom, D., 2000, Migrating sparse-receiver data for AVO analysis at Teal South Field: 70th Ann. Internat. Mtg., Soc. Expl. Geophys., Expanded Abstracts, AVO 4.1, 190–193.

Reilly, J. M., 1995, Comparison of circular "strike" and linear "dip" acquisition geometries for salt diapir imaging: The Leading Edge, **14**, 314–322.

Ritchie, W., 1991, Onshore 3-D acquisition techniques: A retrospective: 61st Ann. Internat. Mtg., Soc. Expl. Geophys., Expanded Abstracts, 750–753.

Schleicher, J., Tygel, M., and Hubral, P., 1993, 3-D true-amplitude finite-offset migration: Geophysics, **58**, 1112–1126.

Smith, J. W., 1997, Simple linear inline field arrays may save the day for 3-D direct-arrival noise rejection: Proc. Soc. Expl. Geophys. Summer Research Workshop.

Starr, J., 2000, Method of creating common-offset/common-azimuth gathers in 3-D seismic surveys and method of conducting reflection attribute variation analysis: U.S. Patent 6 026 059.

Stone, D. G., 1994, Designing seismic surveys in two and three dimensions: Soc. Expl. Geophys.

Stubblefield, S. A., 1990, Marine walkaway vertical seismic profiling: U.S. Patent 4 958 328.

Thomas, J. W., 2000, Method for sorting seismic data: U.S. Patent 6 026 058.

Tura, A., Hanitzsch, C., and Calandra, H., 1998, 3-D AVO migration / inversion of field data: The Leading Edge, **17**, 1578–1583.

Vermeer, G. J. O., 1990, Seismic wavefield sampling: Soc. Expl. Geophys.

————1991, Symmetric sampling: The Leading Edge, **10**, No. 11, 21–27.

————1994, 3D symmetric sampling: 64th Ann. Internat. Mtg., Soc. Expl. Geophys., Expanded Abstracts, 906–909.

————1995, Discussion On: "3-D true-amplitude finite-offset migration," Schleicher, J., Tygel, H., and Hubral, P., authors: Geophysics **60**, 921–923.

————1998a, 3-D symmetric sampling: Geophysics, **63**, 1629–1647.

————1998b, Creating image gathers in the absence of proper common-offset gathers: Exploration Geophysics, **29**, 636–642.

————2000, A strategy for prestack processing of data acquired with crossed-array geometries: Proceedings 20th Mintrop seminar.

Vermeer, G. J. O., den Rooijen, H. P. G. M., and Douma, J., 1995, DMO in arbitrary 3-D geometries: 65th Ann. Internat. Mtg., Soc. Expl. Geophys., Expanded Abstracts, 1445–1448.

Walton, G. G., 1971, Esso's 3-D seismic proves versatile: Oil and Gas J. , **69**, No. 13, 139–141.

————1972, Three-dimensional seismic method: Geophysics, **37**, 417–430.

Wams, J., and Rozemond, J., 1997, Recent developments in 3-D acquisition techniques in Oman: GeoArabia, **2**, 205–216.

Wright, S., and Young, J., 1996, Lodgepole 3-D seismic: Design, acquisition and processing: 66th Ann. Internat. Mtg., Soc. Expl. Geophys., Expanded Abstracts, 397–400.

Zwartjes, P. M., and Duijndam, A. J. W., 2000, Optimizing reconstruction for sparse sampling: 70th Ann. Internat. Mtg., Soc. Expl. Geophys., Expanded Abstracts, SP P7.1, 2162–2165.

Chapter 3
Noise suppression

3.1 Introduction

A 3-D acquisition geometry should be designed such that at the end of the acquisition and processing sequence the desired signal can be reliably interpreted and the noise is suppressed as much as possible. This chapter focuses on noise suppression.

The main types of noise are multiples and low-velocity noise such as ground roll and scattered energy. How much low-velocity noise can be suppressed depends on the choice of field arrays, the stack response (implicitly also on fold) and on various processing steps. One of the reasons to select a *wide* orthogonal geometry is that it allows tackling low-velocity noise by filtering in the shot as well as in the receiver domain. The total amount of multiple suppression depends on the stack response (implicitly also on range of offsets) and on the success of multiple elimination programs, but not on field arrays. At present there is no clear theory on how much noise can be removed in processing. As a consequence, the required noise suppression by field arrays and stacking is relatively unknown, and, to a large extent, the choice of field arrays and fold is dependent on experience.

In this chapter, the effect of field arrays on low-velocity noise and of the stack response on low-velocity noise and multiples is discussed. This chapter begins with a discussion of the properties of the low-velocity noise as essential knowledge for the optimal choice of field arrays (linear or areal, shot and/or receiver arrays). Another very useful piece of knowledge would be a quantitative assessment of the amount of noise (ground roll and scattered energy) relative to the desired primary energy. A potential way of determining this relation is the acquisition of one or more 3-D microspreads. Section 7.2 discusses an example of such a data set.

3.2 Properties of low-velocity noise
3.2.1 "Direct" waves

Usually, the bulk of the energy in the so-called ground-roll cone consists of linear events traveling more or less directly from source to receiver. The linear events along the outside of the cone are usually refracted shear waves (traveling close to the surface), which have a faster velocity than the Rayleigh waves (true ground roll) arriving later. If the near-surface conditions do not vary rapidly, the arrival times of these linear events tend to vary mostly as a function of offset with only minor variation as a function of midpoint position. This in contrast to scattered waves, which also vary as a function of midpoint position due to the fixed position of the scatterer.

In a cross-spread, the midpoints of traces with the same absolute offset are situated on a circle (see Figure 2.6). Therefore a constant-velocity event lies along a circle in each time slice through the cross-spread data, and the 3-D shape of the event is a circular cone. With several linear events, all having slightly different velocities, the cross-spread contains a whole suite of cones. This property is illustrated in Figure 7.4, which shows two time slices through a densely sampled cross-spread.

The apparent velocity of the ground roll in the midpoint domain equals $V/2$ in all directions, V being the ground-roll velocity. The directional apparent velocity in shot and receiver domains varies from ∞ to $V/2$ (cf. Figures 7.2 and 7.3). In a wide geometry, ground roll with apparent velocity close to infinity in one domain will have a small apparent velocity in the other domain. A 3-D circular velocity filter would be most suitable to remove such noise; properly designed cascaded velocity filtering in shot domain and receiver domain would also be suitable.

3.2.2 Scattered waves

In this section, I analyze the properties of a scatterer in the cross-spread. It turns out that these properties are quite special due to decoupling of shot and receiver properties in x and y.

For a cross-spread with its center in (0,0), the traveltime surface of a scatterer with velocity V can be written as follows:

$$t(x_m, y_m, x_d, y_d, z_d) = t_s + t_r = \frac{\sqrt{(2y_m - y_d)^2 + x_d^2 + z_d^2} + \sqrt{(2x_m - x_d)^2 + y_d^2 + z_d^2}}{V}, \quad (3.1)$$

where x_m, y_m are midpoint coordinates, $\mathbf{d} = (x_d, y_d, z_d)$ is the position of the scatterer, and t_s and t_r are traveltimes from source to scatterer and from scatterer to receiver, respectively.

An example of this surface is shown in Figure 3.1a. The figure shows that the traveltime surface is pyramid-shaped with rounded-off edges. The edges run parallel to the x- and y-axes, at $y = y_d / 2$ and $x = x_d / 2$, and the apex is located at $(x_d / 2, y_d / 2)$. For comparison, Figure 3.1a also shows one time contour of the direct wave (the ground-roll cone), with the corresponding time contour of the scatterer as a heavy line. This shows that the traveltime surface of the scatterer lies entirely inside the ground-roll cone. Time slices of modeled data illustrating the same phenomenon are shown in Meunier (1999).

The apex of the traveltime surface of a surface scatterer will always be close to the ground-roll cone. This can be seen by looking for the locus of all apexes with the same traveltime. The diamond shape in Figure 3.1b represents the locus of all apexes with the same traveltime as the circular arrival time of the ground roll. If in a cross-section, e.g., line l in Figure 3.1a, the (local) apex seems to be far away from the ground-roll cone, then the cross-section must cut through the flank of the traveltime surface of the scatterer and the true apex (and the scatterer itself) must be located outside the line of the cross-section.

The apparent velocity of the traveltime surface of the scatterer in the x-direction is

$$V_{app,x}(x_m, \mathbf{d}) = \frac{V\sqrt{(2x_m - x_d)^2 + y_d^2 + z_d^2}}{2(2x_m - x_d)}. \quad (3.2)$$

Note that this velocity is only dependent on the receiver position and on the position of the scatterer $\mathbf{d}$. The apparent velocity tends asymptotically to $V / 2$. The apparent velocity in the y-direction also tends asymptotically to $V / 2$. This means that in any cross-section parallel to one of the axes, the flanks of the traveltime surface always tend to $V / 2$. This explains the predominance of steeply dipping events—all with about the same apparent velocity—inside the ground-roll cone (in case any events can be distinguished). In other words, such events are not necessarily back-scatterers, they may just as well be side-scatterers (depending on the position of the apex). This observation applies to any shot record with receivers along a straight line, not just to cross-spreads. An extensive discussion of the appearance of scatterers pre- and poststack is given in Larner et al. (1983). This paper also shows many illustrative examples of coherent noise.

Figure 3.2 shows the apparent velocity along the gradient of the traveltime surface. This figure may be used to predict what to expect from a radar analysis inside the ground-roll cone of a cross-spread. In a radar analysis

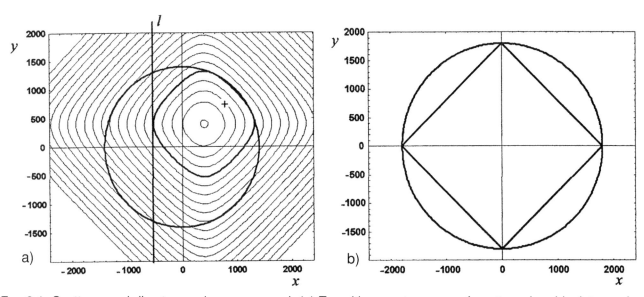

FIG. 3.1. Scatterer and direct wave in cross-spread. (a) Traveltime contour map of scatterer in midpoint x and y. The circle represents the traveltime contour of the direct wave for the same time as the heavy contour of the scattered wave. (b) Locus of all apexes with the same traveltime (diamond shape) fits inside the arrival time of the direct wave (circle).

(Regone, 1997) energy is measured as a function of slowness and azimuth. In the directions of the acquisition lines an apparent slowness of $2/V$ will be found, whereas in all other directions the slowness tends to $2\sqrt{2}/V$. The energy in the flat parts of the traveltime surfaces will show up in the center of the radar plots.

In the (f,k) domain of the scattered energy there will be a maximum apparent velocity V_{max} in either shot or receiver domain for which a velocity filter will stop being effective. Equation (3.2) can be solved to determine for which x_m this V_{max} occurs. This gives

$$x_{m,\text{limit}} = \frac{x_d}{2} \pm \frac{1}{2}\sqrt{(y_d^2 + z_d^2)/(4V_{max}^2/V^2 - 1)} = \frac{x_d}{2} \pm R_{\text{limit}}. \quad (3.3)$$

ticular the ratio of the energy in the apex and the energy in the flanks of that surface. The (surface-wave) energy is first spread according to $1/r_s$ when traveling from source to scatterer, and subsequently according to $1/r_r$ when traveling from scatterer to receiver (r_s, r_r are distance from source to scatterer and from receiver to scatterer, respectively). Hence, the energy is proportional to $1/(r_s r_r)$. The total energy in some range can be described schematically as

$$\text{energy} \approx \iint \frac{1}{r_s r_r}\, dx_m dy_m. \quad (3.4)$$

Because of the separability in x_m and y_m, this integral can be written as the product of two integrals. For instance, integration along the x_m-axis, centered on the apex of the traveltime surface reads

$$\text{energy}_x(y_d, z_d, R) = \int_{x_d/2 - R}^{x_d/2 + R} \frac{1}{\sqrt{(2x_m - x_d)^2 + y_d^2 + z_d^2}}\, dx_m = \sinh^{-1}\left(\frac{2R}{\sqrt{y_d^2 + z_d^2}}\right). \quad (3.5)$$

This equation shows that, for small depth z_d of the scatterer, the width $2R_{\text{limit}}$ of the apex area of the traveltime surface around x_d is proportional to the distance $|y_d|$ of the scatterer from the x-axis. Of course, the corresponding description also applies to a maximum apparent velocity in y.

Let us now investigate the energy as distributed across the traveltime surface of the scatterer and in par-

Substituting R with R_{limit} according to equation (3.3) into equation (3.5) and squaring to take into account the contribution along the y_m-axis shows that the energy in the apex area of the scatterer can be written as

$$\text{energy}(q) = \left(\sinh^{-1}\left(\frac{1}{\sqrt{q^2 - 1}}\right)\right)^2, \quad (3.6)$$

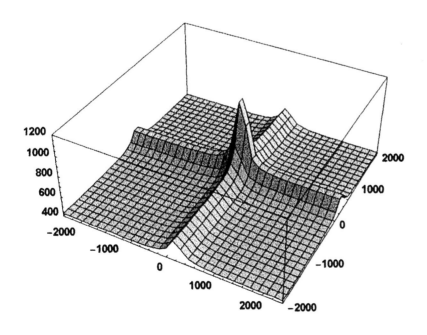

FIG. 3.2. Apparent velocity along the gradient of traveltime surface shown in Figure 3.1.

where $q = 2V_{max}/V$. The significance of equation (3.6) is that the energy in the apex area of the diffraction traveltime surface is a constant value; i.e., the energy that cannot be removed by filtering along the x_m- or y_m-axis is independent of the position of the scatterer with respect to the center of the cross-spread. This is somewhat counterintuitive, because one might expect the energy to become infinite for scatterers close to shot line or receiver line, because there the distance from nearest shot or receiver to the scatterer tends to zero. However, this is compensated by the narrowness of the apex area close to the axes.

The relative amount of energy in the apex area compared to the area of apex plus flank can now be expressed as

$$\text{energy ratio } (x_d/2, y_d/2) = \frac{(\sinh^{-1}(1/\sqrt{q^2-1}))^2}{\sinh^{-1}\left(\frac{2R}{\sqrt{y_d^2 + z_d^2}}\right) \sinh^{-1}\left(\frac{2R}{\sqrt{x_d^2 + z_d^2}}\right)}. \quad (3.7)$$

A reasonable choice of q and of R must be made to get an idea about the relative importance of apex area and flank area. For a ground-roll velocity $V = 1200$ m/s, i.e., velocity in midpoint domain $V/2 = 600$ m/s, it seems safe to assume that energy with apparent velocity above $V_{max} = 1100$ m/s cannot be removed, i.e., assume $q = 1.83$. On the other hand, it might be a reasonable choice to choose $R = 1800$ m (in midpoint domain). This corresponds to 3 s of flank in the center of the cross-spread, and to a smaller time window for scatterers further away from the cross-spread center. Figure 3.3 shows a contour plot of equation (3.7) expressed in dB. Note that the axes of this plot correspond to the x- and y-position of the apexes of the diffraction traveltime surface. Outside this coordinate range the time at the apex of the scatterer is at least 2 s.

3.2.3 Discussion

The analysis in Section 3.2.2 shows that the bulk of the energy of scatterers is concentrated in the flanks of the traveltime curves. Energy in the flanks can be removed by filtering in the shot and receiver domains. Provided sampling is dense enough, filtering in the computer is better than in the field, because of the limited control one has over the response of field arrays (Newman and Mahoney, 1973).

The analysis shows as well that the amount of energy in the flat parts of the traveltime curves is not insignificant. In particular scatterers farther away from the center of the cross-spread have relatively wide flat parts. After the steep parts have been suppressed by 30 dB or more by linear field arrays or by velocity filtering, the flat parts form the dominant part of the scattered noise energy. Areal shot and receiver arrays may have to be used to suppress the energy in those flat parts.

3.3 Shot and receiver arrays in 3-D data acquisition

3.3.1 Introduction

On land the *basic sampling interval* is not normally the same as the basic *signal* sampling interval, because the smallest apparent velocity of the ground roll is always smaller than the smallest apparent velocity of the signal. Only if the maximum frequency of the noise is significantly smaller than that of the signal, might these intervals be nearly the same. For a discussion of sampling and definition of basic sampling interval, see Vermeer (1990) or Section 1.3.

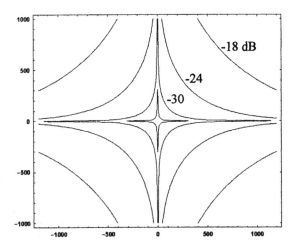

FIG. 3.3. Contour plot of energy in the flat part of the diffraction traveltime surface relative to total energy in a midpoint range of 3600 × 3600 m around the apex as a function of midpoint coordinates of the apex. Cutoff of the apex area is at $V_{app} = 1100$ m/s, the ground-roll velocity is 1200 m/s.

If the seismic data were sampled at the basic sampling interval, velocity filtering could be used to suppress the ground roll. However, the shot and receiver sampling intervals used in land data acquisition are at best equal to the basic *signal* sampling interval, but usually larger (this may change with the advent of high-capacity recording instruments allowing single-sensor acquisition). As a consequence of this coarse sampling, the ground-roll energy tends to be heavily aliased, which means that ground roll cannot be suppressed successfully by velocity filtering. Instead, shot and/or receiver arrays need be used to suppress the ground roll.

For 2-D, linear inline arrays have been discussed extensively in Vermeer (1990) and Section 1.6 provides a summary. It was found that arrays act as crude antialias filters reducing noise. The extension of the theory from 2-D to 3-D involves the use of areal shot and receiver arrays and needs to make a distinction between direct arrival noise and scattered energy. Similarly, as in 2-D, the effect of the arrays on the desired signal should always be taken into account as well.

3.3.2 "Direct" wave noise suppression

Similar to the response of a linear array as given in equation (1.6), the response of an areal array with N array elements located in (x_j, y_j), ($j = 1, ..., N$), can be described by the 2-D discrete spatial Fourier transform

$$p(k_x, k_y) = \frac{\sum_{j=1}^{N}\sum_{l=1}^{N} w_j w_l \exp(2\pi i(k_x x_j + k_y y_l))}{\sum_{j=1}^{N} w_j \sum_{l=1}^{N} w_l}, \quad (3.8)$$

where k_x and k_y are spatial wavenumbers in x and y, respectively, w_j are weights (filter coefficients) for each array element j. Equation (3.8) is normalized to provide $p(0, 0) = 1$. Usually, the absolute value of the array response is plotted. As an example, the contour plot of an array response is shown in Figure 3.4b with its corresponding array elements shown in Figure 3.4a. The contour plot can be interpreted as follows: distances along a radial line correspond to wavenumbers as measured along its particular azimuth and the array response for a plane wave traveling in that direction is obtained by projecting the coordinates of the array elements onto that azimuth. The outer arc of the plot corresponds to the maximum wavenumber one is interested in (e.g., the highest frequency of the slowest event). The central lobe of the array response constitutes the passband of the array, whereas elsewhere all energy should be suppressed as much as possible. Note that it would be sufficient to display only a semicircular plot, because the response for azimuth α is identical to azimuth 180° + α.

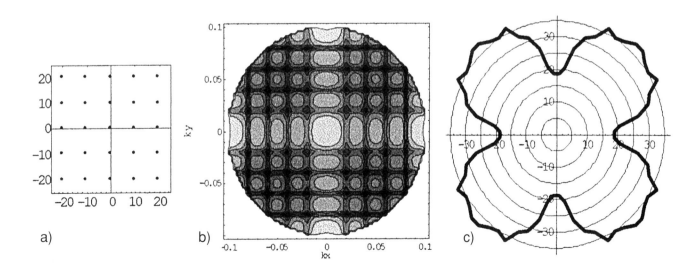

FIG. 3.4. Example of an areal array: (a) position of array elements; (b) contour plot of array response, contour interval is 10 dB; (c) polar plot of average response in range |**k**| = 0.02 till 0.09 m^{-1} expressed in dB suppression. Note relatively poor array response along the two wavenumber axes.

To facilitate the coming discussion on scattered waves, here is more detail regarding performance of arrays for linear events. The amplitude of a linear event can be described as $a = g(d)\, w(t - d/V)$, where $g(d)$ represents the slowly varying geometrical spreading as a function of distance d from shot to receiver, $w(t)$ represents the wavelet as a function of traveltime t and V is velocity. The action of an array is to add the signals received simultaneously at all elements of the array. The amplitudes seen by the different array elements are at any point in time only dependent on the distance

$$d = d(x_s, y_s, x_r, y_r) = \sqrt{(x_s - x_r)^2 + (y_s - y_r)^2}\ .$$

Small variations in distance can be described by

$$\Delta d = \frac{\partial d}{\partial x_s}\Delta x_s + \frac{\partial d}{\partial y_s}\Delta y_s + \frac{\partial d}{\partial x_r}\Delta x_r + \frac{\partial d}{\partial y_r}\Delta y_r,$$

where the various Δs are measured with respect to the nominal position of shot station or receiver station, and $\Delta x_{s,r}, \Delta y_{s,r} \ll d$ (plane-wave assumption).

For a linear receiver array along the receiver line

$$\Delta x_s = \Delta y_s = \Delta y_r = 0.$$

Then for each element i

$$\Delta d_i = \frac{x_r - x_s}{\sqrt{(x_s - x_r)^2 + (y_s - y_r)^2}}\Delta x_{r,i} = \Delta x_{r,i}\cos\varphi,$$

see Figure 3.5. Effectively, for plane waves the length of the array is reduced by a factor $\cos\varphi$.

For an areal receiver array, the same reasoning applies: the projections of all elements of the array on the line SR determine the action of the array for the given shot-receiver azimuth. For an areal shot array, the same formulas apply as for the receiver.

The combination of an M-element linear shot array (along the shot line) and an N-element linear receiver array (along the receiver line) produces $N \times M$ deviations from the nominal shot-receiver distance d. This array is equivalent to an $N \times M$ areal receiver (or shot) array. For equal weights w_j its response $p(k_x, k_y)$ following from equation (3.8) can be written as the product of the responses of the individual arrays (while neglecting phase shifts in x and y)

$$p(k_x, k_y) = \frac{\sin N\pi k_x d_x}{N\sin \pi k_x d_x} \cdot \frac{\sin M\pi k_y d_y}{M\sin \pi k_x d_x}. \quad (3.9)$$

For instance, the 25-element receiver array of Figure 3.4a is equivalent to the combination of a 5-element linear receiver array and a 5-element linear shot array. Most raypaths for the 25 combinations of the two linear arrays are different from the raypaths for a single shot into the equivalent areal receiver array. [Strictly speaking, this should be expressed in equation (3.9): for the combination of a linear receiver array with a linear shot array, k_x corresponds to the varying x-coordinate of the receiver and k_y to the varying y-coordinate of the shot, whereas, for an areal receiver array, both wavenumbers pertain to the receiver coordinates.] Therefore, the underlying assumption is that the linear event does not vary as a function of midpoint (or shot) across the range of the array(s).

Note that the array response also assumes a constant φ for all elements in the array. For small offsets, $\Delta x_{s,r}, \Delta y_{s,r} \ll d$ does not hold, and φ may be different between the different elements and the geometrical spreading $g(d)$ will vary strongly across the array as well. Therefore, for small offsets the array response is no longer equal to the discrete Fourier transform of the position of the array elements.

3.3.3 Scattered-wave noise suppression

For a linear event, the angle with which the raypath leaves the shot is the same as the angle with which the raypath arrives at the receiver; for a scatterer, these two angles are different (see Figure 3.6).

Hence, to investigate the effect of a shot array and a receiver array on a particular scatterer, it is necessary to

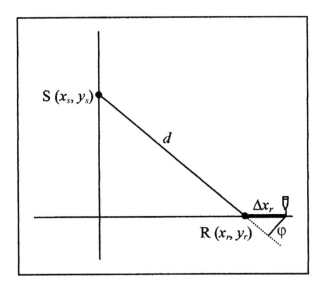

FIG. 3.5. The contribution of an array element to the suppression of a linear event is described by the projection of the element position on SR.

combine the response of the shot array in the φ_1-direction with the response of the receiver array in the φ_2-direction.

Now there is a significant difference between the combination of two linear arrays and an areal receiver array. For instance, for the shot-receiver combination with $x_r = x_d$ and $y_s = y_d$, the angles are $\varphi_1 = 0$ and $\varphi_2 = 90°$. In this case there is no noise suppression by either one of the two linear arrays, whereas the areal geophone array does suppress energy traveling in the φ_2-direction. Linear arrays, if oriented along the corresponding acquisition line, serve as (crude) antialias filters in that direction. However, in the specific example the scatterer's traveltime function is sampled at its apex, where it is horizontal in the shot as well as in the receiver direction (see Figure 3.7). For that situation no antialiasing is required, but suppression of noise energy is required everywhere, also at the apex. Hence, for the best suppression of scattered energy areal arrays have to be considered.

These observations also mean that scatterers in a direction perpendicular to the linear array will not be suppressed, whereas all scatterers inline with the array experience most suppression. It seems that as a consequence, the orientation of a linear array is immaterial for the suppression of scattered energy, if the scatterers are randomly distributed. This is further discussed in Section 3.3.5.

Similar to the direct wave, the array responses for a scatterer assume a constant φ_1 and φ_2 for all elements in the arrays. For small distances of the scatterer to shot or receiver, the plane-wave assumption does not hold.

3.3.4 Analysis of various array combinations

In this section, various array combinations are discussed, some of which are based on actual arrays implemented in the field sometime, somewhere.

Figure 3.4 shows a square array, which can be implemented as such, but which can also be considered as the convolution of two linear arrays. This array response is not quite isotropic: its suppression along the axes is much worse than in between the axes. This is quantified by measuring the average suppression as a function of azimuth over a range of wavenumbers outside the central passband of the array response. Figure 3.4c shows a polar plot for the array of Figure 3.4a.

The reason that the response of the array in Figure 3.4a along the axes is much worse than in between the axes is that the projection of all array elements onto one of the axes effectively reduces the number of elements of the array from 25 to 5. It has been suggested that staggering the array elements would reduce the azimuth dependence of the array response. Figure 3.8a shows the staggered array, Figure 3.8b the array response and Figure 3.8c the corresponding polar plot. Indeed, the array's suppression has improved along the horizontal axis (because there are now effectively 10 elements spaced at 5 m along the horizontal, with weights alternating between 2 and 3), but at the expense of the suppression

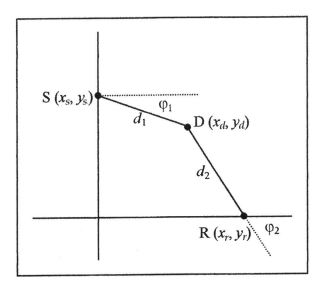

FIG. 3.6. The action of a shot array element on a scatterer is described by the projection of the element position on SD, whereas a receiver array element needs to be projected onto the line DR.

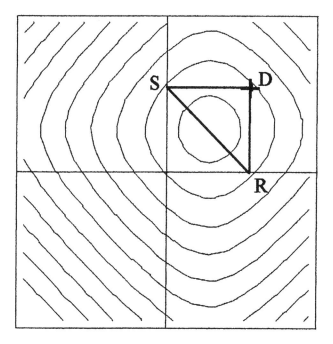

FIG. 3.7. Traveltime surface of scatterer in "+" displayed in midpoint coordinates. Note that the apex lies at the midpoint between S and R.

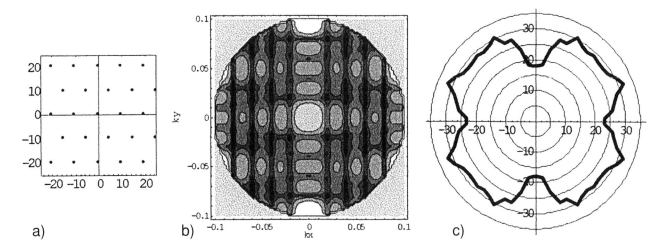

FIG. 3.8. A 25-element staggered array: (a) staggered array, (b) array response, (c) polar plot. As compared to the square array of Figure 3.4, the staggered array shows 5-dB improvement along the horizontal axis, no improvement along the vertical axis, and some 7-dB reduction in suppression along the diagonals.

along the diagonals, whereas the response is still equally bad along the vertical axis.

Arrays with less azimuth-dependency can be constructed by laying out the elements of the array along a number of concentric circles. Figure 3.9 shows two implementations, one with 19 elements, and the other with 28 elements. In both cases the azimuth-dependency is minor, but it is difficult to find an arrangement which has a good response across a wide range of wavenumbers. Yet, it seems to me that this kind of array is to be preferred over the more common square or rectangular arrays. The practical problem of laying out circular arrays might be solvable with some concerted efforts. Arrays consisting of a single circle are being used in some operations (Natali et al., 2001), although the noise suppression by such arrays is not very good. An advantage is that their effective length is larger than the diameter of the circle.

It is always more difficult to use shot arrays than receiver arrays. However, simple shot arrays are often quite feasible, if only because each shotpoint is acquired with three or more vibrators. The combination of a four-point square shot array (in diamond-shape) with a circular receiver array may lead to considerable improvement in noise suppression. This is illustrated in Figure 3.10 for two choices of vibrator point distances: in Figure 3.10a and b for a horizontal and vertical distance of 25 m, in Figure 3.10c and d for a distance of 12.5 m. The imprint of the square shot array is clearly visible on the total array responses.

Departing now from the theoretical considerations, let us have a look at some dilemmas facing the operations geophysicist in practice. For instance, sometimes a series of geophone strings is used, say six geophones per string. Would it make much difference whether the strings were feathered along the receiver line, or whether they were perpendicular to the receiver line (in the latter case, it is easier to maneuver with vibrators from one side of the receiver line to the other).

Figure 3.11 shows results for some different arrays with the same number of geophone strings. Clearly, the zigzag pattern shown in the second row is worse than the other two arrangements shown in the top row and the third row. There is not much difference between feathered strings and perpendicular strings. In both cases there is a direction for which the suppression is not more than 10 dB on average. Adding a fourth string would improve the result, and staggering (alternate geophones in a string on either side of the string) of the geophones also would help. The bottom row illustrates that the shot array can be used to compensate shortcomings of the geophone array. The perpendicular geophone arrangement in the third row has the worst suppression in the inline direction, therefore, orienting the shot array in that direction, with appropriately chosen shot-element interval, improves the response considerably (although the response is still a far cry from the responses shown in Figure 3.10).

If the two arrays are not circularly symmetric there are always scatterers that are located on the loci of worst suppression for both arrays. This is illustrated in Figure 3.12, which shows a feathered geophone array combined with a 3-point shot array. For a given receiver station the scatterers with worst suppression are situated on a line making an angle $\tan^{-1}[-1/3]$ with the positive x-axis and passing through the receiver station, whereas for the shot array all scatterers with worst suppression are situated

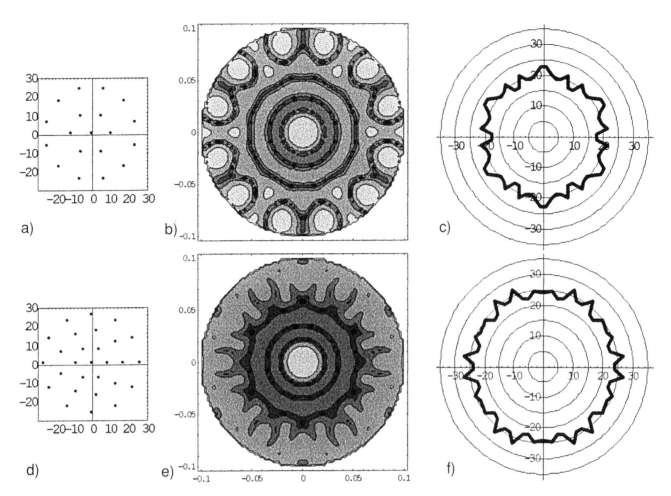

FIG. 3.9. Examples of circular arrays: (a) 19-element array, (b) its array response, (c) its polar plot of average response for wavenumbers 0.02 till 0.09 m^{-1}, (d) – (f) same as (a) – (c) for a 28-element array. The suppression for $|\mathbf{k}| > 0.06$ is not very good in either case, though better for the 28-element array.

along the shot line. A scatterer at the location where the two lines intersect is suppressed by 10 dB only (see Figure 3.12c, d). The best suppression is experienced by scatterers positioned on the intersection of the lines of best suppression for the two arrays (Figure 3.12d). If the shot array is oriented along the shot line, the locations of best and worse suppression change (Figure 3.12e). Note that the responses in Figure 3.12 are only valid for a fixed difference between φ_1 and φ_2 (Figure 3.6). This angle might also be chosen so as to overlay the best direction of one array with the worst direction of the other. In that case another response function would be found.

3.3.5 Discussion

The choice of shot and receiver arrays should depend on the geophysical problem. Knowledge of the amount of scattered energy is essential to come up with the most cost-effective and appropriate solution. Therefore, I recommend acquiring one or more 3-D noise cross-spreads (with very small shot and receiver intervals; see Section 7.2 for an example) for a detailed analysis of the scattered energy. This analysis would provide quantitative information on the energy of the scatterers, the direct waves, and the primaries as a function of wavenumber. Combined with an estimate of the noise suppression in dB by velocity filtering, stacking and migration, the analysis should establish what level of array effort (how many dB suppression across what range of wavenumbers) is required for interpretable data.

The range of wavenumbers to be suppressed determines the spacing between the elements of the arrays and the areal extent of the arrays. The closer the elements the larger the wavenumbers being suppressed and the larger the extent of the array the narrower the central passband. It is generally accepted that the first notch (minimum) of the array response should occur at $2k_N$. Then the central passband extends to $2k_N$, although the noise is already suppressed somewhat between k_N and $2k_N$ by the arrays.

FIG. 3.10. Combination of a four-point diamond-shape shot array with a 28-point circular receiver array: (a) and (b) wide shot array, (c) and (d) narrow shot array.

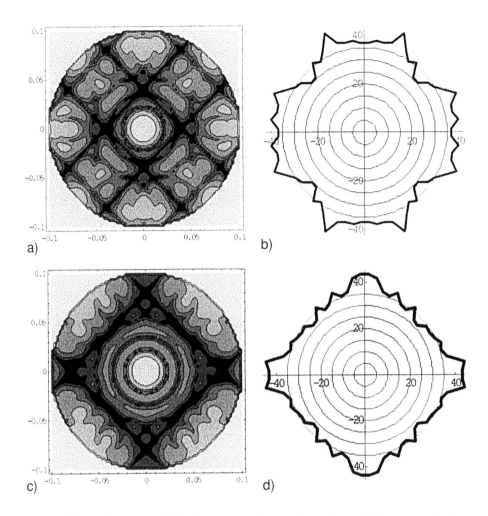

Velocity filtering will also remove some of the noise around k_N (where it does not yet interfere with the signal). For linear arrays the above condition implies that shot and receiver array lengths are equal to station interval. With arrays that are twice as long (fully overlapping arrays) the first notch would occur at k_N, as is customary for antialias filtering, but then the desired signal runs the risk of being affected too much. Shorter arrays are used as well—especially if the station intervals are large—to avoid degradation of the desired signal. In that case the wavefield is undersampled (cf. Figure 1.10), and higher fold may be required for adequate suppression of coherent noise.

If areal geophone and shot arrays are used, it would be sufficient—assuming small variations in the direct wave as a function of midpoint—to have a geophone array producing a notch at $2k_N$ in all directions, and a smaller areal shot array, for instance as in Figure 3.10. Whether the noise is suppressed by the shot array or the receiver array is immaterial, as long as the total suppression is the same. For an equal array response, the noise that remains is the same irrespective of the composition of the two arrays. In practice, there will be some deviations from this, because the linear events will not be strictly a function of offset only. Then a small shot array and a large receiver array will leave more aliased noise in the common receiver than in the common shot. Yet, very often the midpoint-to-midpoint variation of the direct wave will be minimal. In other words, for the direct wave it is sufficient to design an optimal response of the convolution of the two arrays, rather than optimal responses of the individual arrays.

The story is different for scatterers. If an array (shot or receiver) has little suppression for some azimuth, then there will always be scatterer positions that are situated along that azimuth and do not experience much suppression (see Figure 3.12). Hence, if there is a severe scatterer problem, it is important to make both shot and receiver array as azimuth-independent as possible. Yet again, a small areal shot array may be combined with a large areal geophone array.

In case a circular geophone array is combined with a linear shot array, again it is immaterial how the shot array is oriented as far as the linear events are concerned. One

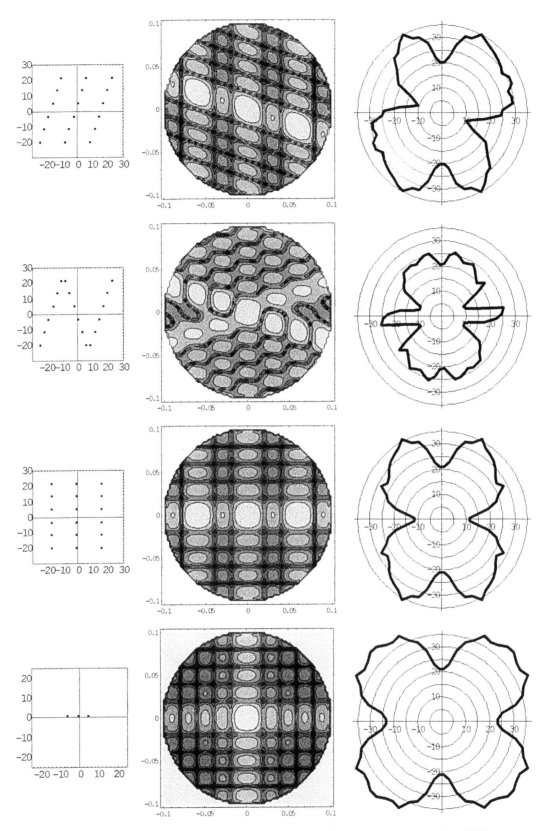

FIG. 3.11. Top three rows describe arrays consisting of three geophone strings each; the middle row array is clearly inferior to the other two. The bottom row describes the convolution of the array above with a three-point shot array.

might have a slight preference for orientation along the shot line, thus reducing any extra aliasing which might occur due to midpoint dependence of the noise. For scatterers, there is an interesting difference. If the shot array is oriented along the shot line, the scatterers experiencing no suppression are situated opposite each shot position (as D versus S in Figure 3.7). On the other hand, if the shot array is oriented perpendicular to the shot line, then all scatterers with bad suppression are situated in the vicinity of the shot line. In other words, in the latter case the same scatterers cause the noise, whereas in the first case scatterers all over the area take turns in contributing to the noise. It is difficult to see which situation is preferable for the final result. Synthetic noise tests might show whether one or the other shot array implementation is to be preferred.

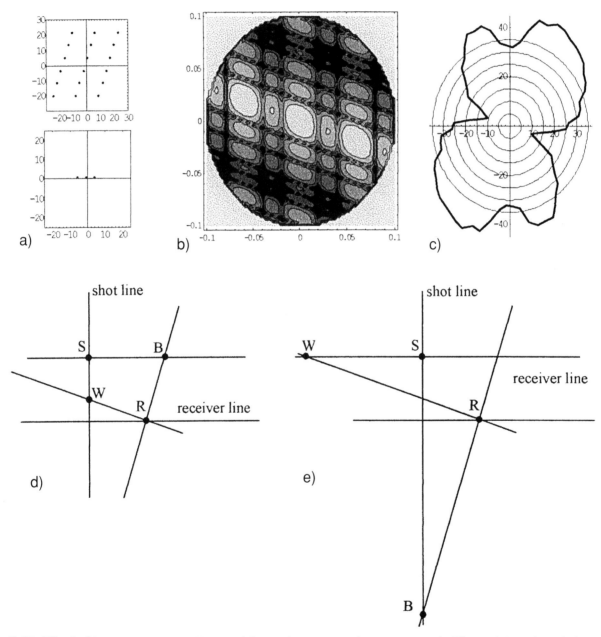

FIG. 3.12. Effect of two arrays on scatterers: (a) geophone array (top, same as in Figure 3.8 top) and shot array (bottom); (b) response of a rotated shot array convolved with geophone array response, rotation selected such that worst directions and best directions of both array responses overlap; (c) polar plot of response shown in (b), in the worst direction the suppression is only 10 dB, in the best direction it is about 60 dB, compare also with polar plot at top of Figure 3.8; (d) location of scatterers with best (B) and worst (W) suppression for shot-receiver combination SR and shot array perpendicular to shot line; (e) same as (d) for shot array oriented along shot line.

If scatterers are not a problem, the easiest way to get good linear-noise suppression is to use a combination of linear shot arrays and linear geophone arrays. This solution requires the smallest number of array elements. Smith (1997) showed with synthetic data tests that these linear arrays (with length equal to station interval) are essential for good noise suppression in case the bandwidth of the noise is large. However, if shot arrays are relatively expensive to implement, areal (preferably circular) geophone arrays are a good alternative. These areal geophone arrays have the added advantage that they do not pass the apexes of the noise cones generated by scatterers.

In very serious noise situations, it might be necessary to make the field arrays wider than long for a better suppression of the apex areas of the diffraction traveltime surfaces (cf. Section 3.2.2). The advent of single-sensor recording (Baeten et al., 2000) would allow the recording of wide receiver lines while postponing side-scatterer suppression to the processing stage.

The choice of arrays may also be related to the acquisition geometry. In a narrow geometry with small maximum crossline offset, most of the linear noise events travel in the inline direction. Then most noise energy may be suppressed if the array response is better in the inline direction than in the crossline direction. If scatterers are important, an areal geophone array may be supplemented by a shot array oriented in the inline direction.

3.4 Stack responses

3.4.1 Introduction

The last part of this chapter is an extension of the discussion in Section 1.6 on stack responses. For 2-D, Section 3.4.2 shows that a regular offset distribution does not necessarily lead to the best stack response, a result which is extended to 3-D in Section 3.4.4. The effect of multiple suppression by stacking is discussed for 2-D data in Section 3.4.3 and is also extended to 3-D in Section 3.4.4.

3.4.2 The 2-D stack response

For a regular geometry and using equal weights for all traces, equation (1.8) describes the 2-D stack response (with $k_i = k_o$):

$$S(k_o) = \frac{\sin N\pi k_o d}{N \sin \pi k_o d}, \quad (3.10)$$

where d is now the constant interval between the traces in the midpoint gather. Figure 3.13a shows the stack response for 48-fold data with $d = 50$ m. The stack response has best suppression around wavenumber $k_o = 1 / (2d)$ and has alias peaks for $k_o = n / d$, n being an integer number. The alias peaks are a consequence of the

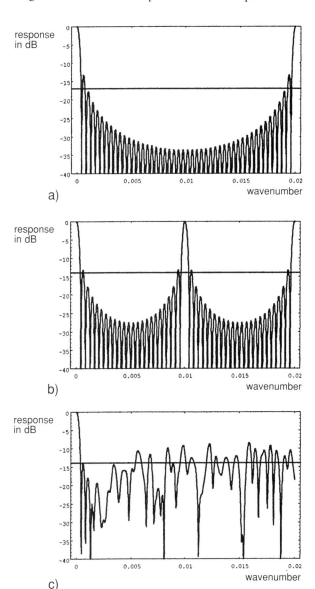

FIG. 3.13. Stack responses of 2-D geometries: (a) regular 48-fold, (b) regular 24-fold, and (c) irregular 24-fold. Horizontal lines indicate the level of random noise suppression. The first alias in the case of the regular 48-fold stack occurs at $k = 0.02$, where no significant coherent noise may be present. However, the first alias of the response of the 24-fold stack, which has double the trace interval of (a), may pass a significant amount of noise. The irregular 24-fold stack, which covers the same offset range as (b) suppresses coherent energy everywhere about as much as it suppresses random noise.

regularity in the offset sampling. In well-sampled 2-D data, the first alias of the stack response ($n = 1$) coincides with the first notch of the field arrays (though only for horizontal events, see Figure 1.15).

For 2-D seismic data, Figure 3.13a illustrates that stacking can suppress coherent noise much better than random noise if the offsets are regularly and densely sampled (Vermeer, 1990). Yet, even for 2-D data, a regular offset distribution is not ideal in general. If the fold-of-coverage is halved by doubling the shot interval, the offset sampling in the CMPs doubles, leading to a first alias peak of the stack response at half the original wavenumber (compare Figures 3.13a and 3.13b). The first alias peak in Figure 3.13b may pass a considerable amount of coherent noise, which was not suppressed by the field arrays either.

For low-fold data, it is better to randomize the offset distribution, as shown in Figure 3.13c. A random offset distribution suppresses coherent noise about as well as it suppresses random noise, whereas a regular offset distribution leads to a periodicity in the offset distribution, which allows the corresponding wavenumbers to escape suppression. Hence, for low-fold data, it is best to have an irregular offset distribution, that is, the CMP should show no periodicities in offset, yet cover the whole range of offsets.

3.4.3 Multiple suppression by stacking

Based on the wave equation, several multiple removal schemes have been introduced using a spatial-temporal filter. These techniques can even be successful in case there is no differential moveout between primaries and multiples. However, here I concentrate on the multiple suppression that is achievable with the stacking process.

3.4.3.1 Multiples with small differential moveout

After NMO-correction for the primary velocity, the moveout Δt of a multiple (its differential moveout) as a function of offset can be approximated by a parabola, provided this differential moveout is small:

$$\Delta t_i = \Delta t_f (x_i / x_f)^2, \quad (3.11)$$

where x_f is some fixed offset and Δt_f the differential moveout for that offset. The stack response for multiples is found by summing the phase-shifts of the traces:

where φ_f is the phase shift in radians at x_f for a frequency f, $\varphi_f = 2\pi f \Delta t_f$.

Figure 3.14 shows displays of the absolute value of the 2-D stack response, as a function of wavenumber (left) and as a function of differential moveout at a fixed offset (right). In the top part of Figure 3.14 equal weights are used, whereas the bottom part has been computed for weights proportional to offset. The offset weighting quasi-linearizes the amplitude of the multiple as a function of offset squared, leading to a stack response with a shape similar to that for equal weighting of linear noise (compare Figure 3.14a and 3.14d).

3.4.3.2 Multiples with large differential moveout

In general, stacking of multiples with large differential moveout should produce residual multiple energy originating from the short offsets only. However, for low-fold, data stacking may also produce residual multiple energy originating from the long offsets.

This is illustrated in Figures 3.15 and 3.16.[1] In particular, in the time window from 2.6 to 2.8 s, Figure 3.15 shows some steeply dipping events, which are not conformable with the overall structure. Figure 3.16 shows the (f,k)- spectrum of the section shown in Figure 3.15. It shows a secondary energy peak around $k_m = k_n / 2$ pointing at a spatial periodicity of four traces. The corresponding acquisition geometry has a shot interval of 100 m, receiver interval of 50 m, and 60 offsets, leading to 15-fold data with 200 m between traces in the CMP, and a spatial offset periodicity of four traces as a function of midpoint. Figure 3.17a shows the (amplitude of the) stack response. It has its first alias peak at $k_m = 1/200$ m^{-1}.

Some algebra leads to the following expression for the apparent velocity V_o of the multiple with velocity V_m after correction for the primary velocity V_p,

$$V_o = \frac{1}{x_o} \bigg/ \left(\frac{1}{t_m V_m^2} - \frac{1}{t_p V_p^2} \right), \quad (3.13)$$

where t_p, t_m are the reflection times at x_o for primary and multiple, respectively. Lines for $f = k_o V_o$ for constant x_o have been drawn in Figure 3.17b. They show that aliasing starts around 40 Hz for the largest offsets. This is confirmed by the secondary peak in Figure 3.16, which

$$S(\varphi_f) = \sum_{j=1}^{N} w_j \exp(i\varphi_j) \bigg/ \sum_{j=1}^{N} w_j = \sum_{j=1}^{N} w_j \exp(2\pi i f \Delta t_j) \bigg/ \sum_{j=1}^{N} w_j = \sum_{j=1}^{N} w_j \exp(i\varphi_f (x_j/x_f)^2) \bigg/ \sum_{j=1}^{N} w_j), \quad (3.12)$$

---

[1] I am indebted to Cees Corsten for the data shown and explanation given in this section.

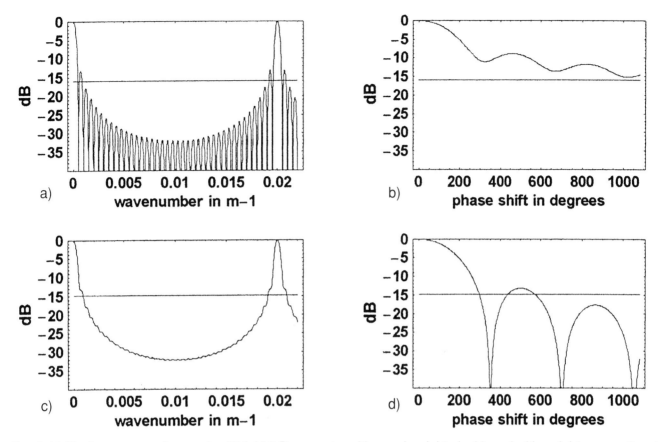

FIG. 3.14 Stack responses for regular 40-fold 2-D geometry with equal weights (a, b), and with weights proportional to offset (c, d). Left: linear noise suppression, right: multiple suppression. Horizontal lines indicate level of random-noise suppression.

also starts around 40 Hz. Lower frequencies do not alias in the CMP for this multiple. Figure 3.17c shows a representative CMP after NMO correction. It shows the steeply dipping multiple, which is only partially suppressed by stacking due to the large distance between the traces. The corresponding 30-fold stack does not show this kind of multiple passed by stacking, because in that case aliasing only starts at 80 Hz.

This little case history clearly shows that peaks in the stack response should be avoided in the wavenumber range where strong noise energy occurs.

3.4.4 3-D stack responses

For 3-D, offset is two-dimensional, hence the stack response is two-dimensional and can be defined as

$$S(\mathbf{k}_o) = \sum_{j=1}^{N} w_j \exp(2\pi \mathbf{k}_o \cdot \mathbf{x}_{oj}) \bigg/ \sum_{j=1}^{N} w_j, \quad (3.14)$$

where $\mathbf{k}_o$ and $\mathbf{x}_{oj}$ are now the two-dimensional offset wavenumber and offset, respectively. For coherent noise that only depends on the shot-to-receiver offset, i.e., is not azimuth-dependent, the stack response can be computed as a function of absolute offset. To a large extent this condition holds for various types of noise, such as the direct arrival ground roll and near-surface multiples in horizontal layering. In the following, I assume this condition holds.

In 3-D parallel geometry (multisource, multistreamer acquisition), the offset distribution is almost as regular as for 2-D lines, except for the sampling of the short offsets along the outer streamers. In other 3-D geometries (orthogonal geometry, zigzag geometry), the offset distribution tends to vary strongly between midpoint gathers.

Table 3.1 lists the main parameters of the geometries for which the stack responses are shown in Figure 3.18. Because the offset distribution varies across the midpoint gathers of the geometry, each gather has its own stack response. Rather than drawing all stack responses of a geometry, the average stack response is plotted, together with the standard deviation in the stack response on both sides of the average.

The observation that the 2-D stack of a regular offset distribution suppresses coherent noise better than random noise (Figure 3.13a) has led to the widespread

64 Chapter 3 Noise Suppression

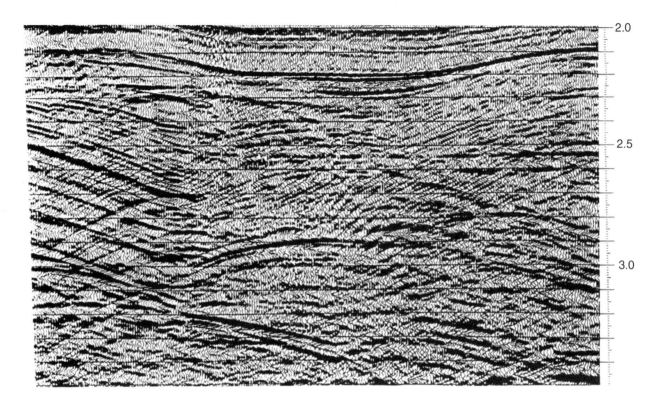

FIG. 3.15. A 15-fold stack showing multiples that were aliased in the CMP in time range 2.6–2.8 s.

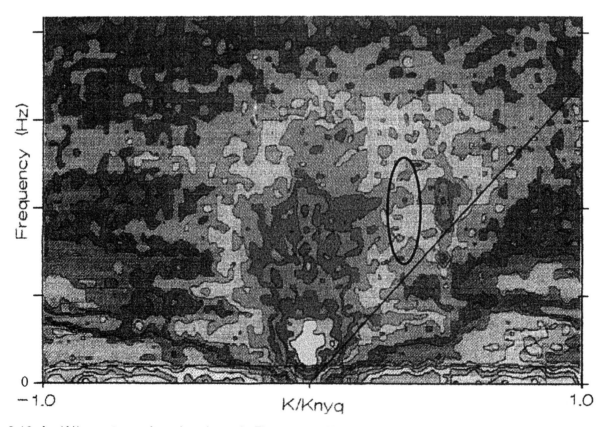

FIG. 3.16. An (f,k)-spectrum of section shown in Figure 3.15. Note (highlighted) band of higher energy at $k = k_N/2$.

Table 3.1. Parameters of 3-D geometries.

Id	Name	Crossline × inline fold	Receiver-line interval	Shot-line interval	Aspect ratio	Figure
1	4-line orthogonal	2 × 15 = 30	200 m	200 m	0.13	3.18a,b
2	4-line brick	2 × 15 = 30	200 m	200 m	0.13	3.18c,d
3	4-line double zigzag	2 × 15 × 2 = 60	200 m	200/2 m	0.13	3.18e,f
4	12-line orthogonal	6 × 6 = 36	450 m	450 m	1.0	3.18g,h

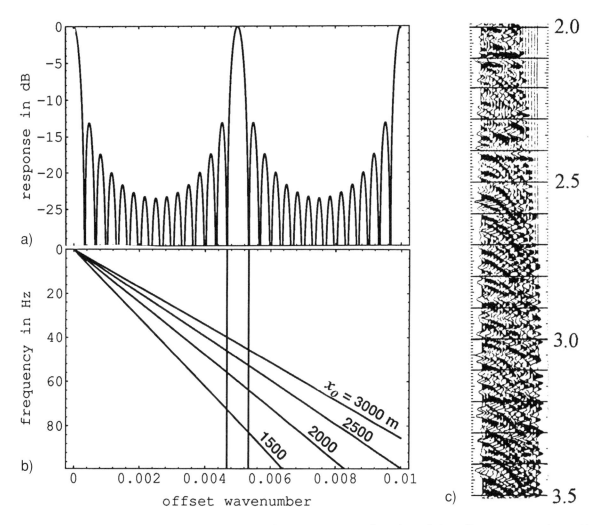

FIG. 3.17. Explanation of multiple aliasing: (a) stack response as a function of the offset wavenumber with alias band at $k_o = 1/200$ m^{-1}; (b) apparent velocity in (f, k_o) of the multiple after NMO correction for various offsets, according to equation (3.13) for zero-offset time 2.6 s, $V_p = 3000$ m/s, and $V_m = 2100$ m/s; (c) CMP, NMO-corrected with primary velocity, showing aliased multiples below 2.4 s.

66   Chapter 3   Noise Suppression

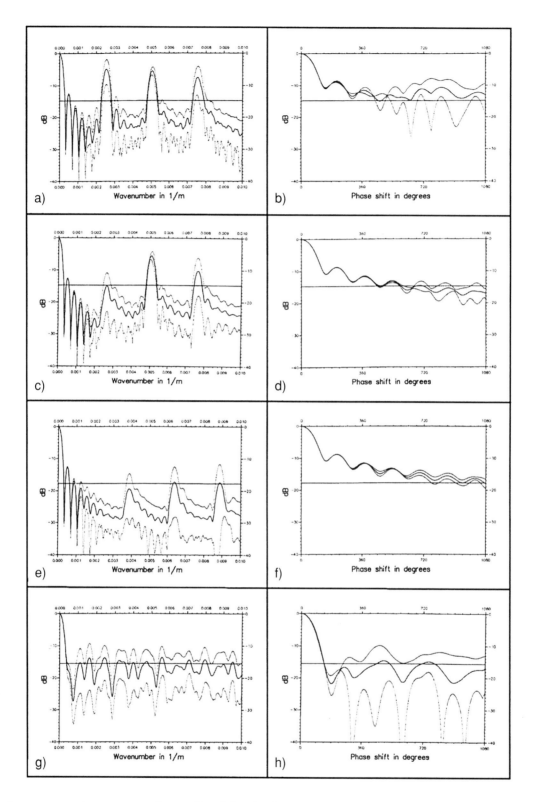

FIG. 3.18. Average stack responses (heavy lines) for the four acquisition geometries listed in Table 3.1. Standard deviations in each average are indicated as well. Left column: amplitude of stack response for linear noise suppression. Right column: amplitude of stack response for suppression of multiples with small differential moveout. (a, b) narrow orthogonal, (c, d) narrow brick, (e, f) double zigzag, (g, h) wide orthogonal.

belief that 3-D survey design should aim for regular offset distributions. However, as shown by the comparison of Figure 3.13b and 3.13c, low-fold data should not have a regular offset distribution for optimum noise suppression.

In 3-D surveys, the fold-of-coverage is often much smaller than in 2-D. If so, a regular offset distribution would produce peaks in the stack response through which coherent noise events could pass. Narrow geometries tend to produce periodicities in the offset distribution, leading to peaks in the stack response. This is illustrated in Figure 3.18a for geometry 1. The brick-wall geometry 2, with the same offsets as geometry 1, has a better stack response than geometry 1, because its first peak in the stack response is much weaker than for geometry 1 (Figure 3.18c). The other peaks of the two stack responses virtually coincide.

Double-zigzag geometry represents a special case. For a small aspect ratio, each CMP in this high-fold geometry has a nearly regular offset distribution, leading to a very good stack response (the first strong peak in the stack response occurs at a high wavenumber due to the high fold). Figure 3.18e shows the stack response of a double-zigzag geometry. It should be realized that in a wide double-zigzag geometry the offsets would be distributed less regularly, leading to a random-noise type suppression.

Selecting a wide orthogonal geometry leads automatically to an irregular offset distribution, making the stack responses of the various CMP gathers as flat as possible on average (Figure 3.18g).

In geometries 1–3, the offsets are distributed quite evenly across the total range of offsets (apart from some periodicities in geometries 1 and 2). This means that their ability to suppress multiples with small differential moveout is about the same, as illustrated by the right side of Figure 3.18. In wide orthogonal geometry, there is a preponderance of long offsets leading to a similar effect as offset-weighting for the 2-D stack response as shown in Figure 3.14d. Figure 3.18h shows that the multiple suppression using a wide geometry is better than the multiple suppression by narrow geometries.

3.4.5 Discussion

It should be emphasized again, that the stack responses of the 3-D geometries were made for absolute offset. This means that they are only valid for events, which are not azimuth-dependent.

In the figures the stack response is always shown as a function of offset wavenumber k_o starting at $k_o = 0$. To judge the effect of stacking, it is also necessary to know the energy distribution as a function of wavenumber of the data. Some 3-D survey design programs incorporate this possibility by allowing the user to specify the range of wavenumbers for which an average stack response must be computed.

The stack response is by no means the one and only criterion by which to judge the quality of an acquisition geometry. For instance, in a wide geometry, dual-domain (shot and receiver) (f, k)-processing or 3-D velocity filtering can take care of much of the ground-roll energy that is not going to be suppressed by the stack. Also migration suppresses much of the energy that does not fit the migration model (Smith and McKinley, 1996). Therefore, the not-so-good stack response of wide geometries (although better than the stack responses of narrow orthogonal and narrow brick geometry) can be compensated to some extent in processing.

References

Baeten, G. J. M. et al., 2000, Acquisition and processing of point receiver measurements in land seismic: 62nd Conf., Eur. Assoc. Geosc. and Eng., Extended Abstracts, Paper B-06.

Larner, K., et al., 1983, Coherent noise in marine seismic data: Geophysics, **48**, 854–886.

Meunier, J., 1999, 3D geometry, velocity filtering and scattered noise: 69th Ann. Internat. Mtg., Soc. Expl. Geophys., Expanded Abstracts, 1216–1219.

Natali, S., Roux, R., Dea, P., and Barrett, F., 2001, Cave Gulch 3-D survey, Wind River Basin, Wyoming: A major gas discovery developed using poststack depth migration: The Leading Edge, **20**, 262–268.

Newman, P., and Mahoney, J. T., 1973, Patterns—With a pinch of salt: Geophys. Prosp., **21**, 197–219.

Regone, C. J., 1997, Measurement and identification of 3-D coherent noise generated from irregular surface carbonates, *in* Palaz, I., and Marfurt, K.J., Ed., Carbonate seismology: Soc. Expl. Geophys., 281–305.

Smith, J. W., 1997, Simple linear inline field arrays may save the day for 3-D direct-arrival noise rejection: Proceedings, SEG Summer Research Workshop.

Smith, J. W., and McKinley, H. J., 1996, Now what's happened to groundroll? A 3D perspective on linear-moveout noise rejection: 66th Ann. Internat. Mtg., Soc. Expl. Geophys., Expanded Abstracts, 72–74.

Vermeer, G. J. O., 1990, Seismic wavefield sampling: Soc. Expl. Geophys.

Chapter 4
Guidelines for design of "land-type" 3-D geometry

4.1 Introduction

In this chapter the symmetric sampling criteria are expanded into guidelines for parameter selection for the survey geometry.

Often, geophysicists dealing with the design of 3-D seismic surveys concentrate on the properties of the bin: offset distribution, azimuth mix, midpoint scatter. In my approach, even more emphasis is put on the spatial properties of a geometry across the bins. These spatial aspects are so important because most seismic processing programs operate in some spatial domain, i.e., combine neighboring traces into new output traces, and because it is the spatial behavior of the 3-D seismic volume which the interpreter has to translate into maps.

These guidelines start with a brief description of the knowledge base, which has to be built to allow a satisfactory choice of all parameters. The first choice to be made is the type of geometry. In general, orthogonal geometry is the geometry of choice for land data acquisition and for marine data acquisition in combination with ocean-bottom cables. Yet, other geometries may also be selected, and a short review outlines pros and cons of various geometries that may be chosen.

This chapter focuses on orthogonal geometry. If 3-D symmetric sampling is taken as a starting point, the choice of parameters for this geometry is simplified considerably. Instead of having to decide on the shot interval and on the receiver interval, a decision need only be made as to the sampling interval. Similarly, the maximum inline and maximum crossline offsets can be made equal. It is also recommended to see what the consequences are of making the shot-line interval and the receiver-line interval the same. Another benefit of symmetric sampling is that the designer does not need to worry about the offset distribution: 3-D symmetric sampling automatically leads to a reasonable offset distribution.

The choice of the various parameters depends on the geophysical requirements, which in turn are often a trade-off between what the interpreter would like to see and what the budget will permit. In my view, the most important geophysical requirements are spatial continuity, resolution, shallowest horizon to be mapped, deepest horizon to be mapped, and the signal-to-noise ratio. These requirements and their consequences for parameter choice are discussed extensively in this chapter.

Although symmetric sampling is a starting point for survey design, there are often good reasons for deviating from it. Various situations are sketched to describe reasons for and consequences of using asymmetric sampling. This chapter is rounded off with a discussion of attribute analysis and model-based survey design.

4.2 Preparations
4.2.1 Objective of survey

The designer of a 3-D seismic survey should be familiar with the objectives of the survey. A rough classification of objectives is

- structural interpretation
- stratigraphic interpretation
- reservoir characterization
  - porosity
  - pore fill
  - fracture orientation
- time lapse

In practice, these objectives need to be refined after obtaining a detailed description of the geological and geophysical problem.

4.2.2 Know your problem

Before the design task can start, some groundwork has to be done to collect and quantify the information that is available for the survey area. Information that will be needed is listed below:

- Time/depth of shallowest event of interest (for statics or for mapping)

- Time/depth of shallowest objective (prospective level)
- Time/depth of deepest objective or main objective
- Required resolution, or maximum frequency at those levels
- Steepest dips at those levels
- Representative velocity function(s) (several may be needed if there is strong lateral variation)
- Representative mute function (might be computed from velocity function)
- Information on data quality problems (multiples, scatterers, ground roll, statics)
- Interpretable survey area
- Interpreted seismic sections
- Raw shots
- Terrain conditions
- For complex geology: model(s) of the structure

Sometimes more information will be useful, e.g., for AVO analysis it would be very helpful to know the main petrophysical parameters around reservoir level.

Much of the information listed above will be referred to explicitly in the design discussion in this chapter. Some other information will be used only implicitly, important is a thorough familiarity with the objectives and the problems so that a survey design can be recommended with confidence in the outcome.

4.3 The choice of geometry

In general, the orthogonal geometry is the geometry of choice for land data acquisition. However, there are situations in which it may be preferable to choose a different geometry. In this section parallel geometry, zigzag geometry, slanted geometry and areal geometry are compared with orthogonal geometry, and target-oriented geometries are discussed. Here, I assume acquisition of P-wave data; a discussion of 3-D survey design and choice of geometry for converted waves is given in Vermeer (1999b) and is expanded into a full discussion in Chapter 6.

4.3.1 Parallel geometry versus orthogonal geometry

A detailed discussion of the pros and cons of streamers (parallel geometry) versus stationary receivers (orthogonal or areal geometry) is given in Vermeer (1997) and reprinted as Chapter 5.

In land data acquisition, parallel geometry would normally be too expensive, because the acquisition line spacings have to be small for good crossline sampling intervals. The close line spacing also requires virtually unlimited access, which is only available in specific environments (deserts, tundras, etc.). Therefore, only in very rare situations, parallel geometry is used on land. Schroeder et al. (1998) use the data of a parallel geometry acquired on land to study the effect of fold and bin size on quality.

Similar to the marine situation, parallel geometry on land is acquired using swaths composed of a few source lines and a few receiver lines. Usually, the crossline fold is 1, which may lead to decoupling of statics in the crossline direction (Wisecup, 1994). Irregular illumination, which is inherent in this configuration (see Section 5.3.2.4), is less severe than in the marine situation in case center-spread acquisition is applied. Moreover, feathering, which is a main disadvantage of parallel geometry using streamers, does not occur on land. Therefore, on land, advantages of the parallel geometry, such as a better stack response than that of orthogonal geometry, can be fully exploited.

In both parallel geometry and orthogonal geometry, common-receiver gathers and common-shot gathers can be sampled if symmetric sampling is applied. This distinguishes these two geometries from other geometries (slanted, zigzag) for which receiver gathers tend to be of variable length. (Here, as usual, I refer to receiver gathers as part of the basic subset of the geometry; i.e., the shots in the gather are located on a single source line.) The main distinction between parallel and orthogonal geometry is that parallel geometry is basically single azimuth, whereas orthogonal geometry is wide azimuth. Usually, this difference has little consequence for the imaging capabilities of the two geometries. Only in very complex geology, some shot-receiver azimuths are not very suitable for illumination (cf. dip/strike decision, Section 5.3.1.1). In those situations, orthogonal geometry may be at an advantage as it will always include shot-receiver azimuths that are most suitable. Parallel geometry is not suitable for investigating azimuth-dependent effects, unless this geometry is acquired in two or more different directions.

Apart from a better stack response for linear noise suppression, parallel geometry has some more distinct advantages over orthogonal geometry. First, for the same fold, parallel geometry has better potential resolution than orthogonal geometry, because of relatively more short offsets, i.e., it suffers less from NMO stretch effects. More short offsets also leads to better imaging of the shallow data (although this advantage may be lost when a wide swath is used). Processing parallel geometry data is much more straightforward than processing orthogonal geometry data. Vermeer (1998b, see also

Chapter 10) discusses the problem of creating common-image gathers from data acquired with orthogonal geometry. AVO analysis using parallel geometry tends to be easier as well, and it can have higher resolution, because it has virtually the same offset distribution (apart from minor variations in the short offsets) in all CMPs (cf. discussion in Section 2.6.9).

Dickinson et al. (1990) discuss a comparison between data acquired with parallel geometry and orthogonal geometry. The final results were not very different, but the CMP gathers of the cross-spread data looked much noisier than those of parallel geometry. This can be attributed to the noise, which may look very incoherent in CMP gathers of orthogonal geometry because of the range of azimuths and also because of the irregular offset sampling.

4.3.2 Zigzag geometry versus orthogonal geometry

Zigzag geometry is most efficient in open areas such as deserts. The distance to be traveled by the vibrators is $\sqrt{2}$ shorter than for an equivalent orthogonal geometry with the added advantage that it is easier for the vibrators to avoid running over the geophones (*equivalent orthogonal geometry*: orthogonal geometry with the same maximum inline and crossline offsets as zigzag geometry, and effectively the same shot-line interval and the same receiver-line interval, i.e., with the same trace density).

All current processing packages are based on binning, and the common perception is that processing works best if all midpoints are located as much as possible in the bin center (yet, in DMO the DMO-correction traces cannot be forced into bin centers anyway). Therefore, the inline move-up of the shots is made equal to the receiver-station interval. Another, perhaps even more compelling, reason for this choice of shot move-up is that it allows center-spread acquisition for each individual shot by moving the active spread together with the shot. As a consequence, the shot interval is $\sqrt{2}$ times the station interval. For alias-free recording of the common-receiver gathers in the zig- and zag-spreads (see Section 2.3.4), the shot interval has to be equal to or less than the basic signal sampling interval, but then the common-shot gathers would be oversampled. The equivalent orthogonal geometry would be oversampled in both shots and receivers, i.e., fewer shots and receivers would be needed in the orthogonal geometry to achieve alias-free sampling of the same maximum frequency. This reasoning suggests that a zigzag geometry is perhaps not as efficient as it seems to be.

Prestack processing of the zig- and zag-spreads has to deal with lower apparent velocities than orthogonal cross-spreads. This can be seen from the contour plots in Figure 2.4. For the same contour interval the contours in the zig-spreads are locally closer than in the cross-spreads. This also means that the prestack migration operator is more likely to alias.

The same sampling disadvantage applies to the double-zigzag geometry (defined in Section 2.3.4). However, the double-zigzag geometry does have the attractive property that its average stack response approaches the stack response of a high-fold 2-D geometry (see Figure 3.18e).

The very good stack response of the double-zigzag geometry is only possible by maintaining a small number of closely spaced receiver lines. Therefore, this geometry has a very high shot density, and it has many short offsets.

In a wide orthogonal geometry the average stack response is not very good, but at least it does not have any peaks (see Figure 3.18g). Suppression of ground roll in that geometry can be achieved mostly by dual-domain (f,k)-filtering or 3-D velocity filtering. In double zigzag geometry, (f,k)-filtering can only be applied satisfactorily in the common-shot gathers, not in common-receiver gathers. In other words, there is no clear reason why suppression of ground roll can be achieved any better in a double-zigzag geometry than in a wide orthogonal geometry.

Multiples with small differential moveout with respect to the primaries are better suppressed by stacking in a wide orthogonal geometry than in the narrow double zigzag. This is caused by the preponderance of long offsets in the wide geometry (compare Figure 3.18f with 3.18h). Similarly for 2-D processing (cf. Figure 3.14), offset weighting in the double-zigzag geometry may lead to better suppression of multiples with small differential moveout.

The energy of multiples with large differential moveout is spread out along the offset wavenumber axis and is best suppressed by a dense equidistant offset sampling in the CMP. For those multiples, if strong, the double-zigzag geometry is at an advantage, unless the multiples could be suppressed satisfactorily by some prestack multiple elimination.

Some multiple elimination programs assume that the multiples have hyperbolic moveout as a function of offset. If such programs are the only ones available for multiple elimination, this may have some consequence for the choice of geometry. Hyperbolic moveout may be assumed for horizontal geologies and for not too complicated geologies provided the azimuth does not vary. For horizontal geologies such as a wide geometry and the double-zigzag geometry, such programs should be equally suitable for multiple elimination. However, for wide geometries and in dipping geologies, moveout of multiples will not vary smoothly as a function of absolute offset, because it will also be dependent on azimuth. For such geologies, the double-zigzag geometry will be at an advantage, as the azimuth variation is very limited in that geometry.

An advantage of double-zigzag geometry and parallel geometry is that the relatively larger number of small offsets leads to better resolution as compared to all wide geometries which have a preponderance of long offsets and suffer more from NMO stretch.

4.3.3 Slanted geometry versus orthogonal geometry

Slanted geometry (sometimes called slash geometry) represents a modification of orthogonal geometry, in particular of brick-wall geometry (for a discussion of brick-wall geometry, see Section 7.3). In this geometry the shot lines are nonorthogonal to the receiver lines. The geometry is an improvement over the brick geometry because the shot line is no longer discontinuous. Instead of cross-spreads, slanted spreads are the basic subsets of the geometry.

For low-fold, and, similarly, brick-wall geometry, slanted geometry tends to have a better distribution of the sparse offsets across the total offset range for each bin, thus reducing the geometry imprint. This advantage is quite irrelevant in areas with some dip, and reduces as well for high fold. Depending on the angle of the shot lines with the receiver lines, LMOS is smaller than in an equivalent orthogonal geometry. (LMOS is defined in Section 2.5.6. The smaller LMOS, the better the shallow coverage, see Section 4.4.3.)

An attribute comparison of slanted geometry with orthogonal geometry is given in Section 4.6.6. The next section compares the subsets of slanted geometry, zigzag geometry and orthogonal geometry in more detail.

4.3.4 Comparison of sampled minimal data sets of crossed-array geometries

Figure 4.1 shows the midpoints of the MDSs of the orthogonal geometry, the zigzag geometry and the slanted geometry for 16 shots recorded in 16 receivers. These numbers have been kept small to avoid clutter of points. Maximum crossline offset is the same as maximum inline offset in all three cases.

In the zig-spread the inline range of midpoints is much larger than in the cross-spread, because the active spread moves with the shots. Whereas in the cross-spread the number of midpoints in the common receiver is constant, the number of points in the common receiver of the zig-spread varies. Moreover, the sampling interval in the common receivers of the zig-spread is $\sqrt{2}$ times that in the common shot. Lines parallel to the edges of the zig-spread represent common-inline-offset gathers. This gather contains a trace from each shot in the zig-spread, just like the common receiver in the cross-spread, but the sampling interval is two midpoints in the inline direction.

The slanted spread has many features in common with the zig-spread. Again the common receivers do not all have the same length, and the sampling interval in the common receiver is larger than in the common shot.

A consequence of the variable length of the common receivers in zig-spread and slanted spread is that dual-domain filtering is not really practical. Hence advantages of spatial continuity cannot be fully exploited in these cases.

The slant of the shot line in zigzag geometry and in slanted geometry also leads to lower apparent velocities of the diffraction traveltime surfaces in the crossline direction. This can be seen by inspection of Figure 2.4a. This may lead to aliasing of the migration operator in these geometries sooner than in orthogonal geometry.

In the cross-spread the shot line and the receiver line split the midpoint area in four quadrants of equal size. In zigzag and slanted geometry the midpoint area is split into unequal areas. Without a special effort to true-amplitude processing, this will lead to geometry effects in prestack-migrated amplitudes.

Another problem with zig-spreads and slanted spreads is that splitting over OVTs is more difficult to arrange. It can be done, however, and it is interesting to note that an OVT in the zig-spread will have the same offset and azimuth range as the equivalent OVT in the cross-spread.

Summarizing, the commonly applied asymmetric sampling in zigzag and slanted geometry leads sooner to aliasing. Dual-domain filtering cannot be carried out. Amplitudes are more difficult to control. On top of this, shot lines are longer in these geometries. Depending on the terrain, this may increase cost. Although the spatial continuity of these geometries is much better than of a brick-wall geometry, they are still inferior to orthogonal geometry.

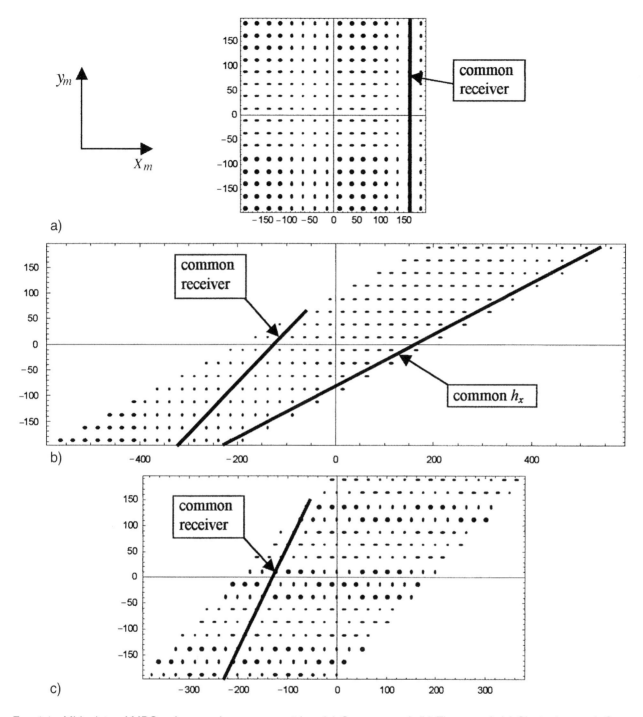

FIG. 4.1. Midpoints of MDSs of crossed-array geometries. (a) Cross-spread, (b) Zig-spread, (c) Slanted spread. Common shots are rows of horizontal midpoints, common receivers are parallel to shot line and are indicated. In zig-spread, each next shot moves a station interval to the right together with the receiver spread, hence two midpoint positions. In slanted spread (with tan α = 2), pairs of shots shoot into the same spread, for the next pair the spread moves one station to the right. Note variable length of common-receiver gathers in zig-spread and in slanted spread. In zig-spread the common inline offset gather (constant h_x) contains a constant number of traces.

4.3.5 Areal geometry

Areal geometries with a very coarse grid of receiver stations and a dense grid of shots are becoming more important, especially in marine data acquisition. It is the geometry of choice for deployment of vertical hydrophone cables. Each vertical hydrophone cable is a self-contained recording system, consisting of 12 or 16 hydrophones at regular intervals along a cable kept vertical in the water. The data are recorded in a floating recording unit (see also Vermeer, 1997 and Section 5.4.2).

Another acquisition technique, using the same geometry, is based on OBS (ocean-bottom seismometer) systems. These systems may be equipped with a hydrophone and a 3-component geophone for recording of the complete wavefield at the sea bottom. Two systems are in use: one type of OBS uses gravity for deployment and buoyancy for retrieval, and a more modern type—based on SUMIC experience (Berg et al., 1994)—uses an ROV (subsea robot) to plant the geophones and to retrieve the equipment from the sea floor.

A similar approach to marine acquisition is used in the Teal South Project (Ebrom et al., 1998). Here 6 four-component receivers are placed at 200 m intervals in four receiver lines with a receiver-line distance of 400 m. Shots are fired every 25 m in both x- and y-directions.

Disadvantages of streamer acquisition such as striping are overcome by this type of acquisition geometry. Moreover, in case of multicomponent recording shear-wave information is becoming available. Similar to the orthogonal geometry used on land and in ocean-bottom cable (OBC) acquisition, the shallow subsurface will not be completely illuminated now, owing to the distance between the recording units.

Another interesting application of areal geometry may develop in combination with shallow VSPs. A number of holes is drilled in which multicomponent sensors are permanently installed well below the weathering zone, but not too deep. Then the area is covered with shots in a dense grid. The data acquired in this way will suffer less from ground roll, and as attenuation of high frequencies takes place mainly in the very shallow subsurface, the data perhaps will contain more usable high frequencies. The technique would be expensive, but might be very appropriate for seismic reservoir monitoring (time-lapse seismic).

A disadvantage of areal geometry is the lower resolution of the 3-D receiver gathers compared to other geometries for the same migration operator radius (Vermeer, 1999a, and Chapter 8). In particular *PS*-data acquired with areal geometry suffer from low resolution (Vermeer, 1999b, and Chapter 6).

4.3.6 Target-oriented geometries

In the remainder of this chapter orthogonal geometry is used as a starting point to show how the geophysical requirements should influence the choice of parameters of that geometry. For instance, in a complex geology it is more important to use small sampling intervals than in a geology which is basically flat. However, there are also situations in which the complexity of the geology requires a local adaptation of the chosen geometry or even a different choice of geometry.

In some situations, the shallow subsurface may vary rapidly locally. In general, this would require a dense acquisition-line spacing such that these variations can be mapped. If the location of the anomaly (such as a shallow top of a salt dome) is known from earlier surveys, it may be sufficient to opt for a locally higher density of acquisition lines.

It may also be necessary to adopt an entirely different geometry. The main example of this situation is a subsurface with reservoirs being truncated (and sealed) by the flanks of more or less circular salt domes. It has been demonstrated that, in that case, a concentric circle shoot leads to better imaging of all flanks than a parallel geometry (Reilly, 1995). There are two reasons for this difference: (a) in circular geometry the raypaths stay outside the salt, whereas in parallel geometry there are many raypaths with one leg through the salt, and (b) the existence of so-called prism waves, consisting of raypaths that are reflected twice, once against the salt flank and once against the clastic sediments, before returning to the surface. These waves can have large energy and travel in a vertical plane more or less perpendicular to the salt flank. The prism waves are difficult to process properly, but are mostly avoided in the circular geometry.

In orthogonal geometry all shot/receiver azimuths are present, so that each CMP also receives contributions from raypaths striking the salt flank. On the other hand there are also shot/receiver combinations that are unfavorable for imaging. Rather than suppressing such traces in processing, one may try to avoid acquiring them. To some extent this may be achieved by what may be called "spider-web geometry" (see Figure 4.2), in which the shots are located along circular lines around the salt dome, and receivers along radial lines (Holland, 2000). The same geometry might be acquired by interchanging locations of shots and receivers; this geometry is not covered in Holland's patent. Constance et al. (1999) describe a real implementation of this geometry, where it is also combined with acquiring 3-D, 3-C VSP data in two well-bores. Bloor et al. (1999) illustrate the benefit of a clever migration-amplitude equalization technique using the surface data described in Constance et al. (1999).

4.4 Design criteria and parameter selection

This section reviews the criteria that have to be satisfied in 3-D survey design and it discusses the consequences of each criterion on the selection of parameters for the nominal geometry. The criteria to be discussed, and the related parameters are listed in Table 4.1:

Though these criteria are discussed separately, they are, of course, also interrelated. Data with a good deal of spatial continuity will, in general, also allow good noise suppression, and the fold resulting from the third and fourth requirement is often large enough for adequate noise suppression.

Table 4.1 suggests that the shallowest horizon to be mapped always determines the line spacing. However, there are exceptions to this general rule, and, therefore, it is prudent to establish the required line interval, maximum offset, station interval, and maximum frequency for a number of different horizons or time levels. This will lead to a number of different values for each of those parameters, from which the designer of the 3-D survey will have to make a judicious selection.

4.4.1 Spatial continuity

Resolution and spatial continuity are important objectives to be met in 3-D geometry design. Resolution is directly dependent on the shot and receiver station intervals, but indirectly also on spatial continuity. Artifacts reducing resolution in the final data may be caused by spatial discontinuities in the acquisition geometry.

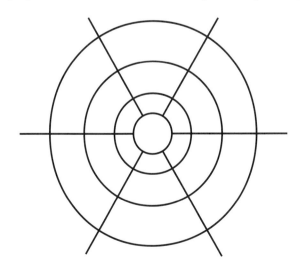

FIG. 4.2. Conceptual spider-web geometry for salt-dome delineation.

Table 4.1. Relation design criteria and parameters of geometry.

Requirement	Parameter
Spatial continuity	Symmetric sampling
Resolution	Shot and receiver station intervals, maximum frequency
Shallowest horizon to be mapped	Line interval
Deepest horizon to be mapped	Maximum offset, spread lengths
Noise suppression	Fold and offset distribution

Therefore, 3-D geometry design should aim to minimize spatial discontinuities.

A problem with orthogonal geometry is that the basic subset of this geometry, the cross-spread, is of limited extent, due to the maximum useful offset at any level. If the data are well-sampled, each cross-spread is spatially continuous, i.e., each point of the underlying continuous wavefield can be reconstructed. However, each cross-spread boundary represents a spatial discontinuity in the 3-D data set. Therefore, it is important to maximize the (useful) extent of each individual cross-spread in order to minimize the overall spatial discontinuity.

Maximizing the extent of a data set that is limited by the maximum offset means that the useful offset has to be maximized in all directions. Basically, this would lead to circular cross-spreads, but for creating regular fold and OVT gathers it is easiest to have square cross-spreads. (For regular fold it is, strictly speaking, sufficient that each cross-spread can be split in the same way into a number of unit-cell sized areas).

Maximizing the spatial extent of each cross-spread optimizes the quality of cross-spread oriented, prestack processing, such as first-break picking, dual-domain filtering, and statics determination. Improved quality also means improved spatial continuity of the final product.

Because symmetric sampling is tantamount to proper sampling of sources and receivers and to maximizing the useful extent of the basic subsets (cross-spreads), spatial continuity is best served by choosing symmetric sampling as a starting point in 3-D geometry design.

4.4.2 Resolution

4.4.2.1 Resolution requirements and maximum frequency

The maximum frequency that can be recorded and processed determines to a large extent the achievable resolution. Often, this maximum frequency is taken for granted, and no effort spent on identifying the resolution requirements. However, the frequency content of the source wavelet can often be influenced (source depth, size of air guns, range of sweep frequencies of vibrators, small or large charges). Therefore, one may try to establish resolution requirements such as what is the minimum layer thickness to be interpreted, or what should be the lateral accuracy of fault positions, and then try to relate those to the maximum frequency that is needed to achieve that resolution. If the maximum required frequency appears not to be achievable, then the resolution requirements will have to be revised or the survey will have to be canceled.

The required maximum frequency depends on the maximum wavenumber that can be achieved. A practical formula to establish the maximum wavenumber is

$$k_{\alpha,\max} = \frac{c}{R_\alpha}, \qquad (4.1)$$

where R is the user-specified minimum resolvable distance, α indicates direction (x, y, or z), c is some constant, and $k_{\alpha,\max}$ is the required maximum wavenumber in direction α. The justification to use this very simple formula is based on the work published in Kallweit and Wood (1982) and is further discussed in Vermeer (1999a) and in Chapter 8.

Resolution is about the resolvability of two events that lie closely together. Figure 4.3 illustrates two events which are just resolved according to the Rayleigh criterion, for which $c = 0.715$ in equation (4.1). In this situation the first negative lobe of one sinc wavelet coincides with the peak of the other wavelet. Figure 4.4 illustrates vertical resolvability using a pinch out, but similar reasoning would apply to horizontal resolution. At the end of this section, the choice of a value for c is discussed further.

Next, the maximum wavenumber must be derived. Other parameters playing a role are the maximum dip angle used in migration, and the type of geometry. (Note that the maximum dip angle used in migration should not be confused with the maximum dip angle assumed to occur in the subsurface.) The type of geometry determines the offset and azimuth mix, which influence resolution. These different parameters can be taken into account in two ways:

1. Determine maximum wavenumber and corresponding required maximum frequency for zero-offset data, followed by a compensation for the loss of resolution caused by NMO stretch as dependent on the mix of offsets.
2. Determine maximum wavenumber and corre-

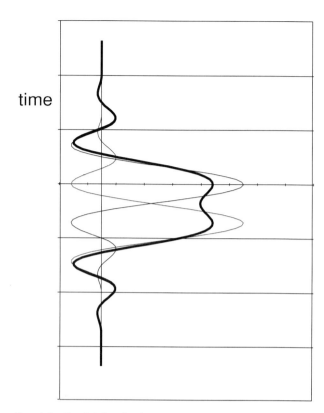

FIG. 4.3. Rayleigh criterion, $c = 0.715$.

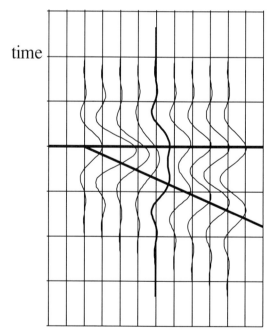

FIG. 4.4. Pinch out model, the thick curve indicates the position where the events are just resolved.

sponding required maximum frequency for the various pseudominimal data sets of an initial guess of the geometry. Then select the maximum frequency, which is the largest of all frequencies found in this way, or select a reasonable average of all maximum frequencies.

Whatever method is chosen, it is important to compensate for the resolution loss due to the NMO stretch effect. In the following I will discuss the derivation of maximum frequency for the situation of a COV gather, followed by a generalization to cross-spreads. A more elaborate discussion of resolution aspects can be found in Chapter 8.

The maximum wavenumber can be derived by raytracing through a representative velocity model, with

$$\mathbf{k} = \mathbf{k}_s + \mathbf{k}_r, \tag{4.2}$$

where $\mathbf{k}$ is the sum of shot and receiver wavenumber vectors. In any subsurface point P, the direction of $\mathbf{k}_s$ and $\mathbf{k}_r$ can be found from the direction in P of the raypath from shot, receiver to P, and the magnitude $|\mathbf{k}|$ of each wavenumber from $|\mathbf{k}| = f/v$, v being the velocity in P. To find the maximum value of $\mathbf{k}$ in any direction α, the corresponding shot/receiver pair has to be found. For the x-direction, k_x is (usually) maximum for the farthest shot/receiver pair that still contributes to the migration result in the output point P. Similarly, in the y-direction. The largest component of k in the z-direction will usually be found for a shot/receiver pair, which is located directly above P on the surface.

This description of finding $k_{\alpha,\max}$ leads to simple formulas. For shot/receiver pairs in a COV gather oriented in the x-direction (see Figure 4.5), the formulas are

$$k_{x,\max} = (2f_{\max} \sin\theta \cos i)/v,$$
$$k_{y,\max} = (2f_{\max} \sin\theta \cos i)/v,$$
$$k_{z,\max} = (2f_{\max} \cos i)/v. \tag{4.3}$$

In these equations, θ is the maximum dip angle being illuminated by the shot/receiver pair, and i is the reflection angle *for the situation which produces the maximum value of the k-component*. This means that θ and i do not have the same value in the three equations. Only for zero-offset sections $i = 0$ in all three directions. Note that equation (4.3) applies to any velocity model for which the raypaths are perpendicular to the wavefronts; it is not just valid for constant velocity as Figure 4.5 might suggest.

In a constant velocity medium the NMO stretch factor $S = t/t_0$ with t reflection time and t_0 normal-incidence time. [For a derivation of the stretch factor taking into account velocity variation, see Vermeer (1990, Section 5.10).] From Figure 4.5b and 4.5c, t/t_0 = RP/MP = 1/$\cos i$, hence $\cos i$ in equation (4.3) represents the NMO stretch effect.

Combining equations (4.1) and (4.3) gives the required maximum frequency as

$$f_{\max} = \frac{c}{2} \cdot \frac{v}{R_\alpha} \cdot \frac{1}{\sin\theta \cos i}, \quad (\alpha = x \text{ or } y),$$

and

$$f_{\max} = \frac{c}{2} \cdot \frac{v}{R_z} \cdot \frac{1}{\cos i}. \tag{4.4}$$

It follows from equation (4.4) that horizontal resolution depends on a processing parameter: θ, the maximum reflection angle (angle of incidence) included in the migration process. Vertical resolution only depends on the acquisition parameters, and is always better than the

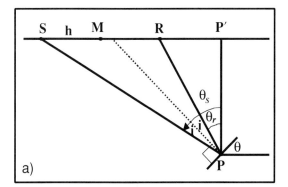
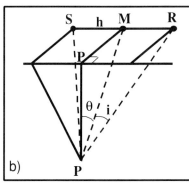
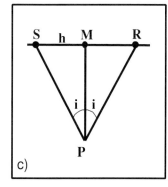

FIG. 4.5. Shot/receiver pair situations for which the wavenumber is largest in P (shot in S, receiver in R and $h_y = 0$). (a) Maximum in x-direction, (b) maximum in y-direction, and (c) maximum in z-direction. Note differences in definitions of θ and i. In x, the shot/receiver pair which is farthest from P determines maximum wavenumber, in z it is the shot/receiver pair closest to P, whereas in y the maximum wavenumber is determined by the shot/receiver pair with the largest y_m and the smallest x_m.

horizontal resolution. (Note that the resolution discussed here is potential resolution; the achievable horizontal and vertical resolution depend on many more processing parameters; see Chapter 8.)

In a COV gather, the determination of i is fairly straightforward, because offset and azimuth of all shot/receiver pairs are constant. If OVT gathers are used as pseudominimal data sets of the orthogonal geometry, offset and azimuth do not vary much. In that case the average offset and azimuth of the pMDS may be used to determine i, and equation (4.4) also applies reasonably well to these OVT gathers.

A choice of $\cos i = 0.9$ would approximately correspond to the maximum-offset-equals-depth criterion, which is often used as a rule of thumb in determining the maximum offset. A migration aperture of $\theta = 30°$ is often a good compromise, it will capture most of the diffracted energy. However, steeper dips require a larger migration aperture.

The quantity v in equations (4.3) and (4.4) is the local interval velocity. To reach the same vertical resolution at deeper levels with higher interval velocities, higher frequencies are required.

In exceptional cases, much better resolution than suggested by $c = 0.715$ in equation (4.1) may be possible. This might be so when extra knowledge about the subsurface may be assumed, e.g., the assumption that all parameters, except thickness, around a horizon of interest hardly vary. One could then attribute any change in horizon attributes to thickness variations. Another such situation may occur with the detection of subtle faults in an otherwise smooth reflector. A factor of $c = 0.25$ (one quarter wavelength resolution) might be used in such cases.

4.4.2.2 Resolution requirements and spatial sampling

The resolution formulas in equation (4.4) assume proper sampling of the minimal data sets of the chosen acquisition geometry. Sampling can be regarded as a means of representing the integrands in the migration formulas, therefore, the migration result depends on the sampling quality and the theoretically best possible resolution can only be obtained with proper sampling of the data to be migrated. A more detailed discussion of the relation between sampling and migration is given in Section 8.3.7. An important conclusion is that proper or alias-free sampling of the input data leads to a well-behaved migration operator response.

On land, alias-free sampling of the total wavefield of a minimal data set is not affordable. As a (first) compromise, alias-free sampling of the *desired* wavefield rather than the *total* wavefield may be chosen as a starting point for a decision on the sampling interval. This requires that the station spacings Δs and Δr should be equal to the basic signal sampling interval (cf. Section 1.3)

$$\Delta s = \Delta r = \frac{V_{r,\min}}{2 f_{\max}}, \qquad (4.5)$$

in which $V_{r,\min}$ is minimum apparent velocity of the P-wave data in the common-shot gather and $f_{\max}$ is maximum frequency. The undesired part of the wavefield, such as ground roll and perhaps converted and shear waves, will be aliased and may have to be tackled by field arrays (see Section 4.4.5.5).

For a modest aim of recording and imaging up to 40 Hz, and with $V_{r,\min} = 2000$ m/s, the shot and receiver station intervals should not be larger than 25 m.

The common-shot gathers used to pick $V_{r,\min}$ are dominated by the NMO effect, whereas the NMO-correction tends to de-alias the steep events. Hence, rather than looking for the minimum apparent velocity in the field data, the minimum apparent velocity is more often determined in the zero-offset domain or stack domain. In these domains the minimum apparent velocity $V_{m,\min}$ is determined by the diffractions. Equation (4.5) is then modified into

$$\Delta m_x = \Delta m_y = \frac{V_{m,\min}}{2 f_{\max}}, \qquad (4.6)$$

where Δm_x and Δm_y are the midpoint intervals in x and y.

For the determination of the sampling intervals from equation (4.6), it is simplest to measure the apparent velocity of diffractions on existing unmigrated stacked data. An alternative is to use a representative velocity distribution as a starting point for the analysis.

Equation (4.6) is quite a stringent requirement, because it looks for the minimum value of the apparent velocity throughout. Usually, the flanks of diffractions will be steepest at shallow levels. As a compromise, one may decide to accept some aliasing at the shallow levels and relax the requirement for alias-free sampling to the levels of interest. A further compromise is to accept some aliasing of the steepest parts of the diffractions and aim only for alias-free sampling of the diffractions included in the migration aperture.

If the overburden may be approximated by horizontal layering, Snell's law may be invoked, and V_m for various levels of interest can be estimated from the interval velocity V_{int} and the departure angle θ (see Figure 4.6), leading to

$$\Delta m_x = \Delta m_y = \frac{V_{\text{int}}}{4 f_{\max} \sin \theta}. \qquad (4.7)$$

The angle θ in equation (4.7) should be interpreted as the largest of maximum dip angle and the migration aperture. As a rule of thumb, a migration aperture of 30° is adequate, because it would use about 95% of the total diffraction energy. Hence, in areas with low geological dip, diffractions dictate the sampling interval, whereas in areas with dips larger than 30° the dip angle determines sampling interval.

If in an area with low geological dip, the dip of the steepest reflection would be used to determine the sampling intervals, rather than the migration aperture, a relatively large sampling interval would result from equation (4.7). This might lead to a lot of migration noise depending on the steepness of the migration operator. To prevent generating this migration noise, the steepness of the migration operator should be taken into account and used in equation (4.7). [Migration noise may also be reduced by the application of antialias filtering to the migration operator (e.g., Lumley et al., 1994)]; however, this would reduce resolution and, if applied at all, it should be restricted to filtering of dips larger than 30°.) In areas of complex geology, it may be necessary to carry out raytracing to find the apparent velocity at various levels.

4.4.2.3 Statics and spatial sampling

Sand dunes in desert areas and mountainous terrain may cause rapid variations in statics. Statics may also vary rapidly in other areas where changes in the near surface occur across small distances. In combination with arrays the intra-array statics may cause loss of high frequencies of the desired wavefield. In such cases the magnitude of the statics may be another criterion to use in the selection of the station spacings. A smaller station spacing than otherwise necessary may have to be used, or it may be that, instead of using field arrays, fold would be increased to suppress noise.

4.4.2.4 Other processing requirements and sampling

Next to migration, (f,k)-filtering is another multitrace process that may suffer from aliasing in the input data. The finer the sampling, the better signal and noise will be separated in the (f,k)-domain, and the more successful the filter will be. Moreover, if the sampling interval is small enough, an (f,k)-filter will remove noise better than field arrays. Therefore, in areas with much low-velocity noise, a small sampling interval may be essential to accommodate the requirements of (f,k)-filtering.

4.4.2.5 Discussion on spatial sampling

It turns out that different areas with a similar velocity distribution may have totally different quality. Often this is related to the complexity of the subsurface. "Fit-for-purpose" sampling intervals do not only depend on the velocity distribution and the corresponding apparent velocities, but also on the energy distribution of the wavenumber spectra. In some areas, heavy faulting may lead to many diffractions with much energy for high wavenumbers. In those areas it is much more important to stick to the rules of alias-free sampling than in more benign geological areas. What may seem overkill in one area, may be just right for another.

It should be realized that the more the requirements are relaxed, the more one relies on fold to suppress the migration noise produced by coarse sampling. There are areas which used to show "no data" zones, that turned into good data zones after reshooting the data with smaller sampling intervals and higher frequencies.

Very powerful interpolation techniques (e.g., Huard and Spitz, 1998) have been developed to compensate for undersampling. These techniques can interpolate "beyond Nyquist," because additional information is provided. Whether such techniques should be relied upon to relax spatial sampling requirements is open to debate.

Ansink et al. (1999) discuss the results of a very interesting high-density, high-resolution survey carried out by Nederlandse Aardolie Maatschappij (NAM) in the Dutch North Sea. A 3-D survey was acquired with 18.75 m crossline sampling. The data set was processed with 18.75 m and with 37.5 m crossline sampling. The steeper dips were better imaged by the 18.75 m; yet, for explo-

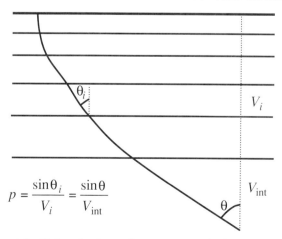

FIG. 4.6. Determination of apparent velocity of event defined by "appropriately" selected departure angle θ, in case of horizontal layering above event.

80 Chapter 4 Guidelines for design of "land-type" 3-D geometry

ration purposes the differences were considered negligible and it was decided to acquire the next surveys with 25 m crossline sampling. The relatively small differences between dense and coarse sampling in this case might be attributed to a number of different causes, such as (1) negative effects of multisource, multistreamer configuration on resolution (see also Section 5.3.2), (2) inability of processing to exploit denser sampling, (3) high signal-to-noise ratio of prestack data (no ground roll, very good sea conditions) allowing nearly perfect interpolation from 37.5 to 18.75 m crossline sampling (the coarsely sampled data were interpolated). Which one, if any, of these causes is most important is difficult to say without further analysis.

4.4.3 Shallowest horizon to be mapped

The smallest offsets in an orthogonal geometry occur at midpoint positions close to the intersections between shot and receiver lines. In the middle of the rectangular area between adjacent shot and receiver lines the smallest offset is about equal to the length of the diagonal of the rectangle. This is the largest minimum offset (LMOS) of the geometry (see Figures 4.7 and 2.23). The larger the distance between the acquisition lines the larger LMOS. As small offsets are needed to illuminate the shallow subsurface, the distance between the acquisition lines determines the shallowest mappable level. Therefore, the seismic interpreter has to identify the shallowest horizon that needs to be fully mappable for an adequate geologic picture of the subsurface.

Experiments have shown (see Section 7.4) that prestack migration of *four-fold data* acquired in Nigeria may already give a tremendous improvement in signal-to-noise ratio, to the extent that it should be possible to map such data. Now this observation would certainly not apply in all imaginable cases, though the quality of the Nigeria data is comparable with that in other areas in the world. This is typically something that could be tested without too much effort, for instance by an exercise such as carried out by Mobil (Lee et al., 1994). (They acquired a long shot line across a number of receiver lines perpendicular to the shot line.)

In my opinion, the criterion of the shallowest-level-to-be-mapped is more objective than "the shallowest level at which complete single-fold coverage should be present." Anyway, the one follows from the other, or vice versa. In some regions of the world, the shallow horizons are steeply dipping. Then the shallow-horizon-criterion cannot be used, and the single-fold criterion has to be used. It is a crucially important design criterion. If the level chosen is too shallow, the distance between acquisition lines chosen might be too small, making the cost of the survey unnecessarily high.

To translate the design criterion of shallowest-level-to-be-mapped into a choice of shot and receiver line intervals, it is necessary to have a representative *mute function* of the survey area. The mute function deter-

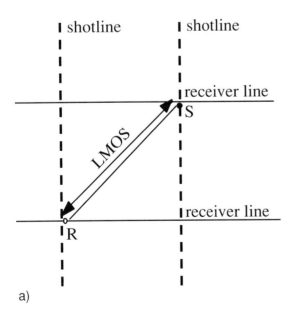
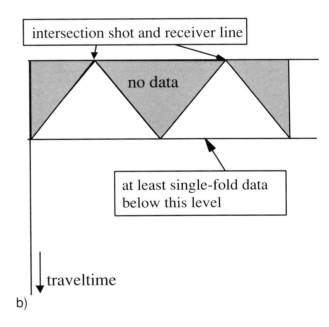

FIG. 4.7. Largest minimum offset is about equal to the length of the diagonal in the rectangle between adjacent pairs of shot and receiver lines (a), and determines the level below which at least complete single-fold data is present (b). The cross-section in (b) is taken along the diagonal of a unit cell (see Section 2.5.2 for definition of unit cell).

mines the maximum offset that contributes to the stacked or migrated section at each traveltime. The mute function should be gleaned from earlier data acquired in the area. If not known, or if it is not completely known (due to small maximum offset in previous surveys), it may be computed from a representative velocity function, assuming some acceptable maximum NMO stretch. The mute function is also important for the next design criterion to be discussed, the deepest-horizon-to-be-mapped.

The procedure to establish the acquisition line interval is described in Table 4.2 and illustrated in Figure 4.8. This procedure assumes the same distance S between shot lines as between receiver lines. In case the shallowest level with complete single-fold coverage t_{min} is used as a criterion, LMOS has to be used. Then LMOS has to be chosen equal to the maximum offset at t_{min}, and $S = \text{LMOS} / \sqrt{2}$. The formula used in 4a of Table 4.2 ensures that the fold at shallow levels equals *at least M*. The formula in 4b is based on *average fold M* at the specified level. This formula may be generalized to any level with mute offset x, provided the maximum inline and crossline offsets are larger than x.

The general procedure described in this section may lead to very small line intervals in case of a need to image shallow salt domes. As a cost-effective alternative, use of nonseismic techniques may be considered for imaging the shallow subsurface. Den Boer et al (2000) describe the use of magnetotellurics for resistivity imaging of shallow salt and show very convincing results. The application of this technique allowed the generation of a better depth model leading to much improved prestack depth migration results.

4.4.4 Deepest horizon to be mapped

The deepest horizon to be mapped provides an upper limit to the maximum offset to be used in the survey. The required maximum offset leads to a choice of receiver spread length L_R and shot spread length L_S (shot spread length is length of shot line being recorded in a single receiver line). For the time being, I will assume symmetric sampling with spread length $L = L_R = L_S$. The mute function provides the maximum offset x_{dp} for the deepest horizon corresponding to the largest time t_{dp} of the horizon (see Figure 4.8). Depending on the requirements of the 3-D survey, there are two different ways of establishing L (Table 4.3).

The first way is used if azimuth-dependent amplitude analysis is to be carried out for the target horizon. Then, for that horizon, the full range of azimuths should be available for the full range of offsets. This means that the absolute offset should be equal to or larger than x_{dp}. This requirement leads to $L_S = L_R = 2\, x_{dp}$. If this procedure is followed, there will be a large collection of traces not contributing to the deepest horizon of interest. These traces are situated in the corners of the cross-spreads, outside the circle with radius $x_{dp} / 2$.

An alternative way of establishing L is to require that there should at least be one trace with offset $x > x_{dp}$ throughout the full-fold area of the survey. This means there should be an OVT gather with all absolute offsets larger than x_{dp}. This requirement is fulfilled if the minimum maximum offset $X_{\text{min max}} = x_{dp}$ (cf. discussion in Section 2.5.6 of Figure 2.23). Figure 2.23 can be used to find an expression for $X_{\text{min max}}$

$$X_{\text{min max}} = \tfrac{1}{2}\sqrt{L_R^2(1-2S/L_R)^2 + L_S^2(1-2R/L_S)^2}\,, \quad (4.8)$$

Equation (4.8) can be used to verify whether a particular choice of L_R and L_S satisfies the requirement for $X_{\text{min max}}$. If this second procedure is followed, not many traces will be muted at the deepest level of interest.

The first procedure is optimal as far as velocity

Table 4.2. Procedure to establish line interval.

	Example
1. Establish shallowest horizon to be mapped. Assume M-fold coverage required	$M = 4$
2. Determine shallowest time t_{sh} of that horizon	t_{sh} = 1000 ms
3. Find maximum offset x_{sh} for t_{sh}	x_{sh} = 2000 m
4a. Line interval $S \approx \dfrac{x_{sh}}{\sqrt{2M}}$ ($M \le 4$)	$S \approx 707$ m
4b. Line interval $S \approx \dfrac{x_{sh}}{2}\sqrt{\dfrac{\pi}{M}}$ ($M > 4$)	
5. Choose S as nearest multiple of Δs	$S = 700$ m

Table 4.3. Procedure to establish spread lengths.

	Example ($S = 700$ m)
1. Establish deepest horizon to be mapped	
2. Determine deepest time t_{dp} of that horizon	t_{dp} = 2200 ms
3. Find maximum offset x_{dp} for t_{dp}	x_{dp} = 3300 m
4a. Choose $L/2$ as nearest multiple of S	$L = 7000$ m
4b. Choose $X_{\text{min max}} = x_{dp}$ and find corresponding L	$L = 5600$ m

FIG. 4.8. The relation between coverage of shallow objective and acquisition line spacing via mute function. (a) Mute function; (b) how to achieve at least four-fold coverage at t_{sh}.

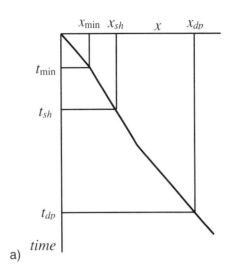
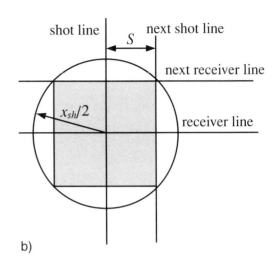

determination is concerned. In particular, if DMO is needed as part of the velocity determination, it will be helpful if the rectangular OVT gathers located around the acquisition lines (see Figure 2.22 and Section 2.6.8) contain all offsets for the target level. On the other hand, from a resolution point of view, it is best to choose the maximum offset as small as possible, because the theoretically best possible resolution is reached for zero-offset and reduces with increasing offset (Chapter 8).

There are, in general, a few other compelling reasons to use long offsets next to velocity-determination requirements:

1. Without some range of offsets it would be difficult to reach sufficient fold for noise reduction. (Fold can also be increased by decreasing line intervals, but that is more expensive than increasing spread lengths.)
2. Long offsets may be needed to create sufficient differential moveout between primaries and multiples for multiple suppression.
3. AVO analysis needs long offsets.

The maximum offset has to be large enough to satisfy all of the applicable requirements. For AVO analysis the required offset should be derived from the desired range of reflection angles, rather than from the mute function as determined from a given NMO stretch factor.

4.4.5 Noise suppression

4.4.5.1 Fold as a dependent or independent parameter

Often the designer of a 3-D survey will have a reasonable idea of the required fold to achieve an adequate signal-to-noise ratio. In that case he may want to use fold as an independent (input) parameter to the design process. His knowledge might even extend farther, allowing a specification of desired fold at all levels of interest.

On the other hand, the survey parameters following from the considerations in Sections 4.4.3 and 4.4.4 will also lead to some fold-of-coverage, because line spacing and maximum offset fully determine the fold of the survey (see Sections 2.5.2 and 4.4.6). Viewed in this way, fold is a dependent parameter. Strictly speaking, fold is always an independent parameter, because when determining line spacing in Section 4.4.3 we had to specify fold for the shallowest level of interest.

It follows from these considerations that another way of arriving at survey parameters would be to specify required fold and maximum offset for each level of interest. This would lead to a line spacing for each level of interest. Selecting the smallest line spacing and the maximum of all maximum offsets would lead to a geometry that satisfies all requirements, but it might lead to a very large full fold of the survey.

4.4.5.2 How to determine desired or required fold

Unfortunately, the important question, how to determine desired or required fold, does not have the clear answer one would like to find in a book on 3-D survey design. Some general remarks are offered instead.

In principle, single-fold data should be adequate for imaging, fold is only necessary to suppress recorded noise. Therefore, 3-D survey design should avoid spatial discontinuities and it should select station intervals that are small enough not to generate migration noise. If sampling is too coarse, fold is also necessary to fight migration artifacts.

The simplest way to determine fold is to consider fold as a dependent parameter (see previous section): select line interval (or LMOS) and maximum offset, and fold follows from these two (or four, in the case of different crossline and inline parameters) parameters.

The pragmatic approach is to base the choice of fold on past experience. Use the success or failure of a given fold in 3-D surveys acquired in similar, perhaps nearby, terrain as a guideline. If no 3-D surveys are available in the area of interest, the fold used for 2-D lines may be used as a guideline. Krey (1987) provides as a rule of thumb that 3-D fold may be taken as 2-D fold times frequency of interest/100 (provided station spacings are the same in 3-D as in 2-D). The more accurate formula (Krey, 1987) is based on a computation of the relative suppression of random noise by the two types of survey. In poststack migration, signal is proportional to the number of traces within the zone of influence (see Section 10.2) and noise is proportional to the square root of the number of traces in the migration aperture. Furthermore, the signal-to-noise ratio improvement by stacking is proportional to the square root of fold. For the same signal-to-noise ratio for 3-D as for 2-D, lower 3-D fold suffices because of the relative difference in number of traces contributing to 3-D migration as compared to 2-D migration.

A weak point in Krey's formula seems to be that it is derived for random noise, whereas usually the main noise problem is shot-generated noise. However, the aliased part of that noise could be considered random, whereas the nonaliased part would normally be suppressed much better than random noise. Therefore, taking Krey's rule of thumb may still form a reasonable starting point for a decision on what fold to use. (This reasoning does not apply to noise that is only dependent on offset in parallel geometry. In a common-offset gather this noise will show up as a horizontal event and will not be suppressed by migration.)

This decision can further be modified by some of the following considerations:

1. Any reduction in the chosen station interval improves the ability to remove noise in processing. This means that the required fold depends implicitly on station interval: better sampling allows lower fold. Ideally, one would like to keep trace density the same for a reduction in spatial sampling interval, i.e., fold proportional to bin area. However, this might be a bit too optimistic.
2. Migration is a very powerful way of reducing noise. Noise can also be suppressed by other pre- or poststack processing steps. In all cases the success in removing noise will be increased with more spatial continuity, i.e., a wide geometry and smooth acquisition lines. This better spatial continuity should also reduce the need for high fold.

Another way of determining required fold is by model-based noise analysis. The analysis can be carried out using model studies with simulated noise or by using 2-D data (Egan, 2000; Savage and Mathewson, 2001). Both approaches have advantages and disadvantages, so it is best to use a combination of the two. For a proper evaluation, it is necessary to simulate the best possible processing sequence. For instance, with wide geometries that are well-sampled it is possible to remove more noise in prestack processing than with narrow geometries in which the receiver gathers are too short for filtering purposes.

Finally, the most thorough way of determining required fold is to acquire high fold 3-D test data and to carry out decimation tests. This is the most expensive method, but it might well be justified in case large areas have to be covered with 3-D seismic data.

4.4.5.3 Fold as an instrument to suppress multiples

Multiple elimination through stacking works best, for a given fold, in a square geometry. This applies to multiples with a small differential moveout with respect to primaries and in general also to multiples with a large differential moveout (see Section 3.4.3). Increasing fold for better multiple suppression needs only to be considered if the geometry is already square. However, there is one exception to this rule: multiples with large differential moveout are best suppressed by a narrow-azimuth high-fold 3-D stack-array geometry (double zigzag, narrow orthogonal, or parallel geometry, of which parallel geometry is best). A high-fold narrow geometry tends to have a more regular offset distribution, thus providing much better linear noise than random noise suppression by stacking. Therefore, in case of severe multiples with large differential moveout, a high-fold narrow geometry may have to be considered. Sections 4.3.1 and 4.3.2 deal with this aspect of geometry choice in some more detail, including the effect of prestack multiple elimination.

4.4.5.4 The importance of regular fold

Sections 4.4.3 and 4.4.4 specified selection of the line interval S as a multiple of station spacing, and selection of maximum inline (crossline) offset X_{max} as a multiple of shot-line (receiver line) interval S. As follows from the formulas in Table 4.4 and Table 4.5, this choice leads to integer values of inline and crossline fold. This means that the nominal geometry leads to constant full fold (apart from the edges of the survey area).

The data set can also be considered as a collection of single-fold cross-spreads. Hence, if fold is constant, this means that where the midpoint area of one cross-spread stops, another cross-spread takes over. In other words, the whole survey area can be covered with single-fold adjacent cross-spreads. Such a collection of minimal data sets is called a pseudominimal data set (pMDS, Section 2.5). Constant fold in a regular geometry (the same maximum crossline offset for all cross-spreads and the same maximum inline offset for all cross-spreads, i.e., identical cross-spreads throughout the survey area, except the edges) also means that the survey area can be covered with a tiling of single-fold offset-vector tiles. Each tile has the size of a unit cell. Each of these tilings is again a pMDS (see Section 2.5.4).

Fold-of-coverage, illumination fold, and image fold are closely tied (for a definition of these terms see Section 2.5.2). Usually, image fold will not be very different from fold-of-coverage (a notable exception is illumination and imaging with *PS*-waves, see Chapter 6). However, at cross-spread edges image fold tends to be irregular because there will be discontinuities in the illumination. The steeper the dips the larger the discontinuities tend to be. The pMDSs formed with OVTs suffer less from discontinuities across the edges of each tile than pMDSs formed from complete cross-spreads (the discontinuities between the OVTs are more abundant but smaller). All this means that regular fold-of-coverage does not guarantee regular image fold, but it certainly helps to minimize irregularities. More about these considerations in Chapters 2 and 10.

Above the level of full fold (the position of the largest offset in the mute function), the mute function takes away the longer offsets. Probably the best way to make fold-of-coverage a constant for each time level would be to select a separate mute function for each offset-vector tile sized 1/4-unit cell as defined in Figure 2.22 and discussed in Section 2.6.4. The mute should be a single step function inside the tile.

The irregularities in image fold may lead to visible acquisition footprints. Once it has been decided to use orthogonal geometry, 3-D symmetric sampling will minimize the acquisition footprint.

4.4.5.5 Shot and receiver arrays

The main purpose of shot and receiver arrays is to suppress ground roll. Smith (1997) demonstrated that aliased ground roll can be reduced considerably by the use of linear shot and receiver arrays located along the acquisition lines. Assuming that linear shot and receiver arrays are used, the survey parameters following from earlier considerations would normally be sufficient for adequate noise suppression.

Table 4.4. Formulas for survey parameters (equal line spacings).

Parameter	Formula	Example
Station spacings	$\Delta s, \Delta r$	25 m
Maximum inline and crossline offset	X_{max}	3500 m
Line interval	S	700 m
Bin size	$b = \Delta s \, \Delta r / 4$	$b = 12.5 \times 12.5$ m
Spread length	$L = 2 X_{max}$	$L = 7000$ m
Inline fold	$M_i = L / 2S$	$M_i = 7000 / 1400 = 5$
Crossline fold	$M_x = M_i$	$M_x = 5$
Total fold	$M = M_i M_x$	$M = 25$
Number of receiver lines	$N_R = 2M_x$	$N_R = 10$
Number of channels per line	$N_{chl} = 2 X_{max} / \Delta r$	$N_{chl} = 280$
Total number of active channels	$N_{tot} = N_{chl} N_R$	$N_{tot} = 2800$
Number of shots / km^2	$S_{dens} = 1\,000\,000 / (\Delta s \, S)$	$S_{dens} = 57.1$/km^2
Distance to build full fold	$D = (X_{max} - S) / 2$	$D = 1400$ m

Table 4.5. Formulas for survey parameters (unequal line spacings).

Parameter	Formula	Example
Station spacings	$\Delta s, \Delta r$	25 m
Maximum inline offset	$X_{max,\,inl}$	3600 m
Maximum crossline offset	$X_{max,\,xl}$	2000 m
Shot-line interval	S	600 m
Receiver-line interval	R	500 m
Bin size	$b = \Delta s\,\Delta r\,/\,4$	$b = 12.5 \times 12.5$ m
Receiver spread length	$L_R = 2\,X_{max,\,inl}$	$L_R = 7200$ m
Shot spread length	$L_S = 2\,X_{max,\,xl}$	$L_S = 4000$ m
Inline fold	$M_i = L_R\,/\,2S$	$M_i = 7200\,/\,1200 = 6$
Crossline fold	$M_x = L_S\,/\,2R$	$M_x = 4$
Total Fold	$M = M_i M_x$	$M = 24$
Number of receiver lines	$N_R = 2M_x$	$N_R = 8$
Number of channels per line	$N_{chl} = 2\,X_{max,\,inl}\,/\,\Delta r$	$N_{chl} = 288$
Total number of active channels	$N_{tot} = N_{chl}\,N_R$	$N_{tot} = 2304$
Number of shots / km²	$S_{dens} = 10^6\,/\,(\Delta s\,S)$	$S_{dens} = 66.7/\mathrm{km}^2$
Inline distance to build to full-fold	$D_{inl} = (X_{max,\,inl} - S)\,/\,2$	$D_{inl} = 1500$ m
Crossline distance to build to full-fold	$D_{xl} = (X_{max,\,xi} - R)\,/\,2$	$D_{xl} = 750$ m

In areas where shot arrays would become unbearably expensive, single shots combined with areal receiver arrays may be considered instead. Also, in case the number of shots per shotpoint has to be small (e.g., because of length of vibrator trucks), this might be compensated by using a larger number of geophones in the crossline direction of an areal geophone array.

Only in areas with a severe noise problem is it necessary to increase efforts for noise suppression. The increased efforts can take the shape of areal (preferably circular) receiver and/or shot arrays; increased fold, i.e., smaller shot and receiver line intervals; or reduced spatial sampling intervals. Which of these options stands the best chance of success may be established with a careful noise test. Chapter 3 gives an extensive discussion of shot and receiver arrays in 3-D data acquisition.

There may be areas where the amount of ground roll is hardly a problem (perhaps because deep shotholes are used). Then, using bunched geophones instead of geophone arrays may be considered. Any aliased noise would have to be suppressed in processing. An advantage of this approach would be that the signal would not be affected by the geophone arrays, in particular intra-array statics would not degrade the signal.

4.4.6 Other survey parameters

Table 4.4 lists various formulas which apply to orthogonal geometry. Note that number of channels and number of shots depend on bin size, whereas fold is independent of bin size. The formula for the distance required to build to full fold is derived in Section 4.5 (Figure 4.11).

The formulas in Table 4.4 demonstrate the importance of a correct choice of S, the shot-line interval, and of X_{max}. If for some reason a smaller S is chosen or needs to be chosen, for instance $S = 500$ m, then fold would soar to 49, and number of channels to 3920. Similarly, if the maximum required offset would be 4900 m instead of 3500 m, fold would also soar to 49, but number of channels would skyrocket to 5488. Fortunately, there are also ways of achieving the same geometry using fewer channels (see Section 4.6.4).

On the other hand, it is interesting to realize that in cases where shallowest level of interest and deepest level of interest coincide, a four-fold geometry might be adequate for mapping the objective level.

Table 4.5 lists the formulas for survey design with unequal line spacings.

4.4.7 The selection of acquisition parameters for areal geometry

The design of the parameters of an areal geometry follows directly from the equivalent orthogonal geometry (Figure 4.9). Equivalence conditions are listed in Table 4.6.

If these conditions are met, the offset distribution of the two geometries is identical (see Figure 4.10). The only difference is the azimuth distribution. With symmetric sampling the shot and receiver grids will be square and the maximum offsets will be the same in inline and crossline direction. However, a more efficient sampling scheme for areal geometry is to use hexagonal sampling for both receiver stations and shot positions as described in Section 2.4.1.

4.5 The survey grid and the survey area

The conventional approach to determine the required extent of a survey is to establish the area that has to be mapped, add the *migration radius* to this area in all directions (the *migration apron*), and then require that full-fold should be acquired in this extended area. With a wide geometry, run-in to full fold is larger (at least in the crossline direction) than in a narrow geometry, hence the requirement of full-fold for migration adds a large area of acquired data to the area to be mapped.

It is helpful to look at the unit cell of a geometry with its contributing shot and receiver stations to get a better insight in the fold build-up. Figure 4.11 shows the unit cell for a 16-fold geometry with all its contributing shots and receivers. In order to acquire full fold, shots and receivers have to be located at most 1.5 line intervals from the edge of the unit cell. For an inline fold M_i, $(M_i - 1)/2$ shot-line intervals would be needed.

This observation leads to the recommendation to use a closed grid of acquisition lines. In this closed grid, the midpoints cover the same area as the acquisition lines, whereas the smallest offsets occur along the perimeter of the survey, i.e., the closed grid maximizes the useful midpoint area.

This recommendation to use a closed grid leads to incomplete cross-spreads around the edges of the survey. However, the smallest cross-spread (in the corners) is still equal to one quadrant, so that reasonable cross-spread oriented processing may be carried out.

As discussed in Section 7.4, prestack migration may lead to good signal-to-noise ratio already for four-fold data. If this is the case, and prestack migration is part of the processing plan, the requirement that the survey area should be large enough to allow full-fold migration into

Table 4.6. Equivalence conditions for orthogonal and areal geometry.

Orthogonal geometry		Areal geometry
Shot-line interval	=	Grid interval receivers in x
Receiver-line interval	=	Grid interval receivers in y
Shot-station spacing	=	Grid interval shots in y
Receiver-station spacing	=	Grid interval shots in x
Maximum inline offset	=	Maximum inline offset
Maximum crossline offset	=	Maximum crossline offset

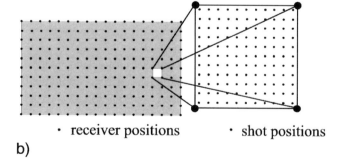

FIG. 4.9. Equivalent geometries. (a) orthogonal geometry, (b) areal geometry. See text for conditions of equivalence.

the area to be mapped, can be relaxed. Four-fold is present everywhere inside the area enclosed by the outer acquisition lines, except in the corners. Because all short offsets are present, this area is the required area for migration.

An alternative is to extend the acquisition lines up to one line interval outside the closed grid. Then four-fold is present anywhere inside the area enclosed by the outer acquisition lines.

The four-fold requirement for adequate imaging is an extreme case. The other end of the scale is that full-fold is purely defined by requirements of signal-to-noise ratio. In other words, all data are required for an acceptable image. In that case the fold-taper zone, given by $(M_{i(x)} - 1) / 2$ shot-line (receiver-line) intervals, should be included to compute the survey area. In intermediate cases a smaller part of the fold-taper zone needs to be included.

Note that the migration radius is the sum of maximum migration distance and the maximum zone of influence (see Section 10.2.2).

4.6 Practical considerations and deviations from symmetric sampling

There may be many practical reasons to use unequal shot and receiver line intervals, or to use unequal maximum inline and crossline offsets. Generally speaking, there are no really valid reasons for selecting different shot- and receiver-station intervals. Exceptions are *PS*-acquisition and oversampling of one of the two intervals (as in receiver sampling in marine streamer surveys). The nominal geometry as decided upon may be implemented in the field in different ways, depending on logistical considerations. Topography and obstacles may require deviations from nominal geometries.

4.6.1 Logistics and terminology

It is not the intention of this book to serve as a manual for dealing with practical aspects of the acquisition of 3-D surveys. The writing of such a manual had better be left to the people with real experience. Yet, some appreciation of practical aspects is essential for the designer of a 3-D survey.

The procedures followed in the field can only be described by using some jargon. Unfortunately, there is no universally established terminology. Therefore, the reader should be aware that my jargon may deviate from somebody else's.

The starting point in a description of the field procedures is the template, which was introduced in Section 2.3.3. The template consists of the collection of active receivers listening to a series of shots. Usually, this series of shots (also called salvo) is located on a single shot line, but not necessarily so. The receivers are located along a number of receiver lines. When all shots of the template

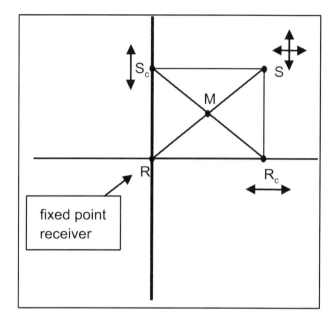

FIG. 4.10. Equivalent offsets between orthogonal and areal geometry. The cross-spread shot/receiver pair (S_c, R_c) has the same offset as the 3-D receiver shot/receiver pair (S, R).

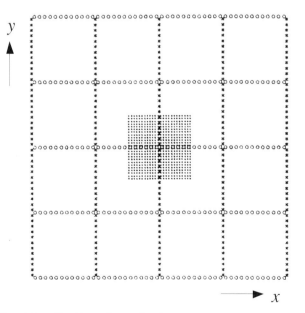

FIG. 4.11. Position of contributing shots and receivers for unit cell of 16-fold square geometry. The unit cell consists of the midpoints indicated by dots, o = receivers, * = shots.

have been fired, the template is rolled (moved) to its next position. Usually, the template is first rolled in the inline direction. The inline roll is equal to the shot-line interval S. inline rolling continues until the width of the survey (or width of zipper, see below) has been covered. The collection of all templates rolled inline is called the swath. Sufficient stations should be available to lay out the whole swath in one go. The required number of channels in the recording instrument equals the number of receivers in the template (times the number of components per receiver in case of multicomponent recording).

After the whole swath has been acquired, a crossline roll is performed. The number of receiver lines picked up in a crossline roll may be one (single-line roll), many (multiline roll) or all (full-swath roll). Pros and cons of the various crossline rolls are discussed in Sections 4.6.4 and 4.6.5.

There may not be enough stations for laying out a swath across the whole width of the survey area. Then the survey area is split into a number of zippers, each zipper consisting of a strip narrow enough to lay out a full swath. At the boundary between two zippers care must be taken that the nominal geometry can be reconstructed from the overlapping parts of the two zippers. This may be implemented in different ways, either by overlapping the receiver lines or by shooting outside the swaths or some combination of these two.

The above description of template and inline roll does not match actual practice of acquiring zigzag and slanted geometry. In these geometries an attempt is made to fire each shot center-spread. In the zigzag geometry, this means that the active stations in the swath move one position for each succeeding shot, and, in the slanted geometry, the active stations move one position each time consecutive shots have moved one station interval in the inline direction.

4.6.2 Harmonizing all requirements

As mentioned in Section 4.4.5.1, the parameters fold, maximum offset, and line interval are interrelated (see also Tables 4.4 and 4.5). The ideal choice for maximum offset and line interval may lead to too large or too low a fold. Virtually always, some compromise has to be found between the "ideal" choices for each individual parameter.

This selection of the best compromise for all acquisition parameters could be viewed as an optimization problem (Liner et al., 1999). Liner et al. (1999) propose to use target values for some main parameters, and to find an acquisition geometry that minimizes a cost function based on the weighted deviations from those target values. Their formulation of the optimization problem might be modified somewhat to ensure even better solutions. In the first place, it would be advisable to also include a measure for the ratio between shot-line interval and receiver-line interval, the optimal ratio being 1.0. A further refinement might be the optimization for different target levels, each level having its own requirements. A proper choice of constraints should lead to an optimal parameter choice. Morrice et al. (2002) minimize survey cost while satisfying certain constraints. My unpublished paper "3D seismic survey design optimization" expands further on this subject and is included on the CD-Rom inserted in this book.

4.6.3. Deviations from symmetric sampling

The use of a wide-azimuth geometry requires more receiver lines than a narrow-azimuth geometry. This requires a rethink of the optimal procedures in the field. A template is square in a geometry where maximum inline offset equals maximum crossline offset. This means that rolling inline or rolling crossline involves the same number of receiver stations per roll, i.e., as many as is present in a single receiver spread. Not only is the number of active stations large, but also the number of required additional stations for rolling.

Acquisition on land is often shot-constrained, i.e., the shot density determines progress and cost of the survey. In that case, a somewhat larger shot-line interval may be compensated by a smaller receiver-line interval. For instance, in the example of Section 4.4.6 (Table 4.4), use 800 m instead of 700 m for the shot-line interval, and use 600 m instead of 700 m for the receiver-line interval. This makes LMOS slightly larger (1000 m instead of 990 m). Choosing 3200 m as maximum inline offset giving inline fold 4, and 3600 m as maximum crossline offset giving crossline fold 6, would lead to a total fold of 24, close to the original 25, and would require 12 receiver lines.

The difference between shot-line interval and receiver-line interval should not become too large, as this would lead to irregular fold at shallow levels. For instance, if the shot-line interval was twice as large as the receiver-line interval, 3-fold data would already be acquired around the line intersections, when single-fold is just reached at the midpoint where LMOS (see Figure 4.7a) is reached. For larger factors between the two line intervals, the variation in fold becomes larger, whereas in a geometry with equal shot and receiver line intervals, the fold variation would be minimal.

In case shots are extremely expensive, using areal geophone arrays may be considered instead of a combination of linear shot and linear receiver arrays.

4.6.4 Different ways of implementing nominal geometry

Because of the necessity of a relatively larger receiver effort for a wide geometry than for a narrow geometry, wide land geometries may be easily receiver-constrained. This was the case for survey A in Nigeria, where all equipment had to be picked up at the end of each day. To reduce planting effort in such a situation, the number of receiver lines can be halved if shots are fired from both sides of the spread (see Figure 4.12a). This means that each shot location has to be visited twice. A further reduction of planting effort can be achieved by halving the spread length, and shooting from all four corners of the remaining lay-out. Then all shot locations have to be visited four times, eventually. This technique could also be used if the number of available channels or units is limited. Another way of making the most of the available number of stations is the WAS technique proposed in Hastings-James et al. (2000). In the WAS technique Figure 4.12a is extended with two series of shots on either side of the spread at a distance equal to the number of receiver lines times the receiver-line interval.

Another constraining factor can be downtime caused by faulty receiver stations. The more receivers, the larger the chance for this kind of downtime. To minimize this downtime, lay out only a limited number of receiver lines in a swath, say six, and then apply a full-swath roll (Figure 4.12b). Now the salvo of shots for a template has to extend far enough outside the swath to allow recording of the required maximum crossline offset. It requires repeated shooting of the same shotpoint (into different swaths), instead of repeated planting of geophones. The full-swath roll can be highly efficient, for dynamite acquisition as well as vibroseis. With dynamite shooting, many shooting crews can work simultaneously on the same shot line; with vibroseis acquisition, the additional advantage is that there are fewer time-consuming moves (including turns) from one shot line to the next. A disadvantage might be that the statics are decoupled for some statics programs.

In the full-swath roll, it tends to be more convenient to let all receivers in the template listen to all shots along the shot line. This leads to asymmetry of the cross-spreads in the crossline direction. The cross-spreads can (should) be made symmetric again in the processing center. As a consequence, *the fold acquired in the field is higher than the nominal fold prescribed by the nominal geometry*. In Figure 4.12b the crossline fold is 8.5, whereas the nominal crossline fold equals 6 ($M_x = 1/2\, L_S / R$). Discarding the redundant traces in the processing center seems to be a waste, yet it avoids strong discontinuities in the attributes of the geometry, which may cause a serious geometry imprint. A typical expression of this irregularity is that it is not possible to split the field data into as many regular OVT gathers as the field fold. Note that the number of receiver lines in this technique can be chosen to fit the available receiver equipment. In a single-line roll, regular fold requires the use of an even number of receiver lines, but in the full-swath roll an odd number of receiver lines may also be chosen. The more receiver lines in the full-swath roll, the smaller the shot repeat factor (number of times the shooting crew has to visit the same shotpoint location). In the example of Figure 4.12b the shot repeat factor is 2.83.

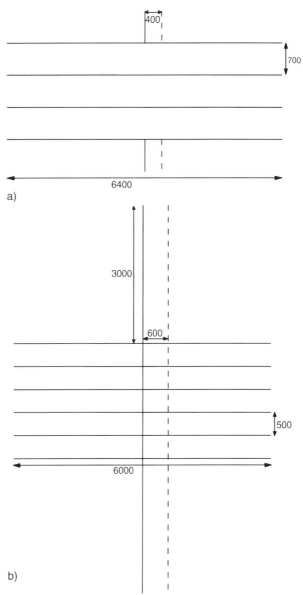

FIG. 4.12. Swaths for cross-spread geometries used in Nigeria; (a) survey A, (b) survey B.

4.6.5 Multiline roll

The multiline roll is a technique that is sometimes used to speed up acquisition. In this technique, the range of shots (shot salvo) fired into each swath extends over several receiver line intervals. As many receiver lines are rolled as there are receiver-line intervals in the shot salvo. This means that each shotpoint location has to be visited only once (unlike in the full-swath roll technique), whereas the number of shots firing into each template is increased. For an equal number of recorded traces, this technique tends to be more efficient than the single-line roll. However, with this acquisition technique, the cross-spreads are asymmetric in the crossline direction.

The difference with the full-swath roll technique is that in that case the cross-spreads are oversized (maximum crossline offset is larger than the nominal maximum crossline offset), whereas in the multiline roll the cross-spreads are undersized (at least one extremum in crossline offset is smaller than the desirable maximum crossline offset; (desirable: for N_R receiver lines the desirable maximum crossline offset equals $N_R / 2$ receiver line intervals). The asymmetry can only be remedied by discarding a large number of nonredundant traces. See next section for a comparison of multiline with one-line roll geometries.

4.6.6 Attribute analysis of one-line roll versus multiline roll geometries and orthogonal versus slanted geometries

In this section deviations from the standard symmetrically sampled orthogonal geometry are investigated. We will look at the slanted geometry and the multiline roll.

Figures 4.13–4.16 each show some major attributes of one of the four geometries being investigated. Figures 4.13b–f show attribute displays for a 20-line geometry, whereas Figure 4.13a shows the template of this geometry, highlighted in a red rectangle. Figure 4.13b shows the "full-fold" display for the whole survey, this display includes all offsets. Figures 4.13c–e show limited-offset displays, Figure 4.13c for offsets up to 800 m, Figure 4.13d for offsets up to 2500 m, and Figure 4.13e for offsets up to 3000 m. These displays clearly show that, even though full-fold is entirely regular across the central part of the survey area, the limited offsets are always irregularly distributed with a range of fold values. The irregularity stems from the fact that fold is built up from overlapping circular areas, each area representing the range of offsets for a given cross-spread. At each time level, the mute function determines the maximum offset used at that level, hence fold is not constant at any time level below the full-fold level (unless the mute function is defined as described in Section 2.6.4).

Figure 4.13f shows two global displays, offset/azimuth and offset density and one bin-oriented display, bin offset distribution. The offset/azimuth display shows for the whole survey the number of traces as a function of offset and azimuth, whereas the offset density function displays the relative abundance of traces as function of offset. The bin offset distribution shows the offset for six rows of bins inside a unit cell. All bins in the left hand division are close to a receiver line, whereas the next divisions move toward the center of a unit cell. The rightmost division, located near the center of the unit cell, shows relatively large white areas in the center. This means that, for those bins, offsets tend to occur in clusters. This is an effect of symmetry inherent in this geometry.

Figure 4.14 shows the attribute displays for a 20-line slanted geometry acquired with a one-line roll. In this example the shot lines make an angle $\tan^{-1}(2)$ with the receiver lines (other angles are used as well). The patterns shown in Figure 4.14c–e look quite different from those in Figure 4.13c–e. Yet, the number of different fold values contributing to each offset range is about the same between the two geometries. The slanted geometry shows more striping (along the receiver lines), whereas the orthogonal geometry shows a grid pattern due to symmetry between shot and receiver lines. The main reason why a slanted geometry is chosen instead of an orthogonal geometry is that it breaks symmetry (if chosen instead of a brick-wall geometry it introduces spatial continuity). This manifests itself in a better offset distribution in the center of the unit cells, as shown in the rightmost division of the bin offset distribution display in Figure 4.14f. There the white areas are not as large as in the corresponding display for the orthogonal geometry. Another reason put forward to choose a slanted geometry is the smaller geometry imprint associated with variations in fold. However, the fold variations between the two geometries may show different patterns, but the range of fold values is about the same.

A slanted geometry has some disadvantages as well: it takes more shot lines to cover the survey area (compare Figure 4.13b with 4.14b), and the slanted spreads have incomplete receiver gathers (the swath moves with the shot line, so that not all shots along the shot line of a slanted spread shoot into the same receivers). See also discussion in Sections 4.3.3 and 4.3.4.

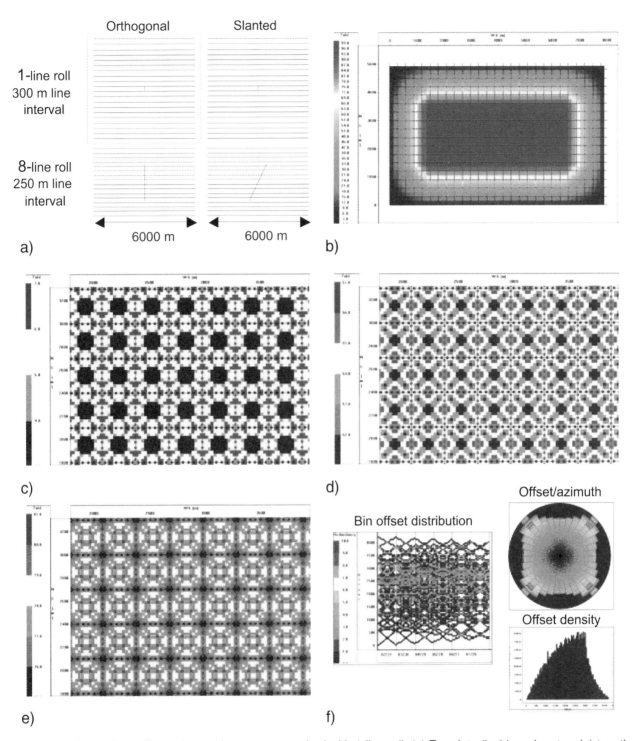

FIG. 4.13. Attributes for 20-line orthogonal geometry acquired with 1-line roll. (a) Template (inside red rectangle) together with templates of Figures 4.14–4.16, (b) full fold of survey, (c) fold for offsets 0–800 m, (d) fold for offsets 0–2500 m, (e) fold for offsets 0–3000 m, (f) other attributes as labeled.

Figure 4.15 illustrates a 16-line orthogonal geometry with an 8-line roll. This means that the next swath in the crossline direction shares only 8 (= 16 – 8) receiver lines with the previous swath. At first sight (Figures 4.15b and c) this change does not make much difference. However, Figures 4.15d and e show a much wider range in fold values, whereas strong discontinuities in fold (blue against yellow or orange) also occur. Such a large variation in fold for any offset range may lead to serious geometry imprints. The bin offset distributions in Figure 4.15f are dependent on the position of the roll boundaries with respect to the bins. In this geometry the concept of unit cell tends to lose its meaning. The shape of the cross-spreads in this geometry depends on their position with respect to the roll. They are all asymmetric with respect to positive and negative crossline offsets. As a consequence, not all offset-vector tiles as defined in Section 2.5.2 are capable of a complete single-fold tiling of the survey area, leading to much larger migration artifacts than in a regular one-line roll geometry. Although this geometry might take less time and effort to acquire in the field, it should be considered inferior as compared to the geometries shown in Figures 4.13 and 4.14.

Finally, Figure 4.16 illustrates a 16-line slanted geometry with an 8-line roll. Fold irregularity for this geometry is similar as for the orthogonal geometry with 8-line roll. An additional disadvantage is that the shots of a slanted spread do not shoot into the same range of receivers.

It would be nice if attribute analysis programs would be able to analyze attributes such as symmetry and suitability for dual-domain filtering of the basic subsets of each geometry.

4.6.7 Conflicting requirements between structural interpretation and AVO

Structural interpretation requires the acquisition of square cross-spreads, because this ensures the same quality and appearance of the data in the crossline direction as in the inline direction. At the same time, the longer the offsets which contribute to the final stacked and migrated data the lower the resolution, because the NMO stretch reduces the potential resolution. Therefore, for structural interpretation, the maximum inline and crossline offsets should not be chosen larger than necessary to achieve a satisfactory signal-to-noise ratio.

On the other hand, AVO analysis tends to be most successful, if the offsets are as large as possible. The maximum useful offset is determined by noise dominating the longer offsets. The structural interpretation requirements and the AVO requirements may be in conflict with each other (not always, e.g. a severe multiple problem may require the use of the largest possible offsets).

If there is a conflict, different methods may be considered to meet AVO requirements in addition to structural requirements:

1. Increase maximum inline offset while maintaining maximum crossline offset.

This is the purist solution, both requirements are met, yet without unduly enlarging maximum crossline offset to maintain symmetry. Maximum inline offset is chosen to suit AVO requirements, maximum crossline offset to suit structural interpretation requirements. In processing for optimal structural interpretation results, the maximum inline offset should be truncated to equal the maximum crossline offset. On the other hand, for AVO processing, the maximum crossline offset may be truncated to reduce the impact of azimuth variation on AVO analysis. AVO processing should take the typical nature of the orthogonal geometry into account as discussed in Section 2.6.9.

2. Increase maximum inline offset while reducing maximum crossline offset.

This is the pragmatic solution. The requirements of structural processing are not fully met, but at least the signal-to-noise ratio should be sufficiently high.

For any reduction in number of receiver lines the efficiency of each shot is reduced, because the number of listening receiver stations is reduced and shot energy is lost. The redundancy in shot and receiver statics will be different. Nevertheless, the impact of a reduction in width of a geometry need not be too dramatic, as long as the smaller common-receiver gathers can still be filtered, i.e., *as long as the two spatial dimensions can be exploited in processing.*

A useful criterion might be the minimal requirement to fully record the ground-roll energy in the two spatial dimensions. Then dual-domain (f,k)-filtering or 3-D velocity filtering can still get rid of most energy traveling with low apparent velocity in all possible directions. If the phase velocity of the fastest noise is, for instance, 750 m/s, and the deepest level of interest is 3 s, then the maximum crossline offset should at least be 2250 m.

3. Acquire a grid of 2-D lines along a subgrid of the 3-D survey.

This solution might be considered when the very long offsets required for AVO are absolutely unnecessary for structural interpretation.

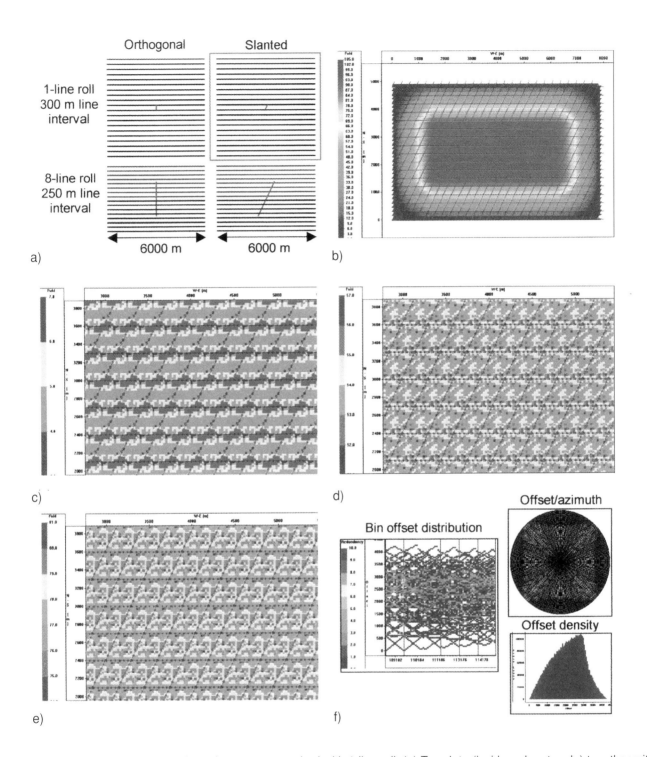

FIG. 4.14. Attributes for 20-line slanted geometry acquired with 1-line roll. (a) Template (inside red rectangle) together with templates of Figures 4.13, 15, and 16; (b) full fold of survey; (c) fold for offsets 0–800 m; (d) fold for offsets 0–2500 m; (e) fold for offsets 0–3000 m; (f) other attributes as labeled.

94 Chapter 4 Guidelines for design of "land-type" 3-D geometry

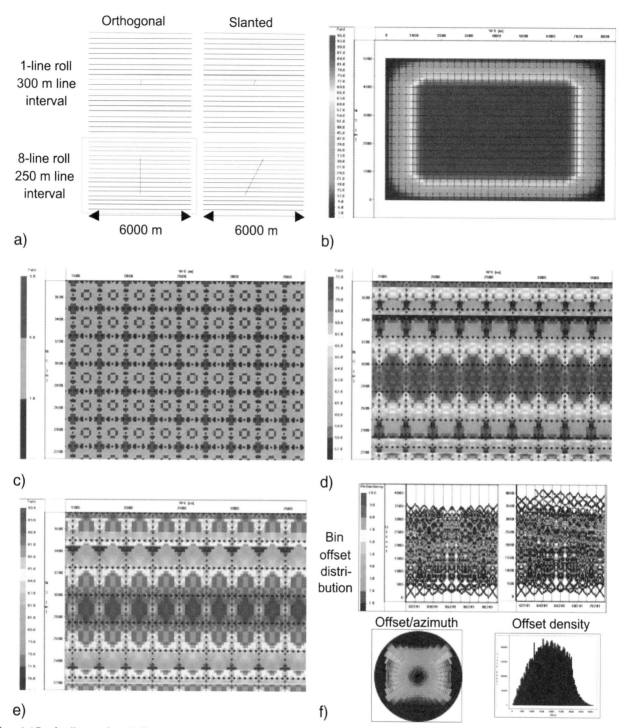

FIG. 4.15. Attributes for 16-line orthogonal geometry acquired with 8-line roll. (a) Template (inside red rectangle) together with templates of Figures 4.13, 14, and 16; (b) full fold of survey; (c) fold for offsets 0–800 m; (d) fold for offsets 0–2500 m; (e) fold for offsets 0–3000 m; (f) other attributes as labeled.

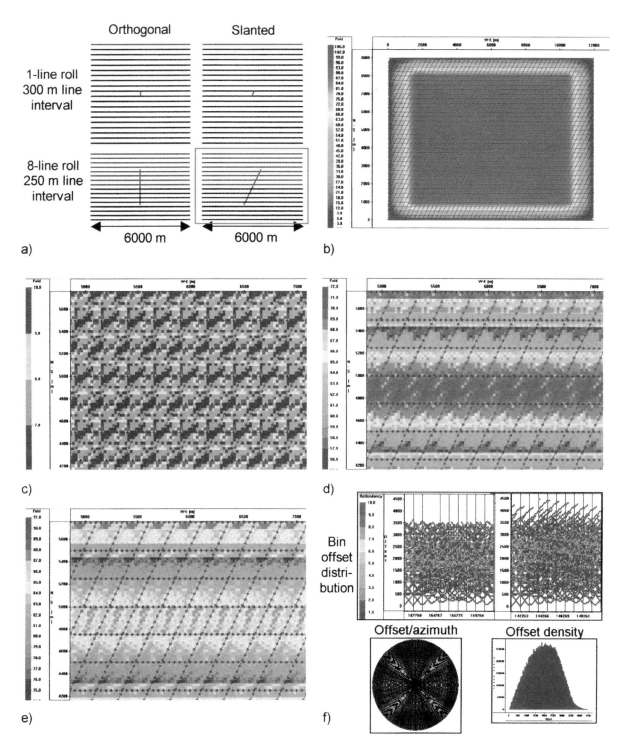

FIG. 4.16. Attributes for 16-line slanted geometry acquired with 8-line roll. (a) Template (inside red rectangle) together with templates of Figures 4.13–15; (b) full fold of survey; (c) fold for offsets 0–800 m; (d) fold for offsets 0–2500 m; (e) fold for offsets 0–3000 m; (f) other attributes as labeled.

4.6.8 Deviations from nominal due to topography and obstacles

Once a nominal geometry has been decided upon, it may not be easy to realize the geometry without modifications. In particular, in heavily built-up and cultivated areas such as in The Netherlands, acquiring 3-D is a daunting task, and sticking to nominal would be impossible. As discussed, spatial continuity of the grid of acquisition lines is of great importance. Therefore, an attempt should be made to acquire common-receiver gathers and common-shot gathers that can be filtered without creating artifacts.

The number of discontinuities in the acquisition lines should be minimized. A first step to achieve this can already be set in the preplanning phase: make use of the natural grain of the area in which data are to be recorded. This involves roads, rivers, and canals, and may vary across the survey. An important consideration is that continuity is more important than having pieces of straight lines. It is perfectly acceptable for filtering purposes to record data along smoothly curved lines. Moreover, sinuous lines have less impact on the environment, because they can wind around large trees (Williams, 1993). The smoothness criterion given in Lindsey (1991) for the acquisition of crooked 2-D lines can be applied in a similar way to 3-D acquisition lines. A similar solution is suggested in Figure 4.17.

In practice, smooth lines are difficult to survey. Figure 4.18 suggests approximating the smooth line by a few straight lines.

The requirement of smoothness can reduce costs in hilly areas. Shot lines may be chosen to follow elevation contours with receiver lines following gradient lines.

Current practice in the presence of *obstacles* is to aim for regular fold as counted in bins (Donze and Crews, 2000). This regularity is achieved most easily by locating shots and receivers as close as possible to the nominal grid point position. If a shot cannot be located at that point, the standard prescription is to move the shot station over an integer number of station intervals to the right or to the left (Figure 4.19a), and to move the receiver spread over an equal number of stations in the opposite direction. This prescription maintains fold and it maintains midpoints in bin centers, but it produces spatial discontinuities in the common-receiver gathers or common-shot gathers.

To achieve a smooth shot line, shots must be shifted

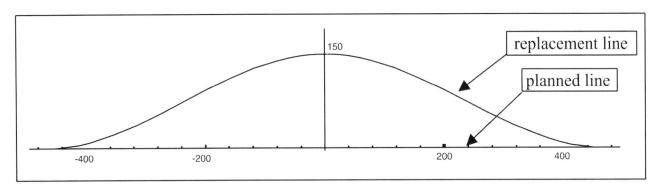

FIG. 4.17. Suggested smoothness criterion for skirting aound obstacles. This figure shows the function $r \cos^2(x/2r)$, with $r = 150$ m. The planned acquisition line runs along the horizontal axis, but an obstacle prevents it being laid out as planned. The obstacle is not shown, but the idea is that the smooth replacement line stays as close as possible to the planned line.

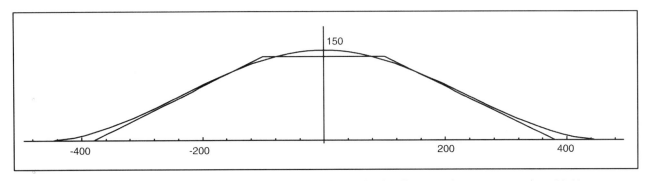

FIG. 4.18. Practical implementation of smoothness criterion. Bends in shot lines can be no greater than 26.6°.

a noninteger number of station intervals to the right or to the left (Figure 4.19b). Figure 4.20 illustrates the benefit of smoothness in an intuitive way. The smoothness criterion may also be formulated as the requirement that the cross-spreads remain minimal data sets. Gesbert (2002) formulates precise criteria for a data set to be a minimal data set. This means that the acquisition lines should not be contorted such that the illumination areas of the cross-spreads are more than single-fold in some places.

Moving some shots to the left also moves the midpoint area of all corresponding cross-spreads to the left. This also has consequences for the cross-spreads adjacent to the affected cross-spreads. The cross-spreads to the left will now have partially overlapping midpoint areas with the shifted cross-spreads, whereas there will be a gap in midpoint coverage between the shifted cross-spreads and the cross-spreads to the right. Yet, it is important to maintain a regular fold (Section 4.4.5.4). The overlap can be taken care of in processing by appropriate weighting, but preferably the gaps should be filled in acquisition.

It is also important that the number of receivers for each shot in a cross-spread stays the same, and that the number of shots for each receiver is constant. This allows filtering in shot and receiver domains. The best way to accomplish regular fold and "rectangular" cross-spreads in a geometry with curved receiver and shot lines, is to *acquire a number of extra stations on either end of the spread or template*. Similarly, curvature of receiver lines may be regularized if some additional shots are fired into each receiver line. (This would be quite expensive in a one-line roll geometry, but easily doable in a full-swath roll.) These extensions of the shot and receiver spreads allow the creation of regular fold in processing (by interpolation to a square bin grid, and discarding or weighting of traces in regions with too high a fold).

The acquisition of redundant receivers and redundant shots might be considered part of the acquisition strategy, even in areas without obstacles. It is good signal-processing practice to try and soften hard discontinuities. With redundant traces outside the nominal edges, (f,k)-filtering across the edges may reduce the discontinuities, in particular those caused by coherent noise. This action would minimize migration artifacts.

A possible solution for skirting along a large obstacle (no obstacle for the receiver lines) is given in Figure 4.21. However, a better solution is to acquire all reciprocity traces by interchanging shot and receiver positions for those shots which cannot be acquired otherwise (Jerry Davis, personal communication). With this solution, only the short offset data inside the no-go-area-for-shots are not acquired.

If it is not possible to avoid an obstacle by skirting around it, and the obstacle would lead to only two or three missing shots, make-up shots should be placed along the planned shot line, just a bit closer together than nominal. This should improve the chances of successful interpolation of shots across the obstacle.

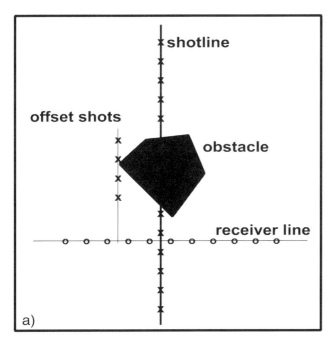
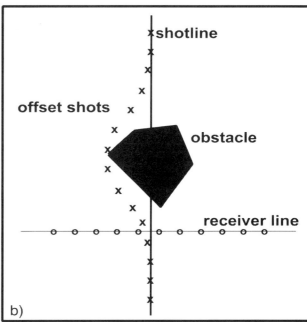

Fig. 4.19. Procedures for dealing with obstacles. (a) Midpoint-centering solution; (b) smooth solution. The smooth solution preserves spatial continuity.

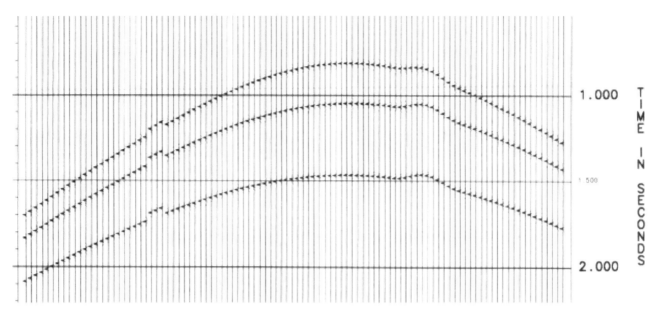

FIG. 4.20. Common-receiver gather illustrating discontinuity in reflection times due to jump in shot positions and effect of smooth positioning of shots.

If it is not possible to avoid an obstacle by skirting around it, and the obstacle would lead to four or more missing shots, then some make-up shots should still be placed along the planned shot line. An additional short shot line might be inserted, depending on the gap to be compensated.

To minimize problems while surveying the planned lines in the field, it is essential to use complete and recent maps that show all obstructions. A 3-D design program should be used that allows the planning of the survey grid on top of satellite maps, topo maps, maps showing power lines, pipelines, and no-access areas, etc.

4.7 Testing

In various places in this document the subject of testing has been touched upon. Testing may be necessary to find out about the minimally required fold at some level to be interpreted (Section 4.4.3); to determine optimal choice of shot and receiver arrays (Section 4.4.5.5), and to develop a better feel for the influence on quality of virtually any other parameter in survey design.

Testing to determine what type of geometry is best is perhaps the most demanding task. Section 7.3 describes the comparison between a brick-wall geometry (the then current technique) and a cross-spread geometry (the proposed technique). It is a fine example of getting very valuable results for a minimum of testing effort.

The acquisition of a number of 3-D microspreads (cross-spreads with small shot and receiver intervals) could be used to measure the amount of scattered ground roll for a definition of field array suppression requirements. A field example of a 3-D microspread is discussed in Section 7.2.

Testing may also be carried out to determine the influence of parameters which are not discussed in this book, such as depth of shot holes, number of sweeps per shotpoint, length of sweeps, etc. In all testing it is important to base judgment on the results of fully processed data. Thomas and Hufford (1998) demonstrate the effect of reducing the number of sweeps; the raw shot records show a large difference between shots with different number of sweeps, whereas after full processing, including stack and migration, the differences become minimal. In the discussed case the reduction in acquisition cost easily outweighed the cost of testing.

Because theory will not always be able to predict the outcome of a particular choice of parameters, experiments are often the only way to obtain a definitive answer.

4.8 Discussion

This chapter outlines a methodology for 3-D seismic survey design. It is based on the recognition of the existence of basic subsets or minimal data sets for all common geometries. In general, application of this methodology should lead to fully satisfactory 3-D survey designs. Other approaches to survey design used in the industry use experience (what worked in the past, should work in the future) often coupled with attribute analysis of proposed geometries, model-based design, optimization techniques (Liner et al., 1999), or focal-beam analysis (Berkhout and Ongkiehong, 1998; Volker et al., 1998;

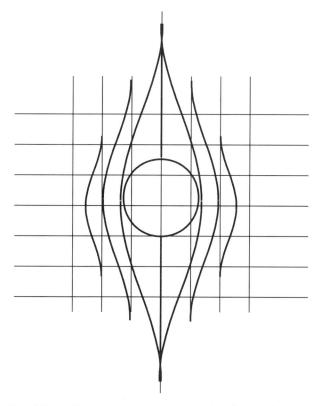

FIG. 4.21. One way to shoot around a large obstacle. Rather than placing many compensation shots close to the obstacle, this spreading out of the lines would produce a better offset distribution. Yet, to maximize coverage under the obstacle, it is necessary to also acquire shots along the interrupted shot line. However, shooting reciprocity traces is a better technique (see text).

Berkhout et al., 2000; Volker et al., 2000). None of these techniques takes into account the basic subsets of the various acquisition geometries and their imaging properties. Section 4.8.1 discusses attribute analysis and Section 4.8.2 model-based design.

4.8.1 Attribute analysis

Currently, none of the existing 3-D design packages (e.g., Mesa, Omni, Reflex) allows the design of 3-D surveys according to geophysical requirements. Instead, they provide the user with means for attribute analysis of a proposed geometry. The attribute analysis is focused on bins rather than on spatial continuity. Examples of attribute analyses are given in Figures 4.13–4.16 (made with Omni). However, attribute analysis of a geometry proposed on the basis of symmetric sampling requirements is hardly necessary as the attributes will just be fine. The attribute analysis may just serve to convince the unconvinced. Attribute analysis is very important for tutorial purposes. In this way, Figures 4.13–4.16 were used to show that a multiline roll produces strong discontinuities in coverage and that slanted geometries have no major advantage over orthogonal geometries. Attribute analysis is also valuable in analyzing changes made in a nominal geometry to allow for obstacles. Most 3-D design packages also allow the preparation of scripts to be followed for the actual recording in the field.

It should be easy to add front ends to existing design packages for the implementation of the design rules given in this chapter. Some companies are making modest attempts in this direction. The actual attribute analysis might be expanded by mapping out the basic subsets of the geometry (similar to the templates which can be shown). In particular, it should be possible to test solutions for obstacles on their robustness with respect to the pseudominimal data sets (see Sections 2.5 and 4.6.8).

4.8.2 Model-based survey design

The objective of the acquisition of 3-D surveys is to get a better picture of the subsurface. In this chapter I have outlined how subsurface information and other geophysical knowledge can be incorporated in the design of 3-D surveys. This information included resolution requirements, maximum dips, velocity distribution, etc. Often, all this information together would allow the construction of a subsurface model. The objective of the survey would then be to illuminate and image that model in an optimal way. Various authors have come up with proposals for model-based survey design (Slawson et al., 1995, Salehi and Kappius, 1998). The approach is to take a proposed geometry and to analyze subsurface illumination (of the model) using raytracing.

In my opinion, these raytracing approaches are not necessary for many geological situations. Usually, it will be sufficient to have a good look at what is necessary for the steepest dips and the shallowest targets in the area. On the other hand, very complex geology may benefit from reassurance about a proposed geometry obtainable with raytracing. For instance, for a locally shallow target, such as around salt domes, one may want to know whether a variable line spacing would be called for, narrow line spacing across the localized shallow target, and a wider line spacing elsewhere.

Up till now, illumination analysis for a whole geometry is very time consuming, because the various procedures are based on raytracing for individual shot/receiver pairs. Illumination is measured in terms of "number of hits per subsurface bin." However, the raytracing approach could benefit a great deal from exploiting the concept of minimal data set and illumination area as discussed in Section 2.5.3.

Rather than counting hits per bin, it is sufficient to count the number of overlapping illumination areas corresponding to the minimal data sets. Each illumination area can be established by tracing a limited number of rays. For mild geologies, it is sufficient to raytrace only for the edges of the minimal data sets, for more complex geologies, some additional shots and receivers may have to be analyzed, until the illumination area can be established with sufficient accuracy.

Often, it is not so much the illumination capability of a survey design that needs to be established, but more its capability of delivering data with adequate signal-to-noise ratio. Modeling the capability of an acquisition geometry to suppress noise is discussed in Section 3.3.5.

4.9 A summary of what to do and not to do in 3-D survey design

There are many ways of acquiring 3-D seismic data, but only parallel geometry and orthogonal geometry can provide optimal spatial continuity by ensuring that the common-receiver gathers have as many shots as the common-shot gathers have receivers. In parallel geometry common-inline-offset gathers ensure spatial continuity in the crossline direction. The properties of areal geometry with sparse receivers listening to a dense grid of shots are about the same as those of orthogonal geometry. The spatial continuity is now ensured by proper 2-D sampling of shots.

On the other hand, I strongly recommend that brick-wall geometry not be used, because its common-receiver gathers are very discontinuous. Slanted geometry is an improvement over brick-wall geometry, but it has no real advantages over orthogonal geometry, whereas its shot interval is larger than in an equivalent orthogonal geometry. Similarly, zigzag geometry has even larger shot intervals and is not recommended.

The multiline roll offers improved efficiency as compared to the single-line roll, but at the expense of greater spatial discontinuities that may manifest themselves as a strong acquisition footprint. Often the full-swath roll will also provide gains in efficiency, but this geometry can be trimmed down to provide the same nominal geometry as the single-line roll.

Wide orthogonal geometry (aspect ratio larger than 0.5) provides similar quality in crossline as in inline direction. Narrow geometry will have more spatial discontinuities in the crossline direction and may show more migration artifacts in that direction.

The single most important parameter currently limiting quality of 3-D seismic is the station spacing. Coarse sampling intervals of 50 or 60 m still seem to be the rule in 3-D acquisition, whereas it has been shown in 2-D data acquisition that much smaller intervals can provide tremendous gain in quality. In 3-D, the same shot and receiver gathers are acquired as in 2-D and the sampling requirements are the same. If finer sampling is used, fold is no longer necessary to compensate for migration artifacts generated in processing, but is only required to separate noise from signal. Although single-sensor recording is becoming technologically feasible, using 20- or 25-m station spacings with correspondingly shorter field arrays would already provide a great improvement over current practice.

The acquisition lines in parallel and orthogonal geometry should preferably be straight, but in the case of obstacles, smooth lines should be selected skirting the obstacles rather than shifting shots an integer number of station spacings to the right or to the left, thus creating discontinuous shot lines. Solitary shots should be avoided as much as possible, it is always important to check the spatial continuity in the receiver gathers.

References

Ansink, I. F., Bourdon, L. M., Corsten, C. J. A., Khalil, A. A., Meiburg, R. M., and Voon, J. W. K., 1999, The effect of fold and crossline spacing on quality and interpretability of 3D marine seismic data in the Dutch North Sea: EAGE Workshop, 3D seismic surveys: Design, tests and experience, Expanded Abstracts.

Berg, E. W., Svenning, B., and Martin, J., 1994, SUMIC — A new strategic tool for exploration and reservoir mapping: 56th Conf., Eur. Assoc. Expl. Geophys., Extended Abstracts, G055.

Berkhout, A. J., and Ongkiehong, L., 1998, Analysis of seismic acquisition geometries by focal beams: 68th Ann. Internat. Mtg., Soc. Expl. Geophys., Expanded Abstracts, ACQ1.8, 82–85.

Berkhout, A. J., Ongkiehong, L., Volker, A. W. F., and Blacquière, G., 2001, Comprehensive assessment of seismic acquisition geometries by focal beams—Part I: Theoretical considerations: Geophysics, **66**, 911–917.

Bloor, R., Albertin, A., Jaramillo, H., and Yingst, D., 1999, Equalised prestack depth migration: 69th Ann. Internat. Mtg., Soc. Expl. Geophys., Expanded Abstracts, SPRO11-3, 1362–1365.

Constance, P. E., et al., 1999, Simultaneous acquisition of 3-D surface seismic data and 3-C, 3-D VSP data: 69th Ann. Internat. Mtg., Soc. Expl. Geophys., Expanded Abstracts, BH/RP4.5, 104–107.

Den Boer, E., Eikelboom, J., van Driel, P., and Watts, D.,

2000, Resistivity imaging of shallow salt with magnetotellurics as an aid to prestack depth migration: First Break, **18**, 19–26.

Dickinson, J. A., Fagin, S. W., and Weisser, G. H., 1990, Comparison of 3-D seismic acquisition techniques on land: 60th Ann. Internat. Mtg., Soc. Expl. Geophys., Expanded Abstracts, 913–916.

Donze, T. W., and Crews, J., 2000, Moving shots on a 3-D seismic survey: The good, the bad and the ugly (or How to shoot seismic without shooting yourself in the foot!): The Leading Edge, **19**, 480, 482, 483.

Ebrom, D., Krail, P., Ridyard, D., and Scott, L., 1998, 4-C/4-D at Teal South: The Leading Edge, **17**, 1450–1453.

Egan, M., 2000, The benefits of seismic modeling when designing the templates for 3-D surveys in the Middle East: GeoArabia, **5**, 86.

Gesbert, S., 2002, From acquisition footprints to true amplitude: Geophysics, **67**, 830–839.

Hastings-James, R., Green, P., Al-Saad, R., and Al-Ali, M., 2000, Wide-azimuth 3-D swath acquisition: GeoArabia, **5**, 103.

Holland, M., 2000, Wide-azimuth, radially-directed seismic data acquisition method: US Patent 6 044 040.

Huard, I., and Spitz, S., 1998, Filling gaps in the coverage of 3-D marine acquisition: The Leading Edge, **17**, 1606–1609.

Kallweit, R. S., and Wood, L. C., 1982, The limits of resolution of zero-phase wavelets: Geophysics, **47**, 1035–1046.

Krey, Th. C., 1987, Attenuation of random noise by 2-D and 3-D CDP stacking and Kirchhoff migration: Geophys. Prosp., **35**, 135–147.

Lee, S. Y. et al., 1994, Pseudo-wavefield study using low-fold 3-D geometry: 64th Ann. Internat. Mtg., Soc. Expl. Geophys., Expanded Abstracts, 926–929.

Lindsey, J. P., 1991, Crooked lines and taboo places: What are the rules that govern good line layout?: The Leading Edge, **10**, No.11, 74–77.

Liner, C. L., Underwood, W. D., and Gobeli, R., 1999, 3-D seismic survey design as an optimization problem: The Leading Edge, **18**, 1054–1060.

Lumley, D. E., Claerbout, J. F., and Bevc, D., 1994, Anti-aliased Kirchhoff 3-D migration: 64th Ann. Internat. Mtg., Soc. Expl. Geophys., Expanded Abstracts, 1282–1285.

Morrice, D. J., Kenyon, A. S., and Beckett, C., 2002, Optimizing operations in 3-D land seismic surveys: Geophysics, **67**, 1818–1826.

Reilly, J. M., 1995, Comparison of circular "strike" and linear "dip" acquisition geometries for salt diapir imaging: The Leading Edge, **14**, 314–322.

Salehi, I., and Kappius, R., 1998, Software optimizes design of survey: The Amer. Oil and Gas Rep., July.

Savage, J. E. G., and Mathewson, J. C., 2001, Prediction of 3-D seismic footprint from existing 2-D data: The Leading Edge, **20**, 464–473.

Schroeder, F. W., Farrington, T. G., Balon, S. G., and Rapp, C. S., 1998, How fold and bin size impact data interpretability: The Leading Edge, **17**, 1274–1284.

Slawson, S., Grove, K., and Monk, D. J., 1995, Model-based 3-D seismic survey design: 57th Conf., Eur. Assoc. Geosc. and Eng., Extended Abstracts, Paper B009.

Smith, J. W., 1997, Simple linear inline field arrays may save the day for 3-D direct-arrival noise rejection: Proceedings SEG Summer Research Workshop.

Thomas, J. W., and Hufford, J. M., 1998, Reducing the cost of acquiring seismic data: 68th Ann. Internat. Mtg., Soc. Expl. Geophys., Expanded Abstracts, ACQ3.4, 125–128.

Vermeer, G. J. O., 1990, Seismic wavefield sampling: Soc. Expl. Geophys.

——— 1997, Streamers versus stationary receivers: 29th Annual Offshore Technology Conference, Paper OTC8314, p. 331–346.

——— 1998a, 3-D symmetric sampling: Geophysics, **63**, 1629–1647.

——— 1998b, Creating image gathers in the absence of proper common-offset gathers: Exploration Geophysics, **29**, 636–642.

——— 1999a, Factors affecting spatial resolution: Geophysics, **64**, 942–953.

——— 1999b, Converted waves: Properties and 3-D seismic survey design, 69th Ann. Internat. Mtg., Soc. Expl. Geophys., Expanded Abstracts, SACQ 2.6, 645–648.

Volker, A. W. F., Blacquière, G., Berkhout, A. J., and Ongkiehong, L., 2001, Comprehensive assessment of seismic acquisition geometries by focal beams – Part II: Practical aspects and examples: Geophysics, **66,** 918–931.

Volker, A. W. F., Blacquière, G., and Ongkiehong, L., 1998, Optimization of 3-D data acquisition geometries: 68th Ann. Internat. Mtg., Soc. Expl. Geophys., Expanded Abstracts, ACQ2.1, 86–89.

Williams, K., 1993, Trends in seismic data acquisition—A perspective: CSEG Recorder, **18**, No. 10, 14–16.

Wisecup, R.D., 1994, The relationship between 3-D acquisition geometry and 3-D static corrections: 64th Ann. Internat. Mtg., Soc. Expl. Geophys., Expanded Abstracts, 930–933.

Chapter 5
Streamers versus stationary receivers

5.1 Introduction

Marine 3-D seismic data acquisition technology is progressing rapidly. On the one hand, there has been a very rapid increase in the number of streamers that can be towed by modern seismic vessels, and on the other hand, the variety of stationary-receiver (seabed) systems is mushrooming. As a consequence, 3-D seismic surveys may be carried out using quite different techniques, and the question of which technique is most appropriate for a given problem needs to be addressed. This chapter reviews pros and cons of the various techniques.

There is a great deal of similarity between a 2-D grid of seismic lines acquired on land and offshore. In both cases sources and receivers are arranged along coinciding straight lines leading to seismic traces, all having the same shot-to-receiver azimuth within one seismic line. The main difference, as far as geometry is concerned, is that in streamer acquisition an end-on geometry is used whereas in land data acquisition a center-spread geometry is possible.

With the advent of 3-D acquisition, marine and land data acquisition geometries started to diverge. In marine acquisition, 3-D was most efficiently achieved by repeating the 2-D geometry, whereas on land, sources and receivers can be decoupled so that other geometries such as orthogonal and zigzag geometries are also feasible, and in fact, more cost-effective.

Acquiring parallel lines in 3-D marine acquisition means that at the start of the survey, the best direction of those lines must be decided. Assuming a dominant dip and strike direction, various authors have discussed the pros and cons of dip or strike acquisition (Larner and Ng, 1984; Manin and Hun, 1992; Arbi et al., 1995).

Considerable gains in efficiency have been reached in marine acquisition with the introduction of multisource, multistreamer (MS/MS) techniques, and even multiboat operations (Davidson and Bandell, 1990; Sandø and Vegeland, 1995; Cramer et al., 1995; Duey, 1996). A record of 10 streamers (Cramer et al., 1995) was soon superseded by a new record of 12 streamers (Duey, 1996) and at present (2001) the record stands at 16 streamers. Basically, these configurations have maintained dominance of the chosen acquisition direction in the shot-to-receiver azimuths, thus maintaining the question of what shooting direction gives the best seismic results. It has been realized that the greater efficiency of MS/MS techniques is achieved at the expense of regular illumination of the subsurface (Vermeer, 1994; Beasley and Mobley, 1995). The presence of obstructions such as production platforms, reduces the efficiency of the MS/MS techniques and requires the use of a two-boat operation (Egan et al., 1991). Uncontrollable feathering forms another reason for irregular illumination of the subsurface.

Bottom cables have been in use for quite some time in transition zone waters. In fact, the industry's very first experiences in shallow water utilized bottom cables. Only after the rediscovery that the combined use of pressure and velocity detectors would allow the necessary removal of the receiver ghost—the dual-sensor technique—could the use of bottom cables be extended into deeper waters (Barr and Sanders, 1989; Barr, 1997). Particularly in areas with many obstructions and in shallow waters, the use of bottom cables (frequently called OBC technique for ocean-bottom cable, though SBC for seabed cable might be more appropriate) is now really taking off (Barr, 1994, 1997; Sanders et al., 1994; Cafarelli, 1995; Meunier et al., 1995; Carvill et al., 1996).

A very special bottom-cable technique was developed by Statoil (Berg et al., 1994; Johansen et al., 1995; Sonneland et al., 1995). In this SUMIC (subsea seismic) technique three-component geophones are attached to the cable and planted in the sea bottom by an ROV. A hydrophone is also part of the system, therefore this kind of acquisition is sometimes referred to as 4-C (four-component). With SUMIC not only P-waves but also S-waves are recorded, and a gas chimney, which would be uninterpretable on a P-wave section, may be resolved in

---

This chapter modified after Vermeer (1997).

the *P-S* section (Granli et al., 1999). The technique is not suitable for 3-D, but investigations are underway to adapt it to 3-D (Berg and Arntsen, 1996).

Other 4-C techniques are also emerging, and are discussed in Section 5.4.4.

An interesting stationary-receiver technique is the vertical hydrophone cable (VHC) (Krail, 1991, 1994). In this technique some 12 to 16 hydrophones are arranged along a vertical cable which is anchored to the sea bottom.

Another stationary-receiver technique is the ocean-bottom seismometer (OBS) which has been used by academia for some twenty years, but now is being considered for use in 3-D seismic data acquisition. An OBS is a self-contained receiving and recording unit residing at the ocean-floor (literally this time: OBSs are even used in waters exceeding 3000 m!) for the duration of the survey. Unless there is a technological breakthrough, VHC and OBS are only suitable for use in an areal geometry in which the receivers are arranged in a widely spaced areal grid, and the shots are arranged in a densely spaced areal grid.

In the following I will expand on the discussion of various marine data acquisition techniques, and compare their relative merits. An important aspect is to what extent the stacked and migrated data are representative of the acoustic impedance of the subsurface. Therefore, the influence the acquisition geometry may have on the final seismic amplitudes is discussed first. Next, streamer acquisition is discussed, the dip/strike question, and the effect of using MS/MS techniques. The stationary-receiver techniques are reviewed with an emphasis on the various geometries that are suitable with those techniques.

5.2 Geometry imprint

Time slices and, in particular, horizon slices of stacked or migrated seismic data often show an amplitude pattern which is typical for the acquisition geometry used in the 3-D seismic survey. This amplitude pattern is often referred to as the geometry imprint or acquisition footprint.

For streamer surveys, the geometry imprint manifests itself as a striping effect: slow variation of amplitude in the inline direction (the shooting direction) and rapid variation in the crossline direction. An example of striping is given in Figure 5.1. In land geometries the shot and receiver line pattern may be visible in the seismic amplitudes. Shallow data, having lower fold than deeper data, tend to have the strongest geometry imprints. These effects of geometry on amplitude are most undesirable, particularly for lithology and porefill prediction, but also for a reliable structural interpretation. Therefore, it is important to choose an acquisition technique and a geometry with which such effects can be minimized.

The geometry imprint is directly related to the offset distribution as a function of CMP position. Systematic variations in offset sampling or periodicities in the offset distribution may create corresponding variations in amplitude. (I will use the term "offset sampling" for the sampling of offsets within a CMP, and the term "offset distribution" for the variation of offset sampling across the CMPs.) The effect is also known from 2-D seismic data; for instance, the odd/even effect in streamer data acquisition with equal shot and receiver station intervals is linked to the fact that the offset sampling of the even CMPs differs from that of the odd CMPs. Why the offset sampling affects the seismic amplitudes might be discussed on the basis of Figure 5.2.

The left side of Figure 5.2 shows an NMO-corrected CMP gather with a very regular offset sampling and virtually constant shot-to-receiver azimuth. The gather shows many events running across the NMO-corrected primaries, and stacking should be able to suppress most of this noise. The right side of Figure 5.2 shows various

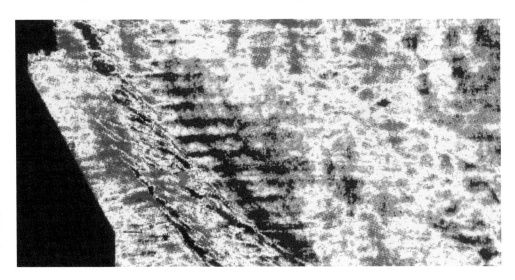

FIG. 5.1. Amplitude striping in 4/4 geometry. The geometry imprint has a periodicity of 16 in the crossline direction.

stacked traces. Splitting the odd and even traces of the CMP over two separate gathers, followed by stacking, leads to two different stacked traces, because the noise events have been sampled at different offsets. A similar reasoning applies to the primaries: as amplitude varies with offset (AVO) and the stack is an average of all sampled offsets, the averaged amplitude of the primary also depends on the offset sampling, even if there were no noise.

Often, the noise events do not change rapidly as a function of CMP position. Hence, if there is a periodicity in the offset distribution, then the noise will be sampled at the same offsets periodically, leading to periodicities in the seismic amplitude. Similarly, if there is a systematic change in the way offsets are sampled from CMP to CMP, then the amplitude effect will be systematic. Yet, in situations where there is no systematic variation in offset sampling, or no periodicity in the offset distribution, the stacked amplitudes are still affected by noise or amplitude variations with offset, even though it is more difficult to recognize the effect.

The ideal way of reducing the geometry imprint to a minimum is by fine and regular sampling of offsets in each CMP. Unfortunately, in streamer acquisition regularity of sampling is not achievable due to uncontrollable feathering of the streamers, and in stationary-receiver techniques, the offset sampling is usually highly irregular (unless parallel geometry is used).

5.3 Streamer acquisition

A main feature in 3-D marine data acquisition using streamers is the decision that has to be made on the shooting direction. Another aspect is that MS/MS configurations produce irregular illumination of the subsurface, whereas uncontrollable feathering compounds the illumination problem. In the following these aspects of streamer acquisition are discussed in some detail.

5.3.1 Shooting direction

The choice of shooting direction is sometimes referred to as the dip/strike decision, but other factors, unrelated to dip or strike, often play a role as well. These other factors could be the presence of a nearby coastline, obstacles in a certain pattern, main current direction, and more. If a rectangular survey area is much longer than it is wide, it is usually more economical to shoot parallel to the long sides of the rectangle than to the short sides. In the latter case it may still be decided to shoot in another direction, if there are good reasons to do so. At any rate, prior to the start of a streamer survey one has to commit to a fixed shooting direction, and many considerations can play a role.

In an area with many obstacles, logistics may dictate the shooting direction. Part of the survey will have to be carried out using an undershoot technique, in which the shooting vessel travels on one side of the obstacle, and

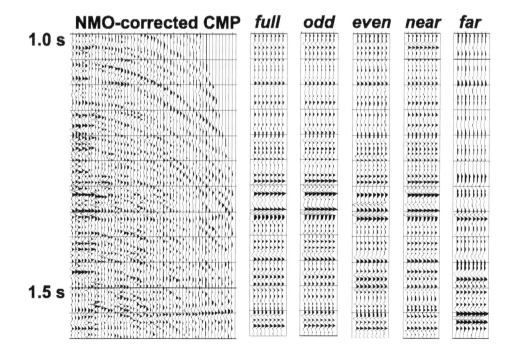

FIG. 5.2. NMO-corrected CMP gather with five different stacks.

the vessel towing the streamer on the other side. The streamer vessel should always remain on the same side (port or starboard) of the shooting vessel (Egan et al., 1991). In that way the shot-to-receiver azimuths all have about the same orientation, which is best for illumination of dipping layers and for DMO.

Often, the undershoot part of the 3-D survey and the regular part are designed to create adjacent midpoint coverage. However, this may lead to illumination gaps in the subsurface, because of the difference in shot-to-receiver azimuths between the two parts. To avoid these gaps, the two parts should have some overlapping midpoint coverage, depending on maximum dip.

To prevent cross-currents from causing differential feathering, i.e., variation in feathering between neighboring streamers, the shooting direction may be chosen to coincide with the main current direction, if any. Unfortunately, a clear and stable current direction does not occur often.

5.3.1.1 Dip/strike decision

To start with, often there is no dip direction that is dominant in the whole survey area. And even if the dipping layers were oriented in some main direction, the fault planes and corresponding diffraction patterns might be mainly oriented at right angles to the dip direction. In all those cases the relevance of the dip/strike decision is reduced.

In case there is a dominant dip direction, there are always some reasons to favor dip shooting, and other reasons to favor strike shooting. The reasons may be truly geophysical, but there may also be reasons related to positioning accuracy and sampling deficiencies.

A geophysical reason to shoot along strike is the imaging of a salt flank. Shooting along strike keeps both legs of the raypath outside the salt dome, making imaging fairly easy, whereas, when shooting dip, one leg of the raypath passes through the salt requiring an accurate estimate of the position of the salt flank for proper imaging. Moreover, much of the energy that should travel through the salt will be reflected before entering the salt, so that less energy is available for reflection against the sedimentary layers. The geometry problem is illustrated in Figure 5.3 which shows a horizon amplitude map around a salt dome. There is a clear correlation between reflection amplitude and shooting direction.

Prism waves (raypaths with a double bounce—against reflector and salt flank before returning to the surface) form another complicating factor in dip shooting (Reilly, 1995). In case the position of the salt dome is fairly well known, a concentric circle shoot survey or a spiral survey

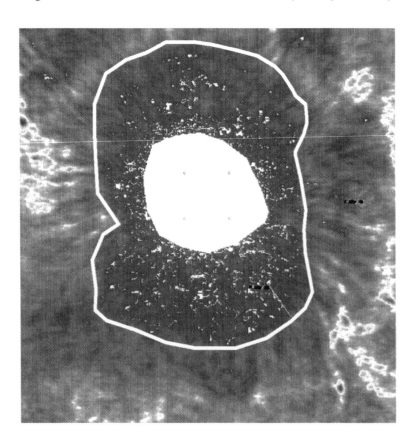

FIG. 5.3. Horizon amplitude map around salt dome. Dark amplitudes are weak. Left and right of the salt dome strike acquisition provides better illumination of the horizon. White line outlines area of weak amplitudes.

can be carried out (Durrani et al., 1987; Marschall, 1990; Hird et al., 1993; Maldonado and Hussein, 1994; Reilly, 1995). With this geometry complicated raypaths are avoided as much as possible.

Another geophysical reason to shoot strike is for AVO analysis. The angle of incidence for a given offset would depend on variations in dip, requiring some correction. When shooting strike this complication can be avoided.

An interesting reason to shoot dip is the existence of a gas chimney along the crest of an elongated anticline. In strike lines along the crest of the anticline, the low-velocity anomaly would create a time delay which is difficult to deal with in processing, whereas in dip lines, undershooting of the anomaly would take place (Sonny Lim, 1992, personal communication).

Larner and Ng (1984) list a number of practical reasons to shoot dip. First, the economics of streamer acquisition favor a finer midpoint sampling in the inline direction than in the crossline direction. It is better to sample finely in the direction where it matters most, i.e., in the dip direction, and if coarse crossline sampling requires interpolation, this can be carried out best in the strike direction. Another reason (which is no longer of great importance due to the increased positioning accuracy of modern streamers) used to be that positioning accuracy tended to be better in the inline direction than in the crossline direction. For strong dips in the crossline direction positioning errors would lead to mis-stacking.

Larner and Ng (1984) also list reasons to shoot strike: Velocity analysis is easier in strike lines, and steeply dipping coherent noise may be removed more easily from sections in which the reflections do not show much dip.

Various authors have investigated the effect of dip versus strike acquisition. The Bullwinkle survey reported in O'Connell et al. (1993) consisted of shooting a survey in two orthogonal directions. A reason to shoot in two different directions was that during 3-D survey design it became clear that no single acquisition direction was optimal. The result confirmed that imaging quality depends on shooting direction, with neither of the two directions being best for all features. Imaging of events was worse when complex raypaths were involved in creating the image, then strike shooting was best. For situations in which such complex raypaths did not play a role, it turned out that steep dips were best imaged with dip shooting. This result might be due to the better sampling in the inline direction, hence better sampling of the fast variations with dip shooting, an argument pro dip shooting also given in Larner and Ng (1984). Whether dip shooting would also be better in case of equal bin size in both directions could not be decided from the Bullwinkle experiment.

Houllevigue et al. (1999) went even a step further than the Bullwinkle experiment. They acquired a marine streamer 3-D survey in four (!) different directions. Their preliminary result shows a time slice through the data set composed from the best parts of the four different data sets, which is better than any of the individual time slices.

Jones et al. (2000) report on a marine 3-D survey, which was acquired in two directions because it was important to get the best image of a graben feature changing its direction inside the survey area. They found that the dip sections produced better images than the strike sections and attribute this result to a finer inline sampling of 12.5 m as compared to the crossline sampling of 18.75 m. In one location, they found that strike produced a better section than dip. However, this could be explained by severe differential feathering in the data acquired in the dip direction.

In a water tank experiment two orthogonal directions were used to find an answer to the dip/strike question for square bins (Arbi et al., 1995). In this experiment "... it was found that the dip survey data produce superior time image results of the target features compared to the strike survey data." Unfortunately, the bin size used in that experiment was very large causing aliasing on input. Aliased input data tend to generate migration noise and incomplete imaging, hence a general conclusion cannot be drawn from that analysis.

Following intuition, I would guess that, in general, the imaging capability of well-sampled common-offset gathers with constant shot-to-receiver azimuth would not depend on azimuth. Only in complex geologies with complex raypaths and azimuth-dependent transmission effects might one expect measurable dependencies on orientation. But then it is best to include all azimuths in the acquisition geometry, because there would not be a clear-cut dip direction.

5.3.2 Multisource, multistreamer acquisition

The first marine 3-D surveys were carried out with the conventional 2-D geometry of a single boat towing one source (array) and one streamer. To increase efficiency in recording 3-D surveys, the industry has seen a gradual increase in the number of midpoint lines (also called bin lines) recorded in one boat pass. The newest vessels can tow eight or even more streamers allowing efficient single-boat operations.

The increase in number of midpoint lines recorded in one boat pass leads to undesirable side effects. This section first describes various MS/MS configurations, followed by a discussion of the undesirable side effects.

5.3.2.1 Multisource, multistreamer configurations

Figure 5.4 provides a schematic display of some common MS/MS configurations. The sources, represented by black circles in this figure, are always kept as close as possible to the boat to minimize the length of the umbilicals (pressure hoses from vessel to airgun arrays). The number of midpoint lines recorded by these geometries equals the product of the number of sources and number of streamers. Very often the distance between midpoint lines is chosen as 25 m. Then the distance between adjacent sources is always 50 m (except in the 4/4 configuration), the distance between streamers is 100 m for configurations with two or four sources, and 150 m for configurations with three sources. Duey (1996) describes a configuration with 24 midpoint lines.

5.3.2.2 Multisource effect on fold

A disadvantage of using several sources is the reduced fold in the individual midpoint lines. This is caused by the time interval needed between successive shots. In that time interval the vessel moves some distance, so that in practice shot intervals smaller than about 18 m are difficult to achieve. The distance between successive shots in a midpoint line is then n times 18 m, n being the number of sources.

Multiples with large differential moveout with respect to the primaries may be severely undersampled, even after NMO-correction, due to the low fold of multisource configurations (Hobson et al., 1992; Wombell and Williams, 1995; Manin and Spitz, 1995; see also Section 3.4.3.2). Various interpolation techniques have been devised to cure this problem (Jakubowicz, 1994; Huard et al., 1996). Impressive examples are shown in Wombell and Williams (1995) and Manin and Spitz (1995). Nevertheless, there is a tendency to prevent the problem by using no more than two sources in modern MS/MS configurations. In poor data quality areas, single-source configurations producing even higher fold are to be considered (Selbekk et al., 2001).

5.3.2.3 Crossline-offset variation

Each midpoint line in an MS/MS configuration is acquired by a unique source/streamer combination having a constant crossline offset (if there is no feathering). The variation in crossline offset between adjacent midpoint lines leads to variation in shot-to-receiver azimuths of traces with the same absolute offset across the survey. Interchanging source and receiver position leads to different azimuths; hence, crossline offset is to be described by a signed value, e.g., receiver x minus shot x for sailing in the y-direction. For example, the crossline offsets of the 3/3 geometry are: (−100, −150, −200, 50, 0, −50, 200, 150, 100 m, for a streamer separation of 150 m).

Figure 5.5 illustrates crossline offset as a function of midpoint line for various MS/MS configurations. Each graph describes the crossline offset for 48 adjacent midpoints, except the graph for the 3/3 configuration which describes 45 adjacent midpoints.

Sailing adjacent boat passes in opposite directions (antiparallel acquisition) instead of in the same direction provides the adjacent midpoints of the two boat passes with exactly opposite shot-to-receiver azimuths (see Figure 5.6), because it reverses the sign of the inline offset between passes. Because of reciprocity, opposite shot-to-receiver azimuths produce exactly the same raypaths. Therefore, antiparallel acquisition may significantly reduce the average azimuth variation. Figure 5.7 charts the variation in crossline offset (defined as the rms of the differences in crossline offset between adjacent midpoint lines) for various geometries. (Strictly speaking, the graphs in Figures 5.5 and 5.7 are not correct for antiparallel acquisition. The crossline offsets of the two adjacent

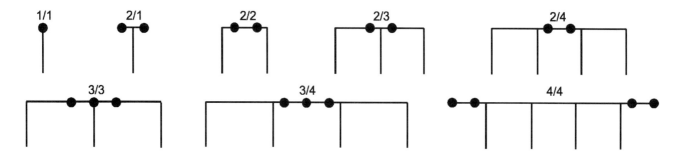

FIG. 5.4. Schematic description of various multisource, multistreamer configurations. Black circles represent sources, vertical lines represent streamers. For 25 m between midpoint lines, pairs of shots always are at a distance of 50 m (except the inner two in 4/4). With two or four sources streamers are 100 m apart, with three sources 150 m.

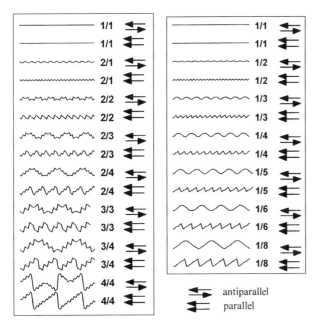

FIG. 5.5. Crossline offset displayed in crossline direction for various multisource, multistreamer configurations. On the right, only single-source configurations are displayed.

midpoints of adjacent boat passes still have opposite sign, but the inline offset also changes sign. In the computations and graphs, the sign of all crossline offsets in every other boat pass has been reversed for antiparallel acquisition.)

Figures 5.5 and 5.7 also illustrate differences between single-source and multisource geometries. The crossline offset in single-source geometries varies smoothly within one boat pass, whereas in multisource geometries it generally shows some rapid jitter. The jitter corresponds to pairs or triplets of sources shooting into the same streamer followed by the same sequence of sources into the next streamer. Note also the large effect antiparallel acquisition has on the variation in crossline offset for the single-source configurations (Figure 5.7).

5.3.2.4 Irregular illumination

The discontinuous behavior of crossline offset leads to irregular illumination of the subsurface. As discussed before in Section 2.4.2.1, this can be illustrated using the illumination pattern of a geometry as in Figure 5.8. Here 48 midpoints adjacent in the crossline direction are selected. The model consists of a plane reflector with 45° dip and a dip direction of 45° with respect to the crossline direction in a constant velocity medium. The depth of the reflector is 2000 m at coordinate (0, 0). For each midpoint the (x, y)-coordinates of the reflection points are plotted for inline offsets ranging from 0 to 3000 m. As expected, for the 1/1 geometry the curves behave in a regular way, whereas for the other geometries, there is a great deal of irregularity. In the 4/4 geometry (Figure 5.8b), there are some areas of the reflector that are never sampled by the large offsets, whereas other areas are sampled more than once. Note that the variations are largest for the large offsets, despite the fact that there the azimuth variations are the smallest. Figures 5.8c and d illustrate that with a smaller number of midpoint lines in one boat pass, the illumination becomes less irregular and that antiparallel acquisition also reduces the irregularities. The case for anti-parallel acquisition is also made by Brink et al., (1997) and Hoffmann (1999).

Differential feathering leads to additional irregularity. In a single boat pass the various streamers usually show about the same feathering, which is quite helpful for keeping the streamers from getting tangled. Significant differential feathering occurs mainly between adjacent boat passes. In the following experiments, a uniform distribution of random feathering angles ranging from $-1.75°$ to $+1.75°$ is used. The feathering angle is assumed constant during a boat pass and the same for all streamers.

Figure 5.9 shows that differential feathering may have a dramatic effect for the 1/1 geometry, whereas in this case feathering hardly affects the subsurface illumination by the 2/4 geometry. With the assumption of constant feathering inside a boat pass, effects of differential feathering are only important between adjacent boat passes. Feathering tends to increase the irregularity of the subsurface illumination in the sense that it reduces irregularity in some places whereas it increases irregularity in other places.

In Figure 5.9, random feathering with zero average has been assumed. But even if feathering tends to be in a single direction (e.g., caused by prevailing cross-currents), then antiparallel acquisition is still recommended as a means to reduce subsurface illumination irregularities. To compensate for irregular coverage caused by random feathering, the technique "steering for coverage" is often applied. This technique, is used to cover midpoints in later boat passes that were missed in earlier boat passes. Steering for coverage is more difficult in antiparallel acquisition. To compensate for irregular coverage and for irregular illumination, parallel boat passes may be used with a little overlap of midpoints between boat passes.

5.3.2.5 Effects of irregular illumination

Due to differential feathering, the offset sampling in the crossline direction can become quite variable,

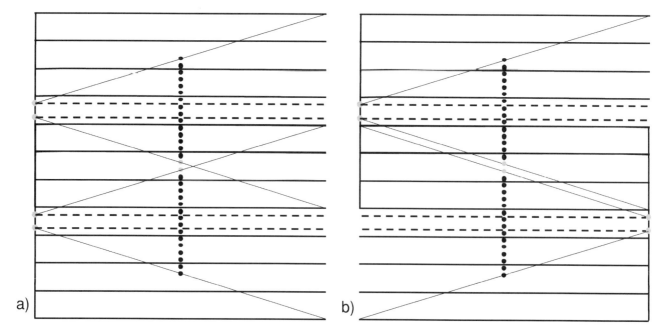

FIG. 5.6. Two adjacent boat passes in 2/8 geometry. Solid lines indicate streamers, stippled lines indicate source tracks. Four of the eight streamers follow the same track in the two boat passes to achieve single crossline fold. Lines connecting sources with the farthest receiver groups indicate the shot-to-receiver offsets for the outer midpoint lines of each boat pass. The vertical row of dots indicates midpoint positions. (a) Parallel acquisition. The adjacent midpoints in the center of the picture have opposite crossline offsets, hence different shot-to-receiver azimuths. (b) Antiparallel acquisition. The adjacent midpoints in the center of the picture have opposite crossline offset and opposite inline offsets, hence identical shot-to-receiver azimuths.

whereas offset sampling in the inline direction will vary much more gradually. Therefore, differential feathering is a major cause of inline striping in streamer data acquisition. Feathering that is not different between neighboring streamers is no cause of striping and is quite acceptable.

Unlike irregular illumination caused by the varying crossline offset in MS/MS acquisition, the irregular illumination caused by differential feathering is not repeatable. As a consequence, the effect of differential feathering on amplitude differs between the baseline survey and repeat surveys in time-lapse studies, making it more difficult to analyze amplitude variations caused by hydrocarbon production. A recent development is the steerable streamer allowing feathering angle corrections up to 3° (Bittleston et al., 2000; Austad et al., 2000). With this technology, it should be possible to get closer to the feathering angles of the baseline survey in the repeat surveys, thus reducing the difference in acquisition imprint between subsequent surveys.

Even without feathering, MS/MS configurations illuminate the subsurface in an irregular way. Figures 5.5 and 5.7 suggest that the effect increases with the width of the geometry and that the largest jumps in crossline off-

set should create the largest effects. Figure 5.10 shows a time slice through the stacked data of a 4/4 geometry (see Figure 5.4). In this time slice, discontinuities occur at the position of the largest jump in crossline offset (between midpoint lines 8 and 9). The discontinuities are largest where the time contours make an angle of 45° with the sailing direction. In that situation adjacent midpoint lines sample different parts of the reflector (cf., Figure 5.8b), leading to sizable differences in stack times. With dip sailing or strike sailing, there would be no differences between the traveltimes of lines 8 and 9.

The time discontinuities, as in Figure 5.10, lead to migration smiles, but the stack itself will not normally give much cause for concern, since every shot/receiver combination contributes reflection energy to the stack. However, after DMO, the situation may change drastically. As DMO moves data back to their normal-incidence point, the illumination gaps discussed for Figure 5.8 will show up as weak seismic amplitudes in the DMO stack. Beasley and Mobley (1995) illustrate this with a synthetic data set. A similar result is shown in Figure 5.11. What has not been illuminated cannot be imaged, therefore DMO equalization techniques (Beasley and Klotz, 1992) cannot solve this problem

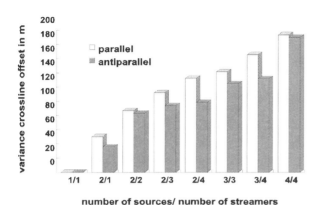

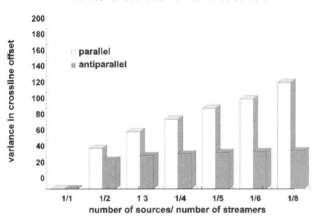

FIG. 5.7. (Left) Variation in crossline offset for various marine 3-D configurations. Top: multisource, multistreamer configurations; bottom: single-source multistreamer configurations. Note that sailing adjacent boat passes in opposite directions often leads to a significant reduction in the crossline offset variation.

FIG. 5.8. (Below) Illumination patterns of some acquisition geometries. Shown are (x, y)-coordinates of reflection points on a dipping interface with 45° dip and azimuth 225°. Each curve represents the reflection points of one midpoint. The curves are shown for 48 adjacent (in the crossline direction) midpoints. Streamer length is 3000 m. Sailing direction from south to north. Reflector depth is 3000 m in origin. (a) Single-source, single-streamer geometry provides regular subsurface illumination; (b) 4/4 geometry (three boat passes), note big gaps in subsurface illumination halfway in each group of 16 midpoints, and oversampling in between; (c) 2/4 geometry (six boat passes) showing smaller gaps, also oversampling in between the gaps; (d) 2/4 geometry, but now acquired in antiparallel mode. In this case antiparallel shooting leads to less irregular subsurface illumination. In b, c, and d, illumination irregularity increases with offset (longest offsets have moved farthest updip into a northeast-direction).

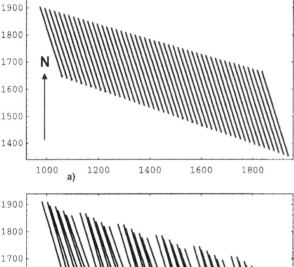

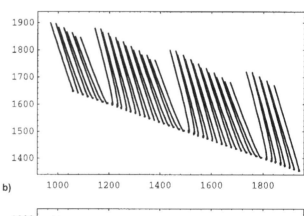

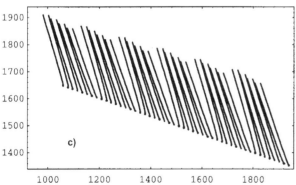

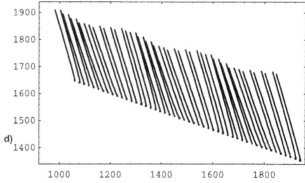

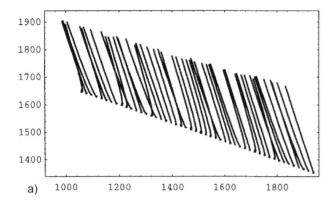
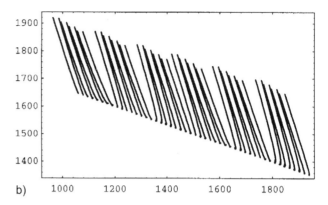

FIG. 5.9. Effect of random differential feathering on the illumination pattern, (a) 1/1 geometry, (b) 2/4 geometry. In this case the curves still correspond to particular source/streamer combinations, but no longer correspond to single midpoints as in Figure 5.8.

FIG. 5.10. Time slice through stacked 3-D data set acquired with 4/4 geometry. Note discontinuities every 16th east-west midpoint line (a horizontal gridline is drawn every tenth midpoint line).

completely, neither can migration correct for the deficiency.

The irregular illumination of the subsurface affects migration and imaging in two ways: first, the images for areas that have not been illuminated by the long offsets will be incomplete, and second, the cancellation of energy along the flanks of the migration operators will be suboptimal leading to migration noise. Both effects cause loss of resolution (see Chapter 8).

The effect of irregular illumination on amplitude is illustrated in Figure 5.12 using horizon amplitude slices of a reflector with 15° dip. Figure 5.12a shows the result for a COV gather. Apart from the edge effects, there is only minor variation in amplitude in this horizon slice. Figure 5.12b shows the result for a 2/4 acquisition geometry. To achieve complete single-fold coverage, a range of offsets had to be used. Such a data set is called a pseudominimal data set (pMDS) and was introduced in Section 2.5. For this configuration, the effect on amplitude is small. In a wider acquisition geometry, the effect on amplitude can be considerably more severe as shown in Figure 5.12c for a 2/8 geometry. The contour interval is 12 amplitude units (twice as large as the color interval). Figure 5.12d shows that, with the same color scale and contour interval as in Figure 5.12c, antiparallel acquisition considerably reduces the severity of the amplitude variations as compared to parallel acquisition. In Figures 5.12a–c sailing is updip, Figure 5.12e shows the result for strike acquisition, again for a 2/8 configuration. In this case there is hardly any effect of the crossline offset variation on amplitude.

Using two sources rather than one leads to a more irregular behavior of the crossline offset as illustrated in Figures 5.5 and 5.7. A zigzag pattern of two midpoint line intervals is superposed on the general trend in crossline offset variation caused by the width of the configuration. Therefore, one might expect that a 1/16 streamer configuration (streamer separation 50 m) would produce better images than a 2/8 configuration (streamer separation 100 m). However, the result for a 1/16 configuration shown in Figure 5.12f is virtually identical to the result for a 2/8 configuration shown in Figure 5.12d. This means that short wavelength (two midpoint line intervals) sampling irregularities have, at least in this case, less effect on the imaging result than the longer wavelength illumination variations caused by the width of the acquisition geometry. Apparently, the irregularity caused by using two sources is evened out by the averaging of amplitudes in the zone of influence (see Section 10.2) around each image point. The main benefit of using a 1/16 configuration would be a doubling of fold as compared to a 2/8 configuration.

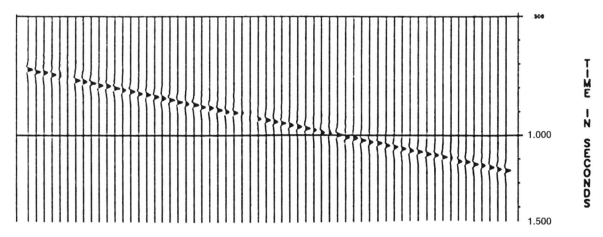

FIG. 5.11. Crossline acquired with 2/4 geometry and feathering after DMO including equalization. The offset range was 1000–1500 m.

The images shown in Figure 5.12 were obtained with a migration program that takes spherical spreading into account and that applies a phase correction. The program does not account for geometry effects. Gesbert (2002) shows that true-amplitude migration taking geometry effects into account would reduce the amplitude variations considerably. The amplitude variation in Figure 5.12a would disappear entirely (apart from the edge effects), because the input data is a true MDS. Cross-sections and amplitude maps in Gesbert (2002) show that there can still be a significant residual footprint for parallel acquisition, whereas the footprint becomes negligible for antiparallel acquisition.

5.3.2.6 Remedies

Illumination by MS/MS configurations is inherently irregular. The wider the configuration, the larger the effects. An obvious remedy to minimize the effects is to limit the width of the geometry. However, a 3-D survey acquired with a narrow geometry is more expensive in general than a survey acquired with a wide acquisition geometry. As an alternative, antiparallel acquisition should be considered. In antiparallel acquisition, the shrinking of the illumination area caused by downdip shooting in one boat pass is compensated by the expansion of the illumination area by updip shooting in the next boat pass. In this way, no serious illumination gaps occur anymore, only areas of higher or lower illumination density. The corresponding amplitude variation in the imaging result as illustrated in Figures 5.12d and f can be corrected for in processing using a true-amplitude migration technique as proposed in Albertin et al. (1999) or in Gesbert (2002). See Section 10.7 for a further discussion of true-amplitude migration.

In general, increasing fold will not do much to reduce illumination irregularities. In particular, illumination gaps from downdip shooting will not disappear. Yet, in a higher-fold geometry, the pMDSs (cf., Section 2.5.1) can be constructed from a smaller range of offsets leading to slightly better sampling of the pMDSs and also to better imaging. One way of increasing fold is to reduce the interval between shots by sailing into the current, if there is a strong predictable current. Using only one source rather than two doubles the fold-of-coverage. It has the disadvantage that the streamers have to be towed very close together. This can be done more safely if steerable streamer configurations are used (Bittleston et al., 2000). Another way to achieve single-source acquisition while ensuring sufficiently small crossline sampling intervals is to use an interleaved acquisition technique (100% overlapping boat passes, i.e., a planned 100% infill). Interleaving doubles trace density (fold), and reduces illumination irregularity on average.

The illumination irregularities are most severe for steeply dipping reflectors while sailing in the updip or downdip direction. Therefore, the irregularities can be minimized by sailing strike to the steepest reflectors (Budd et al., 1995).

The most drastic remedy for irregular illumination caused by crossline-offset variations and uncontrollable feathering is to use a stationary-receiver technique.

5.3.2.7 Operational aspects

There is no doubt that in open waters the MS/MS acquisition technique is highly efficient and cannot be beaten (certainly not in terms of square kilometers per day) by stationary-receiver techniques. On the other hand, the seismic vessels for multistreamer operations

FIG. 5.12. Horizon amplitude slices of migrated pMDSs. All input pMDSs have a regular midpoint grid of 25 × 25 m. Reflector dip is 15°. (a) COV gather with offset 2375 m; (b) 2/4 configuration, shooting downdip, parallel acquisition, inline offsets 2350 and 2400 m; (c) 2/8 configuration, shooting downdip, parallel acquisition, inline offsets 2350 and 2400 m; (d) as (c) with antiparallel acquisition; (e) As (c) shooting strike; (f) 1/16 configuration, antiparallel acquisition, inline offset 2375 m. Displays (a), (b), and (e) have the same color bar, whereas another color bar is used for displays (c), (d) and (f). In all displays contour interval is twice the amplitude step size in the color bar.

must be very powerful, hence are expensive to operate. Towing eight or more streamers is not easy, the outer streamers are especially difficult to control.

A restriction on the production is the amount of time that has to be spent on line turns. In a typical North Sea 3-D survey (an interleaved 1/8 configuration), line turns took about 2.5 hours on average. Deployment of the cables took some 9% of total survey time (see also Figure 5.13). With steerable streamers (Bittleston et al., 2000; Austad et al., 2000) the time needed for line turns can be reduced, because steering can be used to force the streamers into place sooner.

Around obstacles, MS/MS configurations must leave a large gap in the area of coverage as the streamers have to stay away from the obstacles. This needs to be compensated by a special undershooting survey (a two-boat operation), which is time-consuming and expensive. In the above-mentioned survey, 18% of the survey needed undershooting, at 36% of total cost. Again, this problem can be reduced with steerable streamers.

5.4 Stationary-receiver techniques

Figure 5.14 provides a pictorial overview of various stationary-receiver acquisition techniques. A common factor in all of these techniques is that the receivers are referenced in one way or another to the sea bottom. Another common feature is that there is a separate source vessel.

An important distinction between the various stationary-receiver techniques is the geometry that is or may be used. This part of chapter 5 begins with a description of the possible geometries, followed by a description of various stationary-receiver techniques.

5.4.1 Geometries for stationary-receiver techniques

The use of stationary receivers allows the decoupling of the source from the receiver as in land data acquisition. In other words, there is more freedom in the choice of geometry, typical land-type geometries may be used, and there is no physical offset limitation.

The main types of geometry available to the designer of a 3-D marine survey with stationary receivers are parallel geometry, orthogonal geometry, and areal geometry

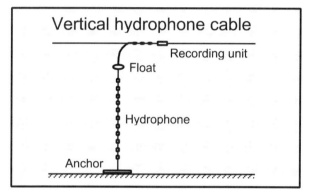

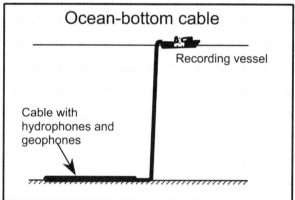

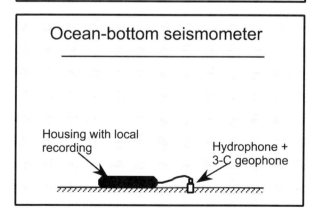

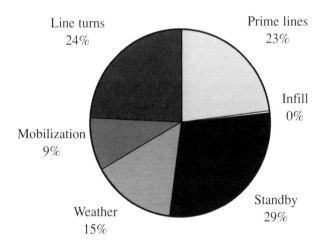

FIG. 5.13. Relative time spent on various activities during interleaved 1/8 survey in the North Sea. Note lack of infill due to interleaving. Downtime due to equipment failure and maintenance not included.

FIG. 5.14. Various stationary-receiver techniques.

(see Section 2.2). In parallel geometry, the source lines and the receiver lines run parallel to each other. The MS/MS configurations described in the first part of this chapter use parallel geometry. With bottom cables similar geometries can be arranged (Sanders et al., 1994).

A main reason to use parallel geometry with stationary-receiver systems is the familiarity with the geometry in marine circumstances and the possibility of tying in to a similar geometry of an adjacent streamer survey. Yet, there are also good geophysical reasons to prefer parallel geometry over orthogonal geometry. The discussion in Section 4.3, comparing various geometries, was intended for land acquisition but applies, to a great extent, to marine acquisition as well. Of course, feathering does not play a role in stationary-receiver systems; moreover, acquisition can be center-spread, alleviating some of the problems associated with the variation in crossline offset in a geometry with parallel shot and receiver lines. One problem might be drifting of the source vessel due to side currents causing gaps in midpoint coverage. Drifting of the source vessel is much less serious in orthogonal geometry.

For techniques employing very expensive receiver units, areal geometry is the preferred geometry, as it requires fewer receiver units for a given survey area. The disadvantage of areal geometry is that it requires a very dense source sampling which is both time-consuming and expensive.

Another disadvantage of areal geometry is the sensitivity to obstacles: where there is an obstacle, there will be a hole in the common-receiver gathers. An interesting opportunity offered by carpeting the survey area with shots is that a short streamer might be towed behind the shooting vessel, thus providing a separate short-offset 3-D cube without much additional cost. Due to the distance between the receiver units in areal geometry, shallow coverage is poor, but with short-offset 3-D, shallow coverage is taken care of, even allowing a larger distance between the receiver units. Carpeting the survey area with shots also allows the simultaneous recording of high-density gravity profiles (N.N., 1996).

Orthogonal geometry and areal geometry do not really commit to a particular shooting direction, all shot-to-receiver azimuths may occur. Hexagonal sampling of areal geometry provides the least dependence of the 3-D survey on direction. The presence of a full range of azimuths also offers the scope for amplitude versus direction (AVD) analysis (MacBeth and Li, 1997).

5.4.2 Vertical hydrophone cable (VHC)

The VHC technique (Figure 5.14 top) was developed and patented by Texaco (Stubblefield, 1990; Krail, 1991, 1994). A vertical cable, along which a string of 12–16 hydrophones is distributed, is anchored to the sea bottom and pulled into a vertical position by a buoyancy sphere. The sphere is kept below the zone of wave action. The signals received by the hydrophones are stored in a storage device located in a recording buoy.

As the patent title (Stubblefield, 1990) suggests, the technique was meant to provide a walkaway VSP without the need of drilling a hole. But it was soon discovered that the technique could also provide an alternative to conventional streamer data acquisition. Because of the expensive nature of the device and the relatively low cost of marine shooting, the use of an areal geometry (Figure 4-9b) is the logical choice for this technology. At the same time, this choice would allow the acquisition of the full range of azimuths which might be helpful for imaging in complex geologies.

The 12 to 16 hydrophones provide as many 3-D common-receiver gathers, each one recording a slightly different signal. VSP-type processing may be applied to separate upgoing and downgoing energy (energy reflected at the sea surface), and may reduce the data set into two representations (up- and downgoing each) of the wavefield at the location of the VHC. This would eliminate the receiver ghost. A high signal-to-noise ratio should be possible with the VHC technique, because (a) the hydrophones are located below the zone of wave action, (b) there are many elements in the hydrophone array, and (c) water-borne noise arrives at all hydrophones at about the same time, hence can be discriminated against easily.

Several full-scale surveys have been carried out with this technique (Havig and Krail, 1996). One of them is the 3-D Strathspey survey in the North Sea in waters of about 145 m. Processing results are very encouraging. Due to the limited number of available systems, the Strathspey survey had to be split over 2×3 adjacent swaths of 3×4 VHCs each. This necessitated considerable overlap of the shot areas between adjacent swaths. For a reasonably sized survey, some 100 to 200 receiver units would be necessary for application of a roll-along technique without repeating shots.

The VHC technique also has a number of shortcomings. First, with the recording buoys, it creates its own obstacles, leading to gaps in the pattern of shots. In a storm, wave action may get hold of the recording buoys and displace the whole system. Unloading tapes and changing batteries have to be carried out while shooting continues, also leading to some missed shots. Another problem is that changing currents will move the cable around, especially the shallowest part, thus violating the assumption of a single receiver position. Improvements in the design should be able to mitigate these problems considerably. However, emerging alternative stationary-receiver techniques are overtaking the VHC technique in importance.

A much cheaper version of the VHC technique is the dual-hydrophone Digiseis (Moldoveanu et al., 1994). In this system only two hydrophones are attached to a vertical cable, allowing immediate radio-transmission of all received data to a recording vessel. It has been used to supplement streamer acquisition in the vicinity of obstacles. In the reported survey, an irregular areal geometry was used with a rectangular grid of 350 m × 320 m for the Digiseis units, and a rectangular grid of 40 m × 25 m for the shots. It is not clear to what extent this technique is capable of removing the receiver ghost.

5.4.3 Dual-sensor OBC

In dual-sensor OBC, acquisition bottom cables are provided with a pressure and a velocity detector at regular intervals. Barr and Sanders (1989) presented a field test of the dual-sensor system. In their paper they argue that the water reverberations have opposite polarity, allowing the suppression of reverberations by summation of the signals of the two sensors in one location. This principle is also explained in Barr (1997). Many papers describe techniques for the combination of the hydrophone and geophone signals (Paffenholz and Barr, 1995; Soubaras, 1996; Ball and Corrigan, 1996).

5.4.3.1 Ghosting

In marine streamer acquisition source and streamer depths must be carefully selected to ensure that the second ghost notch occurs at a high enough frequency (the first notch occurs at zero hertz). For vertically traveling pressure waves, this notch occurs at $f = v_w / (2d)$, where v_w is the wave speed in water, and d is the depth of source or streamer. In OBC acquisition, the depth of the cable equals water depth so that the second ghost notch may occur at a very low frequency and many higher notches will be present in the frequency range of interest.

The ghost phenomenon can be described as a function of frequency:

$$G(f) = 1 + r \exp(-i2\pi f \tau) \quad (5.1)$$

where r is the reflection coefficient at the water surface, and τ is the time difference between primary signal and ghost ($\tau = 2d / v_w$). In the Fourier domain the recorded wavefield equals the product of the wavefield without ghosts, and the source and receiver ghost functions as described by equation (5.1). The dual-sensor technique exploits the fact that hydrophones measure pressure p and geophones measure a component of the particle velocity v. At the surface, the sum of the pressures of up- and downgoing wavefield must be zero, $p = p_u + p_d = 0$, causing $r = -1$, whereas vertical particle velocity for up- and downgoing wavefields must be the same, $v_u = v_d$, causing $r = 1$. This means that the zeros of equation (5.1) for the pressure signal occur at the maxima of equation (5.1) for the vertical particle velocity and vice versa. Figure 5.15 illustrates the ghosts as seen by the geophones and by the hydrophones.

In processing, the two components can be combined to obtain a smooth spectrum without any notches. Because water-bottom reverberations also have opposite polarity (Barr and Sanders, 1989), these can be tackled in the same process.

Although the dual-sensor technique was developed to compensate for the notches in the spectrum for recordings at greater depth, it is also useful for shallow waters. In that case, the geophone signal can be used to compensate for the very weak pressure signal at low frequencies.

5.4.3.2 Geometry

The OBC can be used most efficiently with orthogonal geometry. The implementation of this geometry can be done in various ways. The number and length of the receiver lines which are laid out in one "patch" varies, and shot lines may start beyond or within the reach of the receiver lines. Figure 5.16 shows a patch used by Chevron offshore West Africa (Sanders et al., 1994). A similar patch is reported in Meunier et al. (1995). A very long and narrow patch is described in Carvill et al. (1996). [Here the authors use the word "swath" to describe the patch, whereas elsewhere (Sanders et al., 1994) swath is reserved for acquisition with a parallel geometry. Nomenclature in this field has not yet been settled.] The patches are repeated to generate a full 3-D coverage of the survey area.

Whatever patch is used to maintain a reasonably efficient operation, the recorded cross-spreads will inevitably be asymmetric and different. This may lead to highly variable offset samplings in the CMPs and a noticeable geometry imprint. It is always possible to chop off outside traces in processing in order to create square cross-spreads (or at least rectangular cross-spreads with symmetry around the shot and receiver axes), but to avoid waste, it has to be planned for in the geometry design (cf., discussion of full-swath roll in Section 4.6.4).

During deployment, the cable is launched overboard without much control over where it will go beyond that point. This leads to variations in station spacing. In actual practice, it does not make much difference, since demonstrated in a repeatability experiment reported in Beasley et al. (1997).

5.4.3.3 Logistics

Operating an OBC survey is a complicated matter: four to six vessels are needed for efficiency: a recording boat, a shooting boat, and several cable deployment vessels (Sanders et al., 1994). The shooting boat should not have to wait for the next patch to be ready, but the next patch should not be ready before shooting of the previous patch is completed. Because laying cables is so time consuming, they should be laid out only once in each location; however, this implies repeat visits by the shooting boat to the same locations. The larger the number of stations available, the smaller the shot repeat factor can be.

In the mid 1990s there was still a water depth limitation of some 150 m on the use of conventional dual-sensor cables. The main problem is the retrieval system, not the strength of the cables. Gradually, better retrieval systems are allowing extension to greater water depths.

5.4.4 Four-component marine data acquisition

The advent of four-component (hydrophone plus 3-C geophone) marine acquisition techniques could have a great impact on the E&P business. Application of four-component technology may lead to improved (Johansen et al., 1995; Kristiansen, 1998)

- lithology and pore fill prediction,
- fracture density and fracture orientation determination,
- seismic reservoir monitoring, including compaction analysis,
- imaging inside and below gas chimneys,
- imaging structures with low P-wave contrast and better PS-wave contrasts,
- imaging below salt/basalt, and
- imaging, where there is a strong P-wave multiple.

Until the mid 1990s, only 2-D/4-C experiments had been carried out. Then the results of SUMIC experiments stirred the industry and 3-D/4-C surveys were carried out as well. Before reviewing these developments, coupling issues have to be addressed.

5.4.4.1 Coupling issues

Obviously, proper coupling of the sensors to the medium is of utmost importance. Up to now it has been tacitly assumed in this book that coupling is perfect. Requirements for perfect coupling of vertical geophones

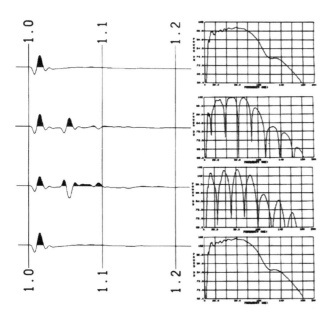

FIG. 5.15. Illustration of dual-sensor principle (copied with permission from Western Geophysical brochure). Wavelets are shown on the left with their corresponding spectra on the right. From above: the source wavelet, the wavelet plus ghost seen by the geophones, the wavelet plus ghost seen by the hydrophone, and the ideal result after combining the hydrophone and geophone signals.

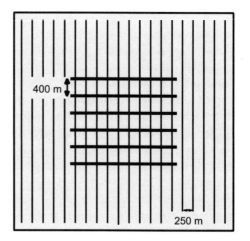

FIG. 5.16. Typical patch used in OBC acquisition with six cables and 20 shot lines. For the next patch cables will be moved to adjacent positions (no receiver overlap), but sources will have to overlap partially.

have been discussed by various authors. Tan (1987) concludes that, for perfect coupling, the density of the measuring device has to be the same as the density of the medium. Vos et al. (1999) suggest that this condition may not be sufficient. They make a distinction between contact coupling, which is perfect if the sensor follows exactly the motions of the surrounding medium (perfect contact between sensor and surroundings), and interaction coupling, which is perfect if the motion of the medium is not influenced by the presence of the sensor (sensor as a scatterer). Interaction coupling not only depends on the density contrast but also on the elastic properties of the whole sensor.

Requirements for perfect coupling are much more difficult to formulate for three-component sensors, particularly if placed at the sea bottom. At the sea bottom, the vertical particle velocity is continuous between seabed and sea. A main complicating factor is that the horizontal particle velocity is not continuous between the two media. Therefore, a three-component geophone preferably should be buried in the sea bed, thus avoiding differential effects between water and bottom. Duennebier and Sutton (1995) provide an extensive discussion of the requirements for proper wavefield measurements at the sea bed. They suggest that the only reliable way of obtaining high-fidelity particle motion data from the seabed is to bury the sensors below the bottom in a package with density close to that of the sediment.

Various aspects of accurate 3-component recording are covered by the term "vector fidelity." A seismic acquisition system exhibits vector fidelity when it accurately records the magnitude and direction of a seismic wave in three dimensions (Tree, 1999). Again, the measuring system should accurately follow the motions of the medium in which it is planted, and it should not have altered the motions of the medium. Tree (1999) reports that the x- and y-components of several tested 3-C geophones in ocean-bottom cables do not respond equally to the ground motion. This "vector infidelity" is probably due to the nature of the cable with its axial symmetry. Gaiser (1998) and Gaiser et al. (2000) developed a technique for correction of vector infidelity of seabed cables under the assumption of perfect coupling by the inline geophone. They assume that the longer contact area in the inline direction allows better pick-up of the ground motion, whereas the crossline and vertical directions may suffer from insufficient contact and rotational effects.

The node systems (Pettenati-Auzière et al., 1997), consisting of single units with 3-C geophones, are planted by a remotely operated vehicle (ROV), and tend to show better vector fidelity than sea-bed cables.

5.4.4.2 SUMIC

Statoil has released results of their experiments with the SUMIC technique (Berg et al., 1994; Johansen et al., 1995; Berg and Arntsen, 1996; Granli et al., 1999). In this technique, a bottom cable is connected to a recording vessel, but unlike conventional OBC, the receiver units are external to the cable, and are planted in the sea bottom using an ROV or underwater robot. The units contain a hydrophone on top, two orthogonal horizontal geophones, and a vertical geophone. In their configuration the receiver units were spaced quite closely along the cable, allowing the recording of high-fold 2-D lines.

Berg et al. (1994) and Granli et al. (1999) show imaging of gas chimneys as the main application of the technique. The *PS*-wave data produced sections suitable for structural interpretation, whereas the *P*-wave sections only produced jumble across the gas chimneys.

Johansen et al. (1995) also show a display of common-receiver records acquired with SUMIC (reproduced here as Figure 5.17). This record and other records shown in presentations show a remarkable quality of the horizontal components, sometimes even better than the hydrophone data.

In early 1997, a survey was carried out in the North Sea using an adapted version of the SUMIC technique. Instead of keeping a small distance between the receiver units, these were spaced at 600-m intervals, and linked via a wired cable to a recording vessel (see Figure 5.18). Sources were fired every 25 m. Hence, this geometry is the same as would be used with a 4-C/3-D OBS survey (see Section 5.4.4.5). Due to logistical problems, only about half of the planned shot lines were acquired.

It is now possible to apply this acquisition technique in 3-D surveys (Pettenati-Auzière et al., 1997).

5.4.4.3 Other 4-C bottom-cable techniques

A somewhat hybrid technique involving six gimbaled geophones from a VSP tool used in OBC mode plus two hydrophones was carried out in 1300-m deep waters offshore Norway (Brink et al., 1996a). Brink et al. (1996b) discuss in detail the coupling conditions of this experiment.

Another technique is the dragged bottom cable. Rather than retrieving the cable between deployments, this cable is made strong enough that it can be dragged to its next position. Perceived advantages of this technique are a better coupling to the sea floor than possible with conventional OBC deployment, and a constant distance between stations (no slack).

Full-scale tests of this technique for 3-D/4-C surveys have been reported for three different surveys acquired in the North Sea (Rognø et al., 1999; McHugo et al., 1999; Rosland et al., 1999). The geometries of these surveys were all different and are reviewed in Chapter 6.

Cafarelli et al. (2000) report results of imaging through a gas cloud in the Gulf of Mexico. The equipment used in this survey was designed to eliminate crossfeed between vertical and horizontal components. Nolte et al. (2000) show some beautiful results of this survey. Originally they found an 11% depth difference between the prestack depth migrated results of the P and the PS data. With an anisotropic velocity model the depths could be matched.

5.4.4.4 4-C acquisition with buried cables

Concern that the required repeatability for time-lapse surveys cannot be reached with streamer acquisition has led to some tests with buried cables. In 1995 BP/Shell acquired data using a buried cable over their Foinaven field (Godfrey et al., 1998). However, in this case the cables were only equipped with hydrophones (densely spaced). The airgun array source was fired in a dense areal grid.

In the 1997 Teal South 4-C/4-D survey, the 4-component receivers were spaced at 200 m, whereas the buried cables were spaced at 400 m (Ebrom et al., 1998). Again an areal geometry was used, with a dense 25×25 m grid of sources.

5.4.4.5 Ocean-bottom seismometers

For more than twenty years, academia has been using OBS units for wide-angle refraction and reflection profiling (WARP). These self-contained units are lowered (by gravity) to the ocean bottom, left there for two weeks or so while shooting takes place, and then retrieved again. The systems usually consist of a glass sphere, which contains a 2-6 channel recording instrument plus batteries, and sometimes one or three geophones; the (external) hydrophone is standard. Unfortunately, even though gravity may firmly plant the whole system on the bottom, internal geophones cannot record the undisturbed seismic wavefield. In particular, the horizontal geophones suffer severely from rotations of the whole system, induced by the height and size of the set-up (Sutton and Duennebier, 1987; Duennebier and Sutton, 1995). For reliable recording of the seismic wavefield, the geophones would have to be separated from the recording unit and be buried (as in Figure 5.14 bottom).

An alternative to the internal geophones would be a gravity-deployed external three-component geophone or a receiver unit (node) planted by ROV as in the SUMIC technique. For applicability to 3-D, the system should be capable of listening during a sufficient length of time, have enough battery power and storage capacity, and the recording fidelity should be state-of-the-art. All these requirements lead to considerable unit cost of such OBSs, necessitating the use of an areal geometry as with VHCs. Moreover, planting of the geophones using an ROV would be time-consuming and expensive. Nevertheless, recent developments in node design (Eivind Berg, 2001,

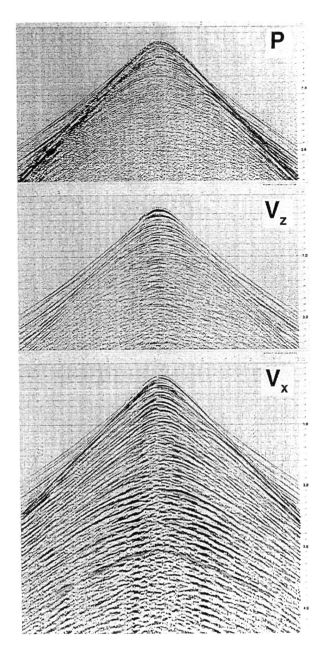

FIG. 5.17. Common receiver panel acquired with SUMIC technique (courtesy Statoil).

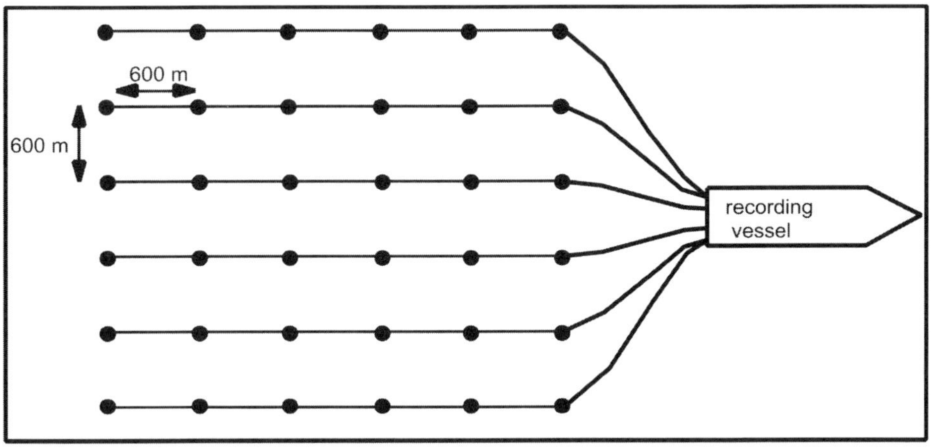

FIG. 5.18. Areal geometry of 4-C receiver units planted by ROVs.

personal communication) would make a 4-C/3-D OBS survey less expensive than a VHC survey, yet easier to handle and with the added benefit of shear-wave data.

5.5 Overview and conclusions

This chapter provides a review of some marine seismic data acquisition techniques. A major observation is that MS/MS acquisition is superior as far as cost and operation in deep waters are concerned, but for the highest quality, it may be worthwhile to consider one of the stationary-receiver techniques. In a comparison of MS/MS techniques to dual-sensor OBC, a similar conclusion is drawn (Barr et al., 1996). Yet, technology in streamer acquisition and seabed acquisition is progressing rapidly, so that by the time this book is published, some observations will have become obsolete.

The advent of 4-C marine recording capabilities opens up a new range of opportunities for the E&P business. SUMIC results have shown that high-quality shear-wave data may be recorded in the marine environment. Further developments and commercialization of those techniques are taking place rapidly.

Processing techniques must also be developed to deal with orthogonal and areal acquisition geometries. Processing of full-azimuth shear-wave data provides yet another challenge. Eventually, the achievements on shear data acquisition and processing that can be anticipated for the marine environment may give a new push to shear-wave recording and processing on land.

References

Albertin, U., et al., 1999, Aspects of true amplitude migration: 69th Ann. Internat. Mtg., Soc. Expl. Geophys., Expanded Abstracts, SPRO 11.2, 1358–1361.

Arbi, N., McDonald, J. A., Zhou, H-W., Ebrom, D. A., and Tatham, R. H., 1995, A comparison of 3-D seismic time imaging results from physical model strike and dip acquisition: 65th Ann. Internat. Mtg., Soc. Expl. Geophys., Expanded Abstracts, 953–956.

Austad, P., Canter, P., Hodnebo, O., and Olafsen, W., 2000, Marine seismic cable steering and computerized control systems: 70th Ann. Internat. Mtg., Soc. Expl. Geophys., Expanded Abstracts, ACQ 3.9, 61–63.

Ball, V. L., and Corrigan, D., 1996, Dual-sensor summation of noisy ocean-bottom data: 66th Ann. Internat. Mtg., Soc. Expl. Geophys., Expanded Abstracts, 28–31.

Barr, F. J., 1994, Ocean-bottom cable use surges for seismic data acquisition: OGJ, Oct. 24, 62–67.

——— 1997, Dual-sensor OBC technology: The Leading Edge, **16**, 45–51.

Barr, F. J., Paffenholz, J., and Rabson, W., 1996, The dual-sensor ocean-bottom cable method: Comparative geophysical attributes, quantitative geophone coupling analysis and other recent advances: 66th Ann. Internat. Mtg., Soc. Expl. Geophys., Expanded Abstracts, 21–23.

Barr, F. J., and Sanders, J. L., 1989, Attenuation of water-column reverberations using pressure and velocity detectors in a water-bottom cable: 59th Ann. Internat. Mtg., Soc. Expl. Geophys., Expanded Abstracts, 653.

Beasley, C. J., et al., 1997, Repeatability of 3-D ocean-bottom cable seismic surveys: The Leading Edge, **16**, 1281–1285.

Beasley, C. J., and Klotz, R., 1992, Equalization of DMO for irregular spatial sampling: 62nd Ann. Internat. Mtg., Soc. Expl. Geophys., Expanded Abstracts, 970–973.

Beasley, C. J., and Mobley, E., 1995, Spatial sampling characteristics of wide-tow marine acquisition: 57th Conf., Eur. Assoc. Geosc. and Eng., Extended Abstracts, B031.

Berg, E., and Arntsen, B., 1996, The potential of subsea seismic: Presented at the EAGE Workshop on Advances in multi-component technology.

Berg, E., Svenning, B., and Martin, J., 1994, SUMIC: Multi-component sea-bottom seismic surveying in the North Sea—Data interpretation and applications: 64th Ann. Internat. Mtg., Soc. Expl. Geophys., Expanded Abstracts, 477–480.

Bittleston, S., Canter, P., Hillesund, O., and Welker, K., 2000, Marine seismic cable steering and control: 62nd Conf., Eur. Assoc. Geosc. and Eng., Extended Abstracts, L-16.

Brink, M., Granger, P. Y., Manin, M., and Spitz, S., 1996a, Seismic methodologies for a 3 components sea floor geophone experiment on a potential flat spot in the Vøring Basin: 58th Conf., Eur. Assoc. Geosc. and Eng., Extended Abstracts, B020.

——1996b, Seismic methodologies for a 3 components sea floor geophone experiment on a potential flat spot in the Vøring Basin: Extended paper for Workshop on advances in multicomponent technology preceding 58th Conf., Eur. Assoc. Geosc. and Eng.

Brink, M., Roberts, G., and Ronen, S., 1997, Wide tow marine seismic surveys: Parallel or opposite sail lines: Offshore Technology Conference, OTC8317.

Budd, A. J. L., Ryan, J. W., Hawkins, K., and Mackewn, A. R., 1995, Understanding the interaction between marine geometries and 3-D imaging methods: 65th Ann. Internat. Mtg., Soc. Expl. Geophys., Expanded Abstracts, 945–948.

Cafarelli, B., 1995, Subsurface imaging with ocean bottom seismic: World Oil, October.

Cafarelli, B., Madtson, E., Krail, P., Nolte, B., and Temple, B., 2000, 3-D gas cloud imaging of the Donald field with converted waves: 70th Ann. Internat. Mtg., Soc. Expl. Geophys., Expanded Abstracts, MC1.7, 1158–1161.

Carvill, C., Faris, N., and Chambers, R. E., 1996, A successful 3-D seismic survey in the "no-data zone," offshore Mississippi delta: Survey design and refraction static correction processing: 66th Ann. Internat. Mtg., Soc. Expl. Geophys., Expanded Abstracts, 9–12.

Cramer, J. E., Ratcliff, D. W., and Weber, D. J., 1995, An innovative split-spread marine 3-D acquisition design for subsalt imaging: 65th Ann. Internat. Mtg., Soc. Expl. Geophys., Expanded Abstracts, 991–994.

Davidson, D. S., and Bandell, A., 1990, A novel 3-D marine acquisition technique: 60th Ann. Internat. Mtg., Soc. Expl. Geophys., Expanded Abstracts, 863–866.

Duennebier, F. K., and Sutton, G. H., 1995, Fidelity of ocean bottom seismic observations: Marine Geophysical Researches, **17**, 535–555.

Duey, R., 1996, Triple-vessel system boosts productivity in the North Sea: Hart's Show Special Edition, SEG Internat. Exposition and 66th Ann. Internat. Mtg.

Durrani, J. A., French, W. S., and Comeaux, L. B., 1987, New directions in 3-D surveys: 57th Ann. Internat. Mtg., Soc. Expl. Geophys., Expanded Abstracts, 177–180.

Ebrom, D., Krail, P., Ridyard, D., and Scott, L., 1998, 4-C/4-D at Teal South: The Leading Edge, **17**, 1450–1453.

Egan, M. S., Dingwall, K., and Kapoor, J., 1991, Shooting direction: A 3-D marine survey design issue: The Leading Edge, **10**, No. 11, 37–41.

Gaiser, J. E., 1998, Compensating OBC data for variations in geophone coupling: 68th Ann. Internat. Mtg., Soc. Expl. Geophys., Expanded Abstracts, SP12.6, 1429–1432.

Gaiser, J. E., Barr, F. J., and Paffenholz, J., 2000, Horizontal and vertical receiver-consistent deconvolution for an ocean bottom cable: US Patent 6 021 090.

Gesbert, S., 2002, From acquisition footprints to true amplitude: Geophysics, **67**, 830–839.

Godfrey, R. J., et al., 1998, Imaging the Foinaven ghost: 68th Ann. Internat. Mtg., Soc. Expl. Geophys., Expanded Abstracts, SP 9.4, 1333–1335.

Granli, J. R., Arntsen, B., Sollid, A., and Hilde, E., 1999, Imaging through gas-filled sediments using marine shear-wave data: Geophysics, **64**, 668–677.

Havig, S., and Krail, P. M., 1996, Vertical cable seismic applications: World Oil, May, 72.

Hird, G. A., Karwatowski, J., Jenkerson, M. R., and Eyres, A., 1993, 3D concentric circle survey—The art of going in circles: 55th Conf., Eur. Assoc. Expl. Geophys., Extended Abstracts, A001.

Hobson, M. R., Arnold, A. J., and Cooper, R. C., 1992, Spatial sampling effects of a multi-source multi-cable recording configuration: Exploration Geophysics, **23**, 163–166.

Hoffmann, J., 1999, Survey planning procedures for wide tow streamer acquisition: Eur. Assoc. Geosc. and Eng., Workshop on 3D seismic surveys: Design, tests and experience, Expanded Abstracts.

Houllevigue, H., Delesalle, H., and de Bazelaire, E., 1999, Enhanced composite 3D cube derived from multi-azimuth 3D marine acquisitions: 61st Conf., Eur. Assoc. Geosc. and Eng., Extended Abstracts, 1-08.

Huard, I., Medina, S., and Spitz, S., 1996, *F-xy* wavefield de-aliasing for acquisition configurations leading to coarse sampling: 58th Mtg., Eur. Assoc. Geosc. and Eng., Extended Abstracts, Paper B039.

Jakubowicz, H., 1994, Wavefield reconstruction: 56th Conf., Eur. Assoc. Geosc. and Eng., Extended Abstracts, Paper H055.

Johansen, B., Holberg, O., and Øvrebø, O. K., 1995, Sub-sea seismic: Impact on exploration and production: 27th Annual Offshore Tech. Conf., Paper OTC 7657, p. ????.

Jones, I. F., et al., 2000, The effect of acquisition direction on preSDM imaging: First Break, **18**, 385–391.

Krail, P. M., 1991, Case history vertical cable 3D acquisition: 53rd Mtg., Eur. Assoc. Expl. Geophys., Abstracts, 204.

——1994, Vertical cable as a subsalt imaging tool: The Leading Edge, **13**, 885–887.

Kristiansen, P., 1998, Application of marine 4C data to the solution of reservoir characterization problems: 68th Ann. Internat. Mtg., Soc. Expl. Geophys., Expanded Abstracts, 908–909.

Larner, K., and Ng, P., 1984, 3-D marine seismic survey direction: Strike or dip?: 54th Ann. Internat. Mtg., Soc. Expl. Geophys., Expanded Abstracts, MAR 1.5, 256–260.

MacBeth, C., and Li, X. Y., 1997, New concepts for marine AVD technology: Presented at the London IIR Conference on Acquisition, Processing and Interpretation of Marine Seismic Data.

Maldonado, B., and Hussein, H. S., 1994, A comparative study between a rectilinear 3-D seismic survey and a concentric-circle 3-D seismic survey: 64th Ann. Internat. Mtg., Soc. Expl. Geophys., Expanded Abstracts, 921–925.

Manin, M., and Hun, F., 1992, Comparison of seismic results after dip and strike acquisition: 53rd Conf., Eur. Assoc. Expl. Geophys., Abstracts, A011.

Manin, M., and Spitz, S., 1995, Wavefield de-aliasing for acquisition configurations leading to coarse sampling: 65th Ann. Internat. Mtg., Soc. Expl. Geophys., Expanded Abstracts, 930–932.

Marschall, R., 1990, Method for recording seismic data: US Patent 4 965 773.

McHugo, S., et al., 1999, Acquisition and processing of 3D multicomponent sea-bed data from Alba field—A case study from the North Sea: 61st Conf., Eur. Assoc. Geosc. and Eng., Extended Abstracts, 6–26.

Meunier, J. J., Musser, J. A., Corre, P. M., and Johnson, P. C., 1995, Two bottom cable 3-D seismic surveys in Indonesia: 65th Ann. Internat. Mtg., Soc. Expl. Geophys., Expanded Abstracts, 976–979.

Moldoveanu, N., et al., 1994, Digiseis-enhanced streamer surveys (DESS) in obstructed area: A case study of the Gulf of Mexico: 64th Ann. Internat. Mtg., Soc. Expl. Geophys., Expanded Abstracts, 872–875.

N. N., 1996, Company profile: ARK Geophysics: First Break, **14**, 470–471.

Nolte, B., et al., 2000, Anisotropic 3D prestack depth imaging of the Donald field with converted waves: 70th Ann. Internat. Mtg., Soc. Expl. Geophys., Expanded Abstracts, MC1.8, 1158–1161.

O'Connell, J. K., Kohli, M., and Amos, S., 1993, Bullwinkle: A unique 3-D experiment: Geophysics, **58**, 167–176.

Paffenholz, J., and Barr, F. J., 1995, An improved method for deriving water-bottom reflectivities for processing dual-sensor ocean-bottom cables: 65th Ann. Internat. Mtg., Soc. Expl. Geophys., Expanded Abstracts, 987–990.

Pettenati-Auzière, C., Debouvry, C., and Berg, E., 1997, Node-based seismic: A new way to reservoir management: 67th Ann. Internat. Mtg., Soc. Expl. Geophys., Expanded Abstracts, 71–74.

Reilly, J. M., 1995, Comparison of circular "strike" and linear "dip" acquisition geometries for salt diapir imaging: The Leading Edge, **14**, 314–319.

Rognø, H., Kristensen, A., and Asmussen, L., 1999, The Statfjord 3-D, 4-C OBC survey: The Leading Edge, **18**, 1301–1305.

Rosland, B., Tree, E. L., and Kristiansen, P., 1999, Acquisition of 3D/4C OBS data at Valhall: 61st Conf., Eur. Assoc. Geosc. and Eng., Extended Abstracts, 6–17.

Sanders, J. I., Starr, J. L., Dale, C. T., and King, W. F., 1994, Ocean bottom cable surveying offshore West Africa: Operational and technical aspects: 26th Annual Offshore Tech. Conf., OTC 7393.

Sandø, I., and Veggeland, T., 1995, Experiences with a super wide 3D marine acquisition in the North Sea: 57th Conf., Eur. Assoc. Geosc. and Eng., Extended Abstracts, P039.

Selbekk, T., Hegna, S., and Krokan, B., 2001, Single source 3D seismic acquisition in a poor data quality area Mid-Norway: 63rd Conf., Eur. Assoc. Geosc. and Eng., Extended Abstracts, F-34.

Sonneland, L., Roed, K., and Navrestad, T., 1995, Subsea seismic: Acquisition and data analysis procedures: 27th Annual Offshore Tech. Conf., OTC 7658.

Soubaras, R., 1996, Ocean bottom hydrophone and geophone processing: 66th Ann. Internat. Mtg., Soc. Expl. Geophys., Expanded Abstracts, 24–27.

Stubblefield, S. A., 1990, Marine walkaway vertical seismic profiling: US Patent 4 958 328.

Sutton, G. H., and Duennebier, F. K., 1987, Optimum design of ocean bottom seismometers: Marine Geophysical Researches, **9**, 47–65.

Tan, T. H., 1987, Reciprocity theorem applied to the geophone-ground coupling problem: Geophysics, **52**, 1715–1717.

Tree, E. L., 1999, The vector infidelity of the ocean bottom multicomponent seismic acquisition system: 61st Conf., Eur. Assoc. Geosc. and Eng., Extended Abstracts, 6–19.

Vermeer, G. J. O., 1994, 3-D symmetric sampling: 64th Ann. Internat. Mtg., Soc. Expl. Geophys., Expanded Abstracts, 906–909.

———1997, Streamers versus stationary receivers: 29th Annual Offshore Tech. Conf., OTC Proceedings, 331–346.

Vos, J., Drijkoningen, G. G., and Fokkema, J. T., 1999, Sensor coupling in acoustic media using reciprocity: J. Acoust. Soc. Am., **105**, 2252–2260.

Wombell, R., and Williams, R. G., 1995, Aliasing and sampling problems for multi-source and multi-cable data acquisition: Exploration Geophysics, **26**, 472–476.

Chapter 6
Converted waves: Properties and 3-D survey design

6.1 Introduction

Multicomponent surface seismic has a long history on land, whereas multicomponent marine data acquisition was virtually unheard of until a few years ago. Then, the interest in multicomponent marine data acquisition received a great stimulus by the pioneering work of Statoil with their SUMIC technique (Section 5.4.4.2; Berg et al., 1994; Johansen et al., 1995; Berg and Arntsen, 1996; Granli et al., 1999). Imaging of gas chimneys was the main application of the technique (Berg et al., 1994; Granli et al., 1999). The PS-wave data produced sections suitable for structural interpretation, whereas the P-wave sections only produced jumble across the gas chimneys.

In SUMIC, ROVs were still used to plant the geophones, but since then a less expensive technique, based on using a dragged bottom cable, was developed. This technique was first tested for 2-D/4-C applications (Kommedal et al., 1997; Kristiansen, 1998). Kristiansen (1998) lists a large number of applications for 4-C (three geophones plus hydrophone) data.

Full-scale tests of the dragged bottom cable technique for 3-D/4-C surveys have been reported for three different surveys acquired in the North Sea (Rognø et al., 1999; McHugo et al., 1999; Rosland et al., 1999). The geometries used in these surveys were all different.

In all marine applications, P-wave energy converted to S-wave energy at the reflecting horizons is the main wave type being analyzed. These PS-waves have asymmetric raypaths leading to special requirements for the survey geometry. Only a few papers seem to have been published on the design of 3-D/3-C seismic surveys. Lawton (1995), and Cordsen and Lawton (1996) deal mainly with binning issues, in association with the asymmetric illumination by PS-waves. In my opinion, binning issues are better left to processing, particularly when spatial interpolation to neighboring bin centers (Herrmann et al., 1997, Beasley and Mobley, 1997) becomes more

This chapter is an expanded version of Vermeer (1999).

generally accepted. Ronen et al. (1999) discussed the irregular illumination by cross-spread 4-C surveys and argued that a careful analysis of this effect is required to plan an optimal survey.

The asymmetric illumination by PS-waves (also often referred to as C-waves, for converted waves) is the major reason that the design of 3-D surveys for converted waves is more complicated than for P- or S-waves. Symmetric sampling requirements (Chapters 1 and 2) no longer apply. To find out what does apply, this chapter looks at some properties of the PS-wavefield in the minimal data sets of various acquisition geometries. The behavior of apparent velocities in the MDSs is discussed to determine sampling requirements. Illumination, resolution, and imaging of converted waves are compared for the different MDSs. In the second part of the chapter, the results of the first part are applied to discuss the suitability of various geometries for PS-wave acquisition. It is found that parallel geometry is most suited for PS-wave acquisition, whereas other geometries tend to have problems with illumination, resolution, or both.

The analyses in this chapter are carried out for a simple isotropic medium with constant P-wave velocity V_p and constant S-wave velocity V_s.

6.2 Properties of the PS-wavefield

6.2.1 Traveltime surfaces and apparent velocity

The difference between V_p and V_s leads to asymmetry between the P- and the S-angles of reflection according to Snell's law. As a consequence, the raypaths are different if shot and receiver are interchanged, and traveltime curves are different in common-shot gathers and common-receiver gathers. For a horizontal reflector, the traveltime curves are still the same, even though the raypaths are different.

Similarly, the diffraction traveltime as a function of offset is different between common-shot diffractions and common-receiver diffractions. In the common shot the diffraction is much steeper because the slow V_s determines the change in traveltime. Figure 6.1a illustrates the traveltime behavior for *PP-* and *PS-*reflections from a horizontal reflector and it shows *PP-* and *PS-*diffraction traveltimes for shot and receiver gathers. The corresponding apparent velocities (as measured in the surface coordinate systems) are plotted in Figure 6.1b. All apparent velocities seem to be controlled by the *P*-wave velocity only, except the *PS-*diffraction in the common shot, which has very low apparent velocities tending towards the *S*-wave velocity.

The asymmetry in *PS-*acquisition becomes more apparent for dipping reflectors. This is illustrated for a reflector with 15° dip in Figure 6.2. Note that the reflec-

tion time curve is steepest for positive offsets in the common shot; there it has an apparent velocity smaller than the *P*-wave velocity.

Figure 6.3 shows contour plots of the diffraction traveltimes for the common receiver, the common shot, the common-offset-vector (COV) gather, and the cross-spread. In the common shot the *S*-wave velocity determines the slopes of the curves, whereas in the common receiver the *P*-wave velocity determines the slopes. The curves in the COV gather have some intermediate slope. This can be understood by realizing that the apparent velocity V_a in the zero-offset gather would tend to $1/V_a = 1/V_p + 1/V_s$ for large distances from the scatterer. Note that, unlike a *PP-*diffraction, the apex of the *PS-*traveltime surface in the COV gather is offset from the dif-

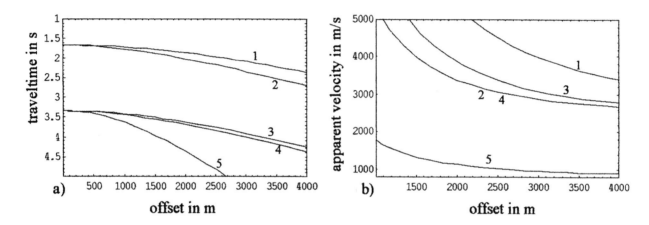

FIG. 6.1. Traveltime curves (a) and apparent velocity (b) for *PP-* and *PS-*reflections and diffractions in constant velocity medium. V_p = 2400 m/s, V_s = 800 m/s, depth of reflector is 2000 m, diffractor at (0, 0, 2000); 1 = *PP-*reflection, 2 = *PP-*diffraction, 3 = *PS-*reflection, 4 = *PS-*diffraction in common receiver, 5 = *PS-*diffraction in common shot.

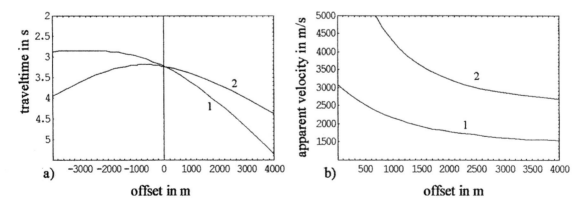

FIG. 6.2. *PS-*reflection in common shot (1) and common receiver (2) for 15° dip. Depth of reflector at position of shot, receiver is 2000 m. (a) Traveltimes. (b) Apparent velocity. The common shot has the steepest traveltime curve and the smallest apparent velocity.

fractor position. The cross-spread shows a mixed behavior: steep flanks in the inline (receiver) direction and gentle slopes in the crossline (source) direction.

6.2.2 Illumination

In *P*-wave acquisition the midpoint coverage is the same as subsurface coverage of horizontal reflectors. Therefore, fold-of-coverage is fairly representative for illumination fold, even for areas with gentle dips. This is quite different for *PS*-wave acquisition due to the asymmetry in the raypaths. For three different minimal data sets, Figure 6.4 shows a comparison of their midpoint area (the same for all three MDSs) with the illumination areas of a horizontal reflector for $V_p / V_s = 1.5$ and $V_p / V_s = 3$. The 2000 × 2000 m square in Figure 6.4 is the midpoint area. This square also represents the *PP*-illumination area for a horizontal reflector. The other curves represent the conversion point curves corresponding to the midpoints along the outline of the square. The cross-spread shows asymmetry: The illumination area is wider in the inline direction and narrower in the crossline direction than the midpoint area. The 3-D shot has the largest illumination area and the 3-D receiver the smallest.

For dipping reflectors the illumination areas will shift updip. The illumination area of a COV gather is not shown in Figure 6.4 to prevent clutter. It would be a square illumination area with the same size as the midpoint area, but shifted towards the receivers.

Often, illumination fold is measured by counting the number of hits per bin. Then, for a bin size equal to the natural bin size of the geometry, illumination fold of a single cross-spread might vary between 0 and 3, whereas there is only a single-fold illumination area. Counting the number of hits per bin neglects the spatial relationship that exists between groups of traces, such as a cross-spread. Counting the number of overlapping illumination areas gives a better measure of imaging capabilities and image fold (see also Figure 6.13 and Section 2.5.3).

6.2.3 Resolution

The wavenumber spectra of different MDSs can be used to compare the resolution that can be reached with those MDSs (Chapter 8). The spectra are composed from the contributions to the spectrum by all individual shot/receiver pairs. For *PP*-wavefields, the contribution of each shot/receiver pair to the resolution in a point *P* is described by

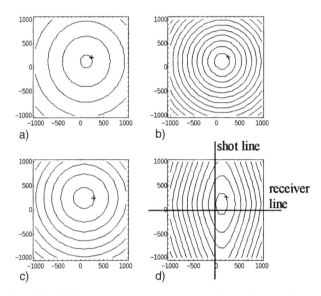

FIG. 6.3. Diffraction traveltime contours for various minimal data sets plotted in midpoint coordinates. Contour interval is 250 ms. Position of diffractor in (250, 250, 500) is indicated by +, $V_p = 2400$ m/s, $V_s = 800$ m/s. (a) Common receiver. (b) Common shot. (c) COV gather (600 m). (d) Cross-spread.

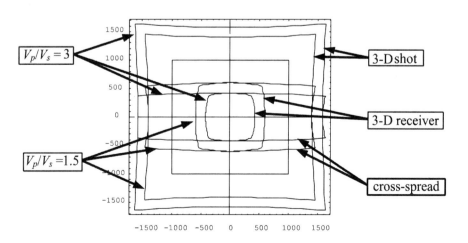

FIG. 6.4. Illumination areas on a horizontal reflector of a 3-D shot, 3-D receiver, and cross-spread for $V_p / V_s = 3$ and $V_p / V_s = 1.5$. The 2000 × 2000 m square represents the midpoint area of the three minimal data sets. The depth of the reflector is 2000 m.

128 Chapter 6 Converted waves: Properties and 3-D survey design

$$\mathbf{k} = \mathbf{k}_s + \mathbf{k}_r = \frac{f}{V_p}(\mathbf{u}_s + \mathbf{u}_r), \quad (6.1)$$

where $\mathbf{k}_s$, $\mathbf{k}_r$ are shot, receiver wavenumber, respectively, f is frequency, and $\mathbf{u}_s$ and $\mathbf{u}_r$ are unit vectors (see Figure 6.5a). For PS-wavefields, a similar relation holds, but now V_s enters the equation as well (see Figure 6.5b and c):

$$\mathbf{k} = \mathbf{k}_s + \mathbf{k}_r = f\left(\frac{\mathbf{u}_s}{V_p} + \frac{\mathbf{u}_r}{V_s}\right), \quad (6.2)$$

It follows from equation (6.2) that for the PS-wavefield $|\mathbf{k}_r|$ is larger than $|\mathbf{k}_s|$. This leads to asymmetry in illumination and resolution, depending on the relative position of shot and receiver as illustrated in Figure 6.5b and c. In Figure 6.5b the vertical component of $\mathbf{k}$ is larger than in Figure 6.5c, whereas it is the other way around for the horizontal component. Another consequence of equation (6.2) is that, for the same frequency, the components of $\mathbf{k}$ for PS-waves are larger than for PP-waves. The collection of all shot/receiver pairs in an MDS illuminate a wide range of dip angles θ, and span a wide range of wavenumbers, indicative of resolution. For different MDSs, the wavenumber spectra are illustrated in Figure 6.6 for a PP-wavefield, and in Figure 6.7 for a PS-wavefield.

Figure 6.6 shows that the wavenumber spectra of the 3-D receiver and the 3-D shot are identical for PP-waves, whereas Figure 6.7 illustrates that the PS-wavenumber spectrum of the 3-D shot spans a much wider range than that of the 3-D receiver. The cross-spread PS-spectrum has a hammock shape, indicative of the asymmetry between inline and crossline direction.

Figure 6.8 shows the projections on the horizontal plane of wavenumber spectra of various minimal data sets for PP- (left) and PS-waves (center and right). Notable is the invariance of the 3-D receiver resolution to V_p/V_s. This is because V_p is kept constant, whereas the V_p-leg of the raypath fully determines the resolution in the 3-D receiver. The asymmetry in the cross-spread

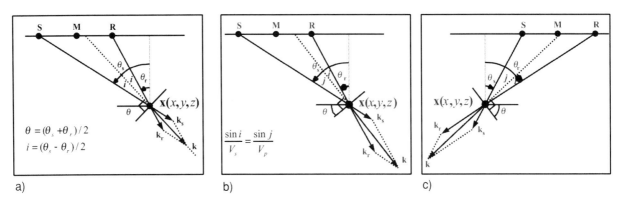

FIG. 6.5. Illumination of subsurface point **x** by single shot/receiver pair (S, R) and corresponding wavenumbers; $\mathbf{k}_s$ and $\mathbf{k}_r$ point in the direction of the raypaths ending in **x**; θ is the dip illuminated by (S, R), and i, j are the corresponding reflection angles. (a) PP-situation. (b) PS-situation with R closest to **x**. (c) PS-situation with S closest to **x**; $V_p/V_s = 2$.

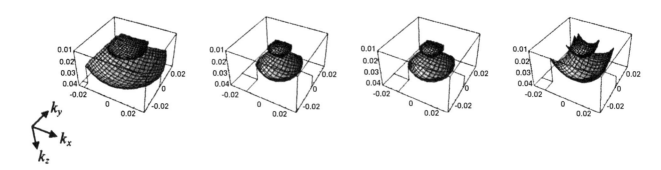

FIG. 6.6. PP-wavenumber spectra for a subsurface point below the center of each one of four minimal data sets. All data sets have the same 1000 × 1000 m midpoint area. The surfaces correspond to constant input frequencies 25 Hz (upper surfaces) and 50 Hz. From left to right: 600 m COV gather, 3-D shot, 3-D receiver, and cross-spread.

leads to less resolution in the crossline (source) direction than in the inline direction. There is also asymmetry in the resolution of the COV gathers. The resolution is best for the downdip shooting part of the wavefield (positive x, y map onto negative k_x, k_y, hence a shot-receiver combination with positive coordinates, source to the left of the receiver, maps to negative k).

Figure 6.8 shows that, except for the 3-D receiver gather, the resolution of *PS*-data is better than the resolution of *PP*-data *for the same frequency* and the same aperture (midpoint range). In practice, *PS*-data tend to have lower maximum frequency than *PP*-data, thus reducing or even losing the relative advantage.

6.2.4 Imaging

In addition to the range of wavenumbers that is available for the imaging process, it is of interest to investigate the imaging process itself, and compare the ability of various MDSs for imaging of the *PS*-wavefield. For this investigation, it is helpful to consider migration as a two-step process (see Figure 6.9), similar to the discussion of the effect of sampling density on the migration result in Section 8.3.7 and Figure 8.14.

In the first step, the seismic section is modified to flatten the diffraction traveltime surface corresponding to the output point. In this process, reflections are turned into bowl-shaped events, with the apex at the point of stationary phase. In the second step, all data of the "diffraction-flattened gather" are summed and provide the image trace. Similar as for forward modeling (Brühl et al., 1996), the zone of influence can be defined (see also Section 10.2.2). The zone of influence encompasses all traces around the point of stationary phase, which contribute, constructively or destructively, to the amplitude of the migrated event at the depth of the image point. The

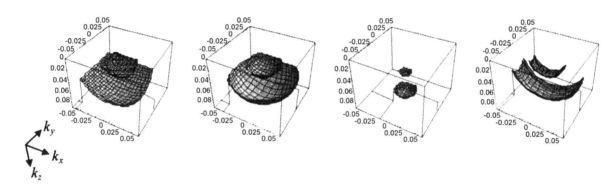

FIG. 6.7. Same as Figure 6.6 for *PS*-situation with $V_p/V_s = 3$. Note that the wavenumber ranges in this figure are larger than in Figure 6.6; the 3-D receiver spectrum has the same size as in Figure 6.6.

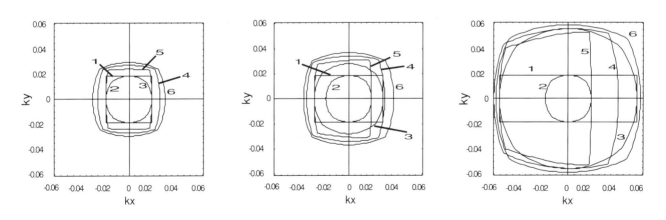

FIG. 6.8. Projection in the horizontal wavenumber plane of the (common-frequency) spectra of six different minimal data sets with the same 1000 × 1000 m midpoint area for *PP* (left), *PS* with $V_s = 1600$ m/s (center) and *PS* with $V_s = 800$ m/s (right). $V_p = 2400$ m/s in all cases; 1 = cross-spread, 2 = 3-D receiver, 3 = 3-D shot, 4 = 600 m COV gather, 5 = 1000 m COV gather, 6 = zero offset. The *PS* zero-offset is hypothetical, as *PS*-waves have zero amplitude for zero-offset.

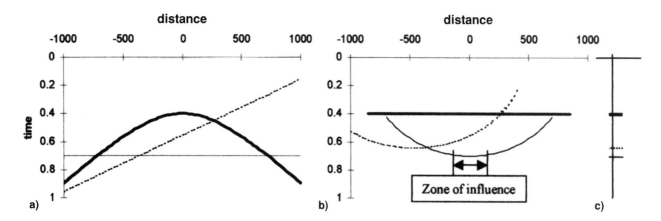

FIG. 6.9. Migration as a two-step process illustrated with a zero-offset section. (a) Input showing diffraction (heavy curve), a dipping reflection (thin dotted curve), and a horizontal reflection (thin drawn curve). (b) Diffraction-flattened gather. In the first step, the input data are realigned according to the diffraction traveltimes in the output point. Shown is the realignment for the output point at $x = 0$, which is the position of the diffractor. Note that the apex of the curve for the horizontal event is located at $x = 0$, whereas the apex for the dipping event is located toward the left. The location of these apexes corresponds to the position of the zero-offset shot/receiver pair, which has illuminated the reflector at $x = 0$. (c) In the second step, the realigned data are summed (stacked) and phase-corrected to form the image trace.

width of the zone of influence depends on the length of the wavelet, on the curvature of the migration-corrected event, and on the domain in which the zone is measured, midpoint domain or subsurface domain (see below). The data outside the zone of influence cannot contribute to the required image and should be canceled in the second step of the migration process.

In 3-D, the migration-corrected reflections become truly bowl-shaped events. To describe the shape of these events in 3-D, contour plots are used in Figures 6.10 and 11. Figure 6.10 shows that for P-wave data the zone of influence is not very different among the various MDSs. (If the output point does not coincide with the center of the minimal data set, the differences become larger.) Figure 6.11 shows that for PS-data large differences exist among the various MDSs.

For proper imaging, it is essential to have complete zones of influence in the migration summation. The elongated shape of the zone of influence in the cross-spread requires more traces in the crossline direction than in the inline direction. This suggests acquisition of asymmetric cross-spreads with much longer shot lines than receiver lines, and it suggests that an asymmetric migration operator range should be used.

The zone of influence contains more data points (traces) in the cross-spread than in the 3-D shot. This would lead to a larger amplitude for cross-spread data than for 3-D shot data. True-amplitude migration should take these effects into account. The zone of influence of the 3-D receiver contains the largest number of traces, giving the 3-D receiver an advantage with respect to noise suppression.

In Figure 6.12 the zones of influence are plotted as a function of reflection point x and y, rather than in terms of midpoint x and y as in Figure 6.11. Figure 6.12 shows that the area of the reflector contributing to the migration amplitude in the output point is about the same in all cases. There are small differences, depending on the off-set mix contributing to the area of the zone of influence. Larger differences would occur if the output point did not coincide with the center of the MDSs.

6.3 3-D survey design for *PS*-waves

6.3.1 Choice of geometry

Very often the choice of geometry will be dictated by circumstances such as available budget. On land, this tends to lead to orthogonal geometry or some derivative thereof (e.g., slanted shot lines), for marine streamer acquisition to parallel geometry and for OBC work to orthogonal geometry. Nevertheless, geophysical requirements should play a role as well, and need to be properly understood. In the first part of this chapter we have seen that illumination depends strongly on which minimal data set is used, hence on acquisition geometry. For equal aperture, resolution between the various MDSs is also very different. In this part I will discuss the consequences of the properties of the *PS*-waves in the various MDSs for 3-D/3-C survey design.

There is a large difference in the properties of parallel geometry as compared to the properties of all other geometries. In its ideal form, the parallel geometry is a *translational* geometry, i.e., its properties do not depend on location, whereas all other geometries are nontranslational, i.e., their properties vary from point-to-point. This difference manifests itself most clearly in the MDSs. The COV gather extends across the whole survey area, whereas the MDSs of all other geometries have limited extent, because the shot/receiver offset increases away from the center of those minimal data sets until being cut off.

6.3.1.1 Orthogonal geometry

For *P*-wave acquisition and processing, the problem of MDSs of limited extent can be solved quite reasonably

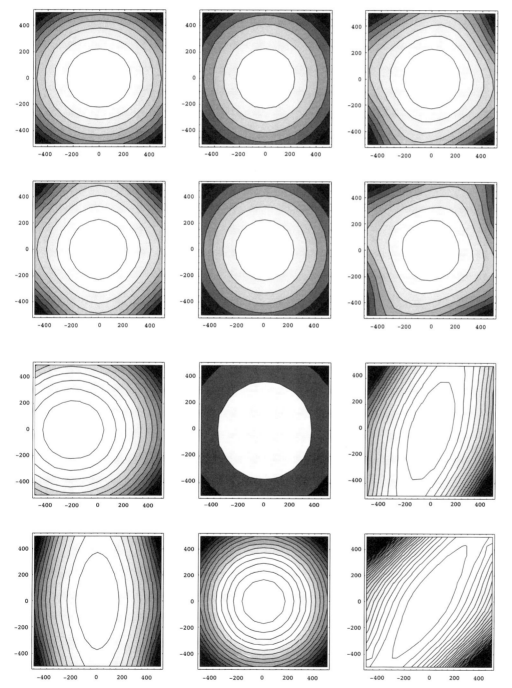

FIG. 6.10. Contour maps in (x_m, y_m) of diffraction-flattened *PP*-traveltimes of horizontal reflector for output point in the center of six different MDSs. Top row: COV gather, 3-D receiver, and slanted spread. Bottom row: Cross-spread, 3-D shot, and zig-spread. The central (light) area in each map may be considered to represent the zone of influence. Depth of reflector 1000 m. V_p = 2400 m/s. Contour interval, 25 m.

FIG. 6.11. As Figure 6.10 for *PS*-reflection with V_p/V_s = 2. Top row: COV gather, 3-D receiver, and slanted spread. Bottom row: cross-spread, 3-D shot, and zig-spread. The zones of influence are very different for the different MDSs. These zones give a representation of the *number* of reflection points being "stacked" into the output point.

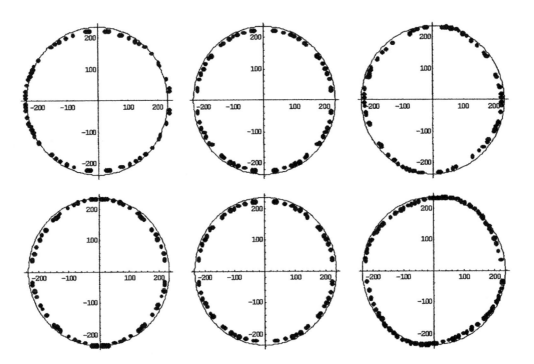

FIG. 6.12. Central contours at 975 m for each graph in Figure 6.11 replotted as the function of reflection point *x* and *y* (dots). Top row: COV gather, 3-D receiver, and slanted spread. Bottom row: cross-spread, 3-D shot, and zig-spread. The drawn circle with the same radius in all plots has been added as a reference. The areas inside the dotted curves give a better representation of the *range* of reflection points being "stacked" into the output point than the zones of influence plotted in Figure 6.11. (The contours are plotted as a series of dots rather than as drawn lines, because of difficulty to compute contours; each dot represents a point in a narrow range around 975 m.)

by the introduction of pseudominimal data sets (pMDS, Section 2.5). For a regular orthogonal geometry, OVT gathers are most suitable as pMDSs (Section 2.5.4). Between the tiles in each OVT gather spatial discontinuities exist, but these discontinuities tend to be of limited significance. If the illumination of a reflector in the subsurface is considered, the illumination by adjacent tiles in an OVT gather will be nearly continuous (depending on the size of the tiles, and the curvature of the reflector), with small overlaps and small gaps (cf. Figure 10.13). This approach tends to work for *P*-wave data because the illumination area of each cross-spread is about equal in size to the midpoint area of the cross-spread; these areas have exactly the same size when a horizontal reflector in a constant-velocity medium is illuminated.

These considerations do not apply to *PS*-wave acquisition and processing. Now the illumination area of each cross-spread is very different from the midpoint area, even for a horizontal reflector in a constant-velocity medium (see Figure 6.4). As a consequence, regular fold-of-coverage does not lead to regular illumination fold. Figure 6.13 illustrates the variation of illumination fold for a 16-fold square orthogonal geometry and $V_p/V_s = 2$. For larger V_p/V_s, the variation in illumination fold would be even larger. It is still possible to construct single-fold tilings (100% cubes) across the whole survey area by taking rectangles from the same location in each cross-spread, but their illumination areas are strongly discontinuous as shown in Figure 6.14. Similar reasoning applies to all other nontranslational geometries.

As a consequence, it is impossible to obtain a regular *PS*-illumination of the subsurface using one of the non-translational geometries. Although the MDSs may be sampled properly, edge effects (migration artifacts) on the prestack (time or depth) migrated results are inevitable and are much more serious than in *P*-wave imaging. The oral version of a paper by Buia et al. (2001) mentioned severe migration artifacts in the crossline direction in a 3-D/4-C survey acquired with orthogonal geometry. Likely, the artifacts are stronger in the crossline direction than in the inline direction, because the zone of influence will be incomplete more often in the crossline direction (see Figure 6.11, bottom left). The problem is only mitigated by using a high fold-of-coverage, and may be further reduced by careful processing.

In orthogonal geometry, resolution in the receiver-line direction is determined by the *S*-wave velocity and in the shot-line direction by the *P*-wave velocity. Therefore, as

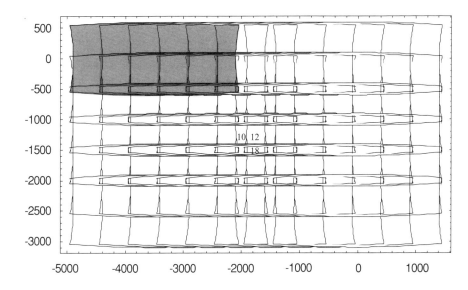

FIG. 6.13. Overlapping illumination areas for 16-fold orthogonal geometry. $V_p/V_s = 2$. Gray area indicates illumination area of one cross-spread. In narrow horizontal strip PS-illumination fold varies between 15 and 18, whereas in broader horizontal strip illumination fold varies between 10 and 12.

shown in Figure 6.8, resolution is much better in the receiver-line direction than in the shot-line direction. In case of a dominant dip direction, it is advisable to orient the receiver lines in that direction.

6.3.1.2 Parallel geometry

The ideal parallel geometry consists of a collection of pure COV gathers in which each COV gather has indeed constant inline offset and constant crossline offset (constant absolute offset and constant azimuth). The ideal geometry can only be acquired by acquisition of each midpoint line separately (repeated 2-D). Acquiring the seismic data in this way is highly expensive; therefore, parallel geometry is always acquired with a number of shot and receiver lines in one pass. In streamer acquisition, the configuration often consists of two shot lines (produced by two source arrays towed behind the vessel) and 4 to 12 streamers. In this way 8 to 24 midpoint lines are acquired in one boat pass.

Because laying cables is time-consuming and shooting sources is relatively cheap, OBC acquisition tends to be carried out the other way around: there the configuration (a swath) is formed by a few receiver lines (often two cables) and many shot lines. Figure 6.15 shows an arrangement in which the cables are laid out at a distance of 210 m from each other. If the next two cables are shifted over 420 m, the total width covered by the shot lines must be 840 m in order to produce regular midpoint coverage in the crossline direction. This leads to 315 m width of shot lines outside each cable.

The geometry of the swath was selected such that the midpoint lines of one cable can be interleaved with those of the other cable. In our example, setting the origin

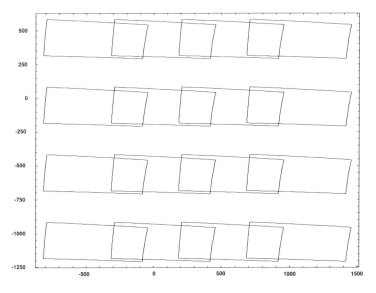

FIG. 6.14. PS-illumination of horizontal reflector by single-fold tiling of offset-vector tiles for $V_p/V_s = 2$. The offset-vector tiles correspond to the upper right-hand corners of 16 cross-spreads similar to those described in Figure 2.21. Even though fold-of-coverage is exactly 1 throughout, illumination fold varies between 0 and 2.

halfway between the two cables, the cables are at locations ± 105 m, and the shot lines should be at ± 30 m, ± 90 m, etc, until ± 390 m, assuming a 60-m shot-line interval. This leads to 840 / 60 = 14 shot lines and 14 × 2 = 28 midpoint lines with 420 / 28 = 15-m crossline interval. In this way crossline fold is always one, and the total fold only depends on the inline parameters. (Note that the arithmetic requires setting the width of the geometry at 840 m, which is number of shot lines times shot-line interval, rather than taking the distance between the two outside shot lines which is 780 m.)

Figure 6.15 illustrates that *PS*-illumination with this acquisition geometry is no longer regular, because the illumination ranges for each cable are narrower than the 420-m midpoint range. For large V_p/V_s values, there are even illumination gaps. The gaps are largest for large inline offsets (as can be understood by inspection of the illumination area of a cross-spread as shown in Figure 6.4). For the situation of Figure 6.15 ($V_p/V_s = 3$, depth of reflector is 2000 m), the illumination gap equals ≈53 m for zero inline offset, larger for larger inline offsets. By adding a few shot lines on either side of the swath, complete illumination can be achieved. Yet, the remaining density variation in illumination will lead to amplitude striping unless very careful processing is carried out.

It tends to be faster to roll only one cable at a time, rather than both cables as suggested in Figure 6.15. However, with this kind of shot-line configuration, some shot lines would have to be repeated for the cable that is not rolled. A more efficient technique, requiring a smaller total number of shot lines, is to acquire the survey in two passes: first acquire cable positions 1, 3, 5, etc., followed by acquiring cable positions 2, 4, 6, etc.

6.3.1.3 Areal geometry

The areal geometry is also, a nontranslational geometry. It is the geometry used in the Teal South time-lapse experiment (Ridyard et al., 1998). The use of areal geometry tends to be practical only with 3-D receiver gathers and not with 3-D shot gathers. Unfortunately, the illumination area of a 3-D receiver is relatively small, whereas resolution tends to be lower than achievable with *PP*-data. This requires a relatively high density of 4-C receivers. An advantage of this geometry is that it is most suitable for analysis of azimuth-dependent effects.

Figure 6.16 illustrates the illumination fold of a 16-fold areal geometry for $V_p/V_s = 2$. The geometry is equivalent to the orthogonal geometry used to produce Figure 6.13, i.e., the distance between receiver units in *x* and *y* is equal to the acquisition line intervals of the orthogonal geometry, and the maximum inline and crossline offsets are also equal to those of the orthogonal geometry. Illumination fold varies between 4 and 9 and fold would be even smaller for larger V_p/V_s. The distance between the receiver units would have to be reduced considerably to reach illumination folds of 16 on average. Note that the relatively low illumination fold of this geometry may be compensated by the larger number of traces in the zone of influence (see Figure 6.11) leading to better signal-to-noise ration of each image.

Figure 6.17 illustrates the illumination by the top-right corner of each common-receiver gather for the geometry illustrated in Figure 6.16. As in Figure 6.14, the midpoints of this data set form a continuous coverage of the survey area. Figure 6.17 shows that illumination is far from continuous.

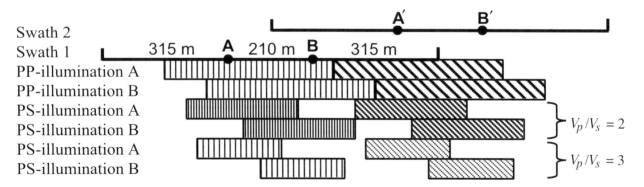

FIG. 6.15. Illumination with swath acquisition. In swath 1, two cables A and B are laid out on the sea bottom at 210 m distance. The range of shot lines equals 840 m. Swath 2 is shifted over 420 m with respect to swath 1. Illumination of a horizontal reflector for swath 1 is indicated with vertical shading, for swath 2 with diagonal shading. This geometry ensures regular midpoint coverage, i.e., regular *PP*-illumination of a horizontal reflector. *PS*-illumination is not regular and even shows gaps in the case of large V_p/V_s.

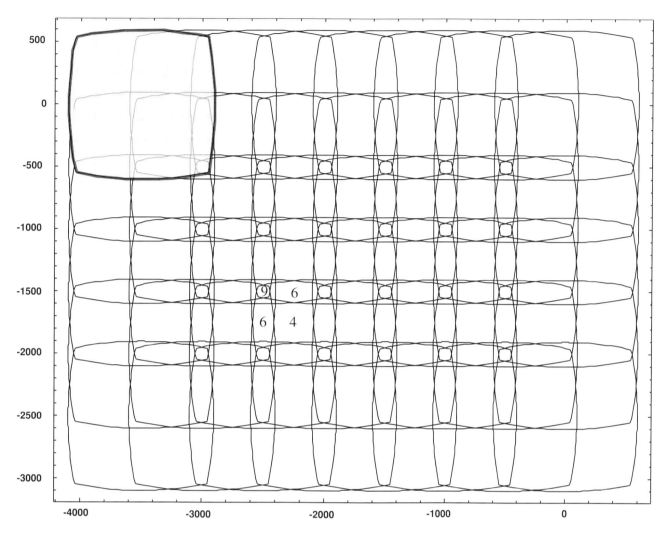

FIG. 6.16. Overlapping illumination areas for 16-fold areal geometry. $V_p/V_s = 2$. Gray area indicates illumination area of one common receiver. In narrow horizontal strip, PS-illumination fold varies between 6 and 9, whereas in broader horizontal strip, illumination fold varies between 4 and 6.

6.3.1.4 Parallel versus orthogonal geometry and areal geometry

Illumination appears to be the most important property determining which geometry is to be preferred for PS-acquisition. Irregular illumination cannot be avoided by either parallel (except repeated 2-D) or orthogonal geometry. Yet, it appears that taking the irregularity into account in processing is easier with parallel geometry than with orthogonal geometry. The reason is that in parallel geometry common-inline-offset gathers are continuous in the inline direction and can be made to have some overlapping illumination in the crossline direction. Hence, it can be attempted to regularize the illumination areas of each common-inline-offset gather by removing overlaps (interpolation might be more difficult). In orthogonal geometry or areal geometry it is more difficult to create (continuous) single-fold illumination gathers from the data.

For resolution, parallel geometry is preferred over orthogonal and areal geometry as well. In parallel geometry, horizontal resolution is better for downdip than for updip shooting, whereas vertical resolution is better for updip than for downdip shooting (cf., Figures 6.5b and c). This asymmetry can be taken care of by center-spread acquisition. On the other hand, in orthogonal geometry, resolution of crossline dips is inferior to inline dips. This problem might be alleviated by using an asymmetric migration operator radius, with a considerably larger radius in the crossline direction than in the inline direc-

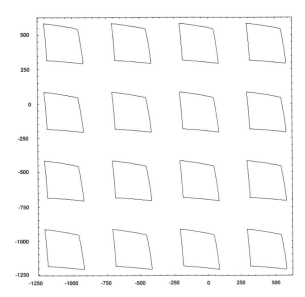

FIG. 6.17. *PS*-illumination of horizontal reflector by single-fold tiling of offset-vector tiles for $V_p/V_s = 2$. The offset-vector tiles correspond to the upper right-hand corners of 16 common-receiver gathers similar to those described for cross-spread geometry in Figure 2.21. Even though fold-of-coverage is exactly one throughout, illumination fold varies between 0 and 1.

tion. However, resolution is also affected by the large spatial discontinuities between the illumination areas of separate cross-spreads. These lead to migration artifacts.

Hence, parallel geometry tends to be better for *PS*-acquisition than orthogonal or any other crossed-array geometry. Apart from a cost benefit, the only advantage of orthogonal geometry is that it allows a more robust analysis of azimuth-dependent effects such as fracture orientation. Determining the principal directions of the fast and the slow shear modes will not be as accurate with parallel geometry (Thomsen, 1999). However, if anisotropy is only a nuisance, making life of the processor difficult, then parallel geometry will suffer least from its presence. An advantage of areal geometry over orthogonal geometry might be that the irregularities in areal geometry are symmetric, including more symmetry in azimuth-dependent effects.

Areal geometry can be sampled more efficiently using a hexagonal sampling grid, both for the receiver units and the shotpoints (see Section 2.4.1).

For all geometries suffering from irregular illumination, the imaging result should benefit from application of the migration-equalization technique proposed in Albertin et al. (1999). See also Section 10.7.

6.3.2 Sampling

The sampling interval in any spatial domain is determined by the smallest apparent velocity and the largest frequency. This means that for equal maximum frequency the sampling of the receivers in a common shot depends on the *S*-wave velocity requiring denser sampling than *P*-wave acquisition, whereas the sampling of the shots in a common receiver can be the same as for *P*-wave acquisition. This leads to asymmetric sampling requirements.

Sampling parallel geometry is of special interest. Here again, proper sampling of the field data requires a smaller sampling interval for the receivers than for the shots. The required midpoint sampling of the COV gather (see Figure 6.3) depends on the harmonic average of *P*-wave and *S*-wave velocities, hence it seems to be less strict than in the 3-D shot gather. However, to realize the required midpoint sampling for each offset, shot and receiver sampling intervals would have to be equal to the required midpoint sampling interval, because each offset only occurs at every other midpoint. Therefore, proper sampling of COV gathers can best be achieved by interpolation of properly sampled shot and receiver gathers.

6.3.3 Other considerations

Garotta and Granger (2001) discuss a number of differences between *P*-wave and *PS*-wave acquisition, all leading to the requirement of extra effort in *PS*-wave acquisition. The differences they discuss are: (1) *PS*-reflectivity is in general lower than *P*-reflectivity when compared as a function of offset, (2) The conversion of *P* to *S* moves the energy to larger wavenumbers, (3) due to the higher wavenumbers in *PS*, absorption affects the *PS* data more than the *P* data, (4) intra-array statics cause a loss of higher wavenumbers in the *PS* data. They conclude that field effort may have to be doubled compared to the effort required for getting the *P*-wave data just right.

Garotta and Granger (2001) assumed equal quality factors for *P* or *S* mode propagation. However, the usually very low cohesion of the sediments on the sea floor may cause considerable absorption of the shear-wave energy in the first few meters or decimeters of the sea bed. This may necessitate burial of the sensors to capture shear-wave energy before it is lost.

6.4 Discussion

In this chapter, some theoretical considerations have been given on the design of 3-D/3-C seismic surveys. A very simple model was used. It will be interesting to see whether these theoretical considerations can be confirmed by analysis of actual 3-D/3-C surveys.

Chevron's Alba survey was acquired with parallel geometry (McHugo et al., 1999). The crossline midpoint range of their geometry was 1050 m (42 source lines times 50 m interval / 2) for a crossline shift between swaths of 800 m, giving a 250-m midpoint overlap between adjacent swaths. This geometry ensures full *PS*-illumination in the crossline direction of horizontal reflectors for V_p/V_s ratios up to three. Illumination-density variations in the crossline direction are inevitable and need be addressed in processing. The geometry has large crossline offsets, leading to gaps in the shallow illumination. The authors report that "the converted wave processing gave an excellent image of the target zone."

Amoco's Valhall survey was acquired with orthogonal geometry (Rosland et al., 1999). One reason to use orthogonal geometry was that 2-D tests in the area showed weak but non-negligible crossline energy, possibly caused by azimuthal anisotropy (Thomsen et al., 1997). The initial 3-D processing results were obtained without exploiting this crossline energy (Brzostowski et al., 1999).

Statoil's Statfjord survey was acquired with a very dense 50 × 50 m coverage of shots across eight 4-C cables of 5 km each and 300 m between the cables (Rognø et al., 1999). This survey lends itself to simulation of orthogonal geometry and areal geometry. Parallel geometry might also be simulated.

The preliminary results on these three 3-D/4-C North Sea surveys reported in 1999 were supplemented by a number of follow-up papers, most and quickest on the Alba survey. The Alba survey had a clear objective (image top reservoir using the larger *PS*-acoustic impedance contrasts) and it was acquired with parallel geometry which is easier to process. So, MacLeod et al. (1999) could quickly publish extensively on the results of the survey, including new insights on the shape of the reservoir and results of two wells drilled on the basis of the new interpretation. Hanson et al. (2000) note that "interpretation of converted wave seismic data radically changed the structural picture of the reservoir and allowed placement of development wells in previously undrained locations." Other noteworthy papers on the Alba field are Duranti et al., (2000), and van Riel et al. (2000).

The Valhall survey has not seen many follow-up papers yet. Granger and Bonnot (2001) report the interesting result that resolution of the fast shear-wave data are better than that of the mere radial component; in other words, taking shear-wave splitting into account improved resolution.

The large volume of data acquired in the 3-D/4-C Statfjord survey seems to form an impediment to fast and reliable results. Ettrich et al. (2000) state that first results show significant potential to improve imaging of the target.

The experience with the North Sea surveys seems to suggest that parallel geometry (Alba) is easiest to process and leads to quickest results. Buia et al. (2001) reported migration artifacts in data acquired with orthogonal geometry. On the other hand, Nolte et al. (2000) and Cafarelli et al. (2001) show some very good imaging through a gas cloud in the Gulf of Mexico Donald field using orthogonal geometry without any obvious processing problems.

6.5 Conclusions and recommendations

In multicomponent data acquisition the converted waves, i.e., *PS*-waves, have asymmetric raypaths leading to asymmetry in the requirements for optimum parameters. Some conclusions are

- Receiver sampling is determined by *S*-wave velocities, and shot sampling by *P*-wave velocities
- Illumination, even of horizontal reflectors, is asymmetrical with cross-spreads, whereas common-offset gathers with constant azimuth have regular illumination
- Horizontal resolution in cross-spreads is much better in the receiver-line direction than in the shot-line direction (for maximum crossline offset equals maximum inline offset)
- Common-offset gathers have better resolution for downdip shooting than for updip shooting. Therefore, the parallel geometry should be acquired with center-spread shooting, so that negative as well as positive offsets are acquired.

Obviously, *PS*-acquisition is much more complicated than *P*-acquisition. Some general guidelines are

- If possible, notably in OBC work, choose parallel geometry rather than orthogonal or areal geometry
- Areal geometry might be best for azimuth-dependent analysis. The strong asymmetry of orthogonal geometry might make azimuth-dependent analysis particularly difficult
- Harmonize requirements of *P*-wave acquisition with those of *PS*-wave acquisition (after all, they are acquired at the same time)
- Parallel geometry:
  ◦ best to use center-spread

- Orthogonal geometry:
  - receiver lines to be oriented in the dip direction
  - receiver-line interval to be smaller than shot-line interval
  - use illumination plots for typical targets to verify illumination and imaging capability of geometry
  - use illumination plots also in processing to regularize illumination fold

References

Albertin, U., et al., 1999, Aspects of true amplitude migration: 69th Ann. Internat. Mtg., Soc. Expl. Geophys., Expanded Abstracts, SPRO 11.2, 1358–1361.

Beasley, C. J., and Mobley, E., 1997, Spatial dealiasing of 3-D DMO: 67th Ann. Internat. Mtg., Soc. Expl. Geophys., Expanded Abstracts, SP3.7, 1119–1122.

Berg, E., and Arntsen, B., 1996, The potential of subsea seismic: Presented at Eur. Assoc. Geosc. and Eng. Workshop on Advances in Multi-Component Technology.

Berg, E., Svenning, B., and Martin, J., 1994, SUMIC: Multi-component sea-bottom seismic surveying in the North Sea—Data interpretation and applications: 64th Ann. Internat. Mtg., Soc. Explor. Geophys., Expanded Abstracts, 477–480.

Brühl, M., Vermeer, G. J. O., and Kiehn, M., 1996, Fresnel zones for broadband data: Geophysics, **61**, 600–604.

Brzostowski, M., et al., 1999, 3D converted-wave processing over the Valhall field: 61st Conf., Eur. Assoc. Geosc. and Eng., Extended Abstracts, Paper 6-43.

Buia, M., Delaney, D., de Tomasi, V., and Vetri, L., 2001, 3D-4C acquisition design for fractures characterization in carbonate reservoir—A case history in offshore Adriatic, the Emilio field: 63rd Conf., Eur. Assoc. Geosc. and Eng., Extended Abstracts, F-33.

Cafarelli, B., Madtson, E., and Nahm, J., 2001, Imaging through gas clouds with converted waves: World Oil, **222,** No. 6, 61–67.

Cordsen, A., and Lawton, D. C., 1996, Designing 3-component 3-D seismic surveys: 66th Ann. Internat. Mtg., Soc. Expl. Geophys., Expanded Abstracts, ACQ3.7, 81–83.

Duranti, D., et al., 2000, Reservoir characterisation of a remobilised sand-rich turbidite reservoir—The Alba field: 62nd Conf., Eur. Assoc. Geosc. and Eng., Extended Abstracts, X-05.

Ettrich, N., et al., 2000, PS pre-stack depth migration case study of the 3-D 4-C OBC survey over the Statfjord field: 70th Ann. Internat. Mtg., Soc. Expl. Geophys., Expanded Abstracts, MC1.6, 1154–1157.

Garotta, R., and Granger, P-Y., 2001, Some *PS* mode acquisition requirements: 63rd Conf., Eur. Assoc. Geosc. and Eng., Extended Abstracts, L-09.

Granger, P-Y., and Bonnot, J-M., 2001, C-wave resolution enhancement through birefringence compensation at the Valhall field: 63rd Conf., Eur. Assoc. Geosc. and Eng., Extended Abstracts, P118.

Granli, J. R., Arntsen, B., Sollid, A., and Hilde, E., 1999, Imaging through gas-filled sediments using marine shear-wave data: Geophysics, **64**, 668–677.

Hanson, R., et al., 2000, 4C seismic data and reservoir modeling at Alba field, North Sea: 70th Ann. Internat. Mtg., Soc. Expl. Geophys., Expanded Abstracts, RC3.1, 1453–1455.

Herrmann, P., David, B., and Suaudeau, E., 1997, DMO weighting and interpolation: 59th Conf., Eur. Assoc. Geosc. and Eng., Extended Abstracts, A050.

Johansen, B., Holberg, O., and Øvrebø, O. K., 1995, Sub-sea seismic: Impact on exploration and production: 27th Annual Offshore Technology Conference, OTC 7657.

Kommedal, J. H., Barkved, O. I., and Thomsen, L. A., 1997, Acquisition of 4 component OBS data—A case study from the Valhall field: 59th Conf., Eur. Assoc. Geosc. and Eng., Extended Abstracts, B047.

Kristiansen, P., 1998, Application of marine 4C data to the solution of reservoir characterization problems: 68th Ann. Internat. Mtg., Soc. Explor. Geophys., Expanded Abstracts, RC1.8, 908–909.

Lawton, D., 1995, Converted-wave 3-D surveys: Design strategies and pitfalls: Can. Soc. Expl. Geophys., Ann. Mtg. Abstracts, 69–70.

McHugo, S., et al., 1999, Acquisition and processing of 3D multicomponent seabed data from Alba field—A case study from the North Sea: 61st Conf., Eur. Assoc. Geosc. and Eng., Extended Abstracts, 6-26.

MacLeod, M. K., Hanson, R. A., Bell, C. R., and McHugo, S., 1999, The Alba field ocean bottom cable survey: Impact on development: The Leading Edge, **18**, 1306–1312.

Nolte, B., et al., 2000, Anisotropic 3D prestack depth imaging of the Donald field with converted waves: 70th Ann. Internat. Mtg., Soc. Expl. Geophys., Expanded Abstracts, MC1.8, 1162–1165.

Ridyard, D., Maxwell, P., Fisseler, G., and Roche, S., 1998, A novel approach to cost effective 4C/4D— The Teal South experiment: 60th Conf., Eur. Assoc. Geosc. and Eng., Extended Abstracts, Paper 2-01.

Rognø, H., Kristensen, A., and Amundsen, L., 1999, The

Statfjord 3-D, 4-C OBC survey: The Leading Edge, **18**, 1301–1305.

Ronen, S., Bagaini, C., Bale, R., Caprioli, P., and Keggin, J., 1999, Coverage and illumination of sea bed 4C surveys: 61st Conf., Eur. Assoc. Geosc. and Eng., Extended Abstracts, 6-18.

Rosland, B., Tree, E. L., and Kristiansen, P., 1999, Acquisition of 3D/4C OBS data at Valhall: 61st Conf., Eur. Assoc. Geosc. and Eng., Extended Abstracts, 6-17.

Thomsen, L., 1999, Converted-wave reflection seismology over inhomogeneous, anisotropic media, Geophysics, **64**, 678–690.

Thomsen, L. A., et al., 1997, Converted-wave imaging of Valhall reservoir: 59th Conf., Eur. Assoc. Geosc. and Eng., Extended Abstracts, B048.

van Riel, P., Leggett, M., Besson-Huerlimann, A., and Hanson, R. A., 2000, Integrated reservoir characterisation of the Alba field from multi-volume seismic data: 62nd Conf., Eur. Assoc. Geosc. and Eng., Extended Abstracts, A-07.

Vermeer, G. J. O., 1999, Converted waves: Properties and 3D survey design: 69th Annual Internat. Mtg., Soc. Expl. Geophys., Expanded Abstracts, SACQ2.6, 645–648.

Chapter 7
Examples of 3-D symmetric sampling

7.1 Introduction

In this chapter some field data examples are discussed. In Section 7.2, results are shown for a 3-D microspread, i.e., a single cross-spread acquired with small shot and receiver station intervals. Section 7.3 discusses the first test of symmetric sampling which was carried out in Nigeria in 1992. Finally, Section 7.4 illustrates some low-fold migration results using data from the Nigeria test.

7.2 3-D microspread
7.2.1 Introduction

Traditionally, a microspread or noise spread has been the tool for detailed investigation of the properties of the wavefield to be recorded in 2-D seismic data acquisition. Ideally, the noisespread consists of a single shot recorded by a receiver spread with receiver station intervals that are small enough for alias-free recording of the total wavefield, including ground roll. Such a data set allows analysis of the noise in relation to the signal and serves as a tool for the design of field arrays. Examples of noise spreads are shown in Figure 4.16 of Vermeer (1990).

In 1979, a noise spread with extremely fine receiver station sampling was acquired in the Paris basin by the field crew of Shell's E&P Lab. With its 0.25-m sampling interval, it was appropriately called the nanospread. Berni and Roever (1989) used this data set to illustrate the effect of statics variation across field arrays (intra-array statics). This paper showed an important application of noise spreads: the investigation of recording effects which cannot be analyzed after acquisition with usual spatial sampling intervals and arrays.

A disadvantage of the 2-D noisespread is that it does not allow the investigation of 3-D effects. Instead, one would need a 3-D microspread. Therefore, in the context of Shell's research project, "Fundamentals of 3D seismic data acquisition," it was decided to acquire such a data set. Originally, the idea was to acquire a 3-D shot with a dense coverage of geophone stations around it. This would involve months of acquisition time. Then we discovered the cross-spread as the basic subset of the orthogonal geometry, and we realized that such a cross-spread would be far easier to acquire, while still providing insight into the 3-D effects.

First, some pilot 2-D microspreads were acquired to find a location where sampling requirements would not be too demanding for the number of available channels. We also wanted to acquire the data along the roadside to avoid permitting problems. Hence we needed two roads crossing each other at straight angles. Finally, an appropriate location was found in the Noordoostpolder, one of the polders in the former Zuiderzee in The Netherlands.

The 3-D microspread was acquired in May 1992. Processing was carried out by Justus Rozemond. In the following, I will discuss the acquisition parameters of the survey and show some preliminary processing results. Unfortunately, processing of the data set was never fully completed.

7.2.2 Acquisition parameters of 3-D microspread

The acquisition lay-out is sketched in Figure 7.1. There were 960 shots fired into 960 receivers with the shot and receiver intervals being 2 m. The nearest shots to the receiver line had a crossline offset of 1 m. As the number of available channels was only 480, the data set was acquired in two passes along the shot line: first the left half of the receiver spread was acquired, followed by the right half. Maximum inline offset was 959 m. This configuration allowed for the recording of the full ground-roll cone up to 3 s. A vibrator was used as the source. The source signature consisted of a single 27-s linear sweep ranging from 8 to 60 Hz. Each receiver station consisted of a single geophone.

The data quality of the acquired 3-D microspread was excellent. There were no missing shots nor any missing receivers.

7.2.3 Cross-sections and time slices

Figure 7.2 shows a shot gather for a shot with 1-m crossline offset. Note that noise that might look random with a coarser sampling interval turns out to be coherent virtually everywhere with this fine sampling. The noise cone, which is conveniently called ground roll, does not consist only of Rayleigh waves; the least steep events are not surface waves but body waves, i.e., refracted shear waves. The Rayleigh wave velocity is about 180 m/s, whereas the apparent velocity of the shear waves ranges from 250 to 420 m/s. Given the 2-m spatial sampling interval and a maximum frequency of 60 Hz, the Rayleigh waves are still aliased above 45 Hz, but the shear waves are sampled without aliasing. The first break comes in at a velocity of 1670 m/s.

Figure 7.3 shows a shot gather for a shot with 901-m crossline offset. The refracted shear wave now appears at 2.5 s and has much higher apparent velocity in this cross-section. Note the clean P-wave data above the noise cone.

Figure 7.4 shows two time slices through the 3-D microspread. The slice at 596 ms cuts mostly through P-wave energy. These events are relatively flat, hence the time slice shows long apparent wavelengths. On the other hand, the slice at 3596 ms cuts mostly through the ground-roll cone with much steeper events, hence this slice shows much energy with short apparent wavelengths. The inside of the ground-roll cone contains many events extending only across a small distance. These events tend to look more continuous in the cross-section of Figure 7.2.

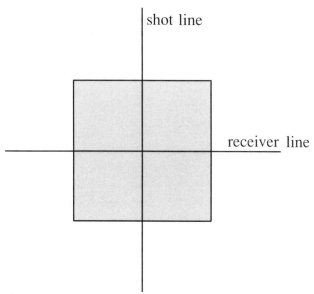

FIG. 7.1. Lay-out of a 3-D microspread; 960 shots were fired into 960 receivers. Shot and receiver station intervals were 2 m.

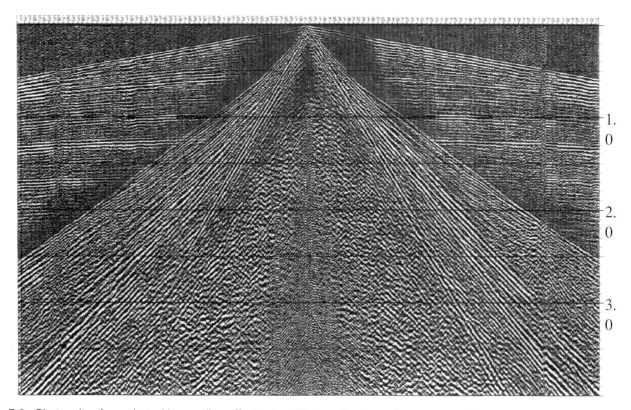

FIG. 7.2. Shot gather for a shot with crossline offset = 1 m. This section is analogous to a 2-D microspread.

The most striking features of these time slices are the circular behavior of many events, and the spatial continuity of the data. Of course, the circular behavior is caused by the traveltimes of many events being a function of offset only, whereas constant offset is represented by a circle in the time slice (cf., Figure 2.6). The shot-to-shot variation is minimal; continuity in the horizontal direction (the common-receiver direction) is as good as in the vertical direction. A rather large discontinuity can be seen in the upper part of Figure 7.4a at shotpoint 540,

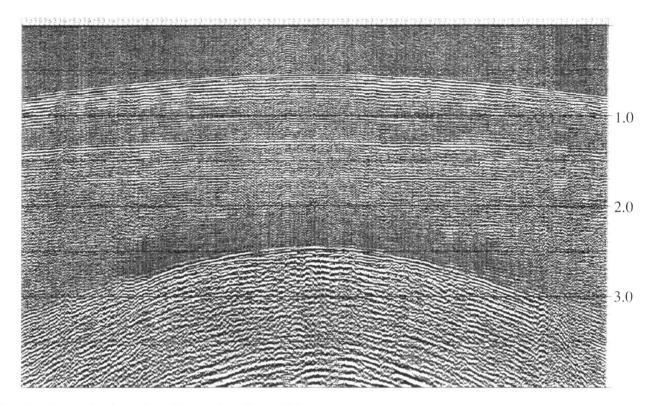

FIG. 7.3. Shot gather for a shot with crossline offset = 901 m.

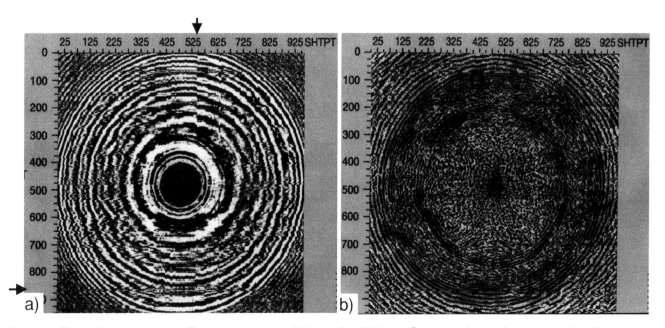

FIG. 7.4. Time slices through a 3-D microspread, (a) 596 ms, (b) 3596 ms. Common-shot gathers are vertical lines, common-receiver gathers are horizontal. Arrows in (a) indicate discontinuities.

144 Chapter 7 Examples of 3-D symmetric sampling

and another strong discontinuity at receiver 870. These discontinuities may be caused by positioning errors or statics.

7.2.4 *(f,k)*-filtering results

Figure 7.5 shows a receiver gather for a receiver at 1-m inline offset. Note again, the near perfect continuity across this gather, even though each trace is the result from a different physical experiment. The section looks very similar to Figure 7.2, except for a mysterious coherent event with an apparent velocity between that of the *P*-wave first break and the shear-wave first break. Unfortunately, I have no explanation for this event.

Figure 7.6 shows the same receiver gather as in Figure 7.5, but now after *(f,k)*-filtering in the shot domain. The parameters of the *(f,k)*-filter were chosen to reject all steeply dipping energy and to pass the reflection energy. At first sight nothing has changed, but closer inspection reveals that inside the ground-roll cone, the steeply dipping events are more abundant. In the unfiltered receiver gather, there are many interfering scattered events. By *(f,k)*-filtering in the shot domain, the steep events perpendicular to the receiver section have been removed, leaving behind the components that dip steepest in the common receiver. This emphasizes again the 3-D nature of all events and in particular that of the scatterers.

Finally, Figure 7.7 shows the receiver gather of Figure 7.5 after *(f,k)*-filtering in the common-receiver domain. Now all steep events in the receiver gather have been removed. The strong smearing of the data indicates the inability of the *(f,k)*-filter to carry out the surgical action which is really desired: removing the steep events while not affecting the less steep events.

7.2.5 Discussion

The figures of the 3-D microspread shown here illustrate that much of the energy that may be considered random in conventional acquisition, appears to be coherent upon fine sampling. Also, shot-to-shot variations were minimal in this case, so that the common-receiver gathers in this data set look as nice as the common-shot gathers. Even though only small maximum inline and crossline offsets were used in this experiment, it still suggests the reasonableness of the recommendation to acquire data with equal maximum offset in both directions. There is no preference for one or the other direction.

3-D microspreads could be useful in making 3-D survey design decisions. First, it should be possible to measure the relative strengths of the noise and the signal.

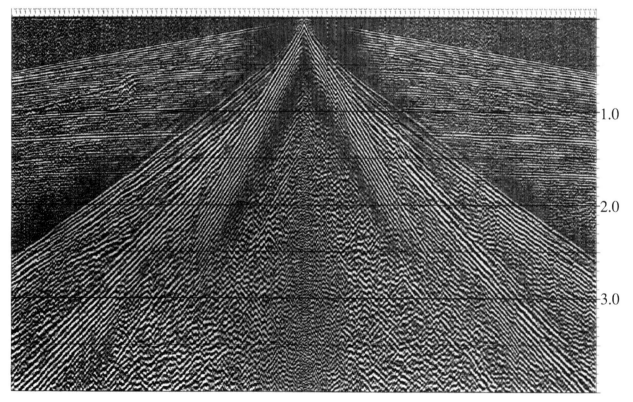

FIG. 7.5. Receiver gather for a receiver at inline offset = 1 m.

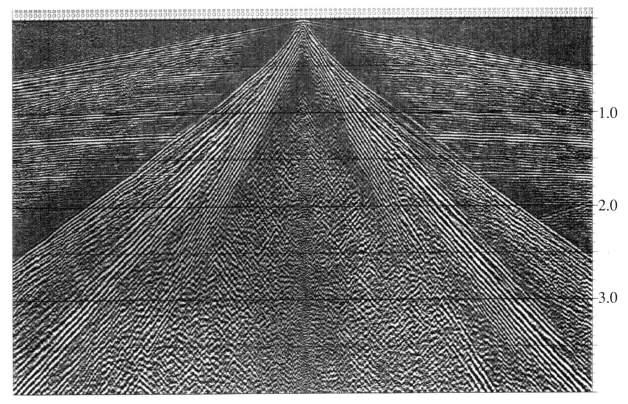

FIG. 7.6. Receiver gather of Figure 7.5 after (f,k)-filtering in the shot domain.

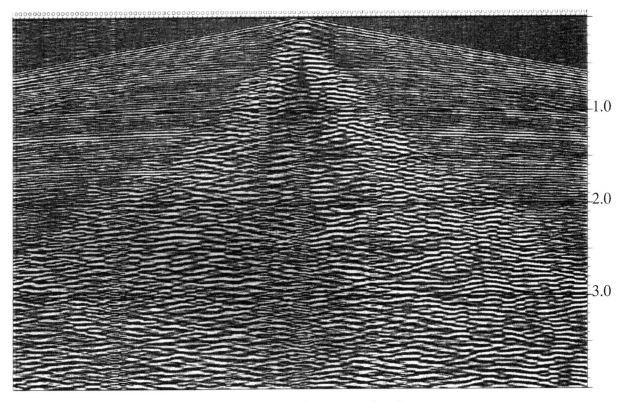

Fig. 7.7. Receiver gather of Figure 7.5 after (f,k)-filtering in the receiver domain.

Then, together with an estimate of the noise suppression by stacking and migration, it should be possible to define the required amount of noise suppression for the combination of shot and receiver arrays for a given station spacing. This is clearly an area where very useful research could be carried out, and could supplement the work published in Krey (1987).

Another application of 3-D microspreads is as an analysis tool of scattered energy to answer the question whether areal arrays are necessary for adequate suppression of the scatterers. In this particular case, it is clear from Figure 7.6 that most of the energy remaining after removal of steeply dipping events in the common-shot gathers consists of steeply dipping events in the common-receiver gathers. In other words, the amount of energy concentrated in the apexes of the scatterers is relatively small compared to the amount of energy in the flanks of the scattered events. Therefore, in this case, it would be sufficient to have a combination of linear shot and receiver arrays, perhaps followed by a 3-D velocity filter (Smith, 1997).

The advantage of the 3-D microspread is that it is fully representative of the noise that is going to be encountered when using orthogonal geometry (apart from variations in character through the survey area). Up till now, the box test (Regone, 1997) is commonly used for this purpose. However, the box test is quite a special survey, not fully representative of the noise that is going to be acquired; moreover, it uses small arrays which already attenuate some of the noise one wants to know the details of.

7.3 Nigeria 3-D test geometry results
7.3.1 Introduction

In brick or brick-wall geometry, the shot lines are staggered such that the pattern of shot lines and receiver lines resembles the pattern of bricks in a brick wall (Figure 7.8b). This geometry was introduced in the late 1980s. According to Wright and Young (1996), brick geometry is "one easy way to insure superior offset sampling." With the staggering, the offsets of the traces in each bin are changed into a pattern which is more evenly spread across the offset range than in the equivalent continuous shot-line geometry (Wright and Young, 1996). This expresses itself also in a smaller LMOS for the brick than for the continuous geometry as illustrated in Figure 7.8. Moreover, as Figure 3.18 shows, (narrow) brick geometry has a better stack response than narrow continuous geometry.

However, with an increasing width of acquisition geometry (larger maximum crossline offset), most of these advantages disappear (except the LMOS advantage), whereas the advantage of using continuous shot lines increases with the width of the survey geometry. With brick geometry, the common-receiver gathers are broken into small segments, whereas in the continuous shot-line geometry, the receiver gathers are continuous. In fact, they are not any different from the common-shot gathers (see Figure 7.9). One of the objectives of 3-D survey design should be to keep the number of spatial discontinuities in the acquired data to a minimum. In brick geometry, the character of the inline data will be different from the character in the crossline direction, as is clearly shown in Figure 7.10, copied from Moldoveanu et al. (1999). The difference is mainly due to discontinuities in the recorded noise, leading to discontinuities in the stacked crossline section. These discontinuities will turn into extra migration smiles after migration.

At the 1992 Shell Geophysical Conference, I presented a paper (with co-author Justus Rozemond) with the message "Don't use brick." In the audience was Paul Wood, then head of acquisition in Shell Petroleum

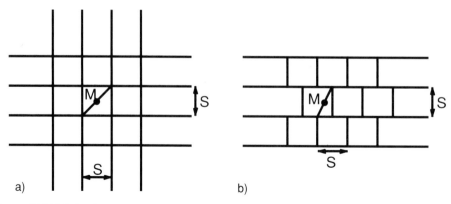

FIG. 7.8. Comparison of LMOS for (a) continuous shot-line geometry and (b) brick-wall geometry. For the same line intervals, LMOS is larger in (a) than in (b).

Nigeria 3-D test geometry results   147

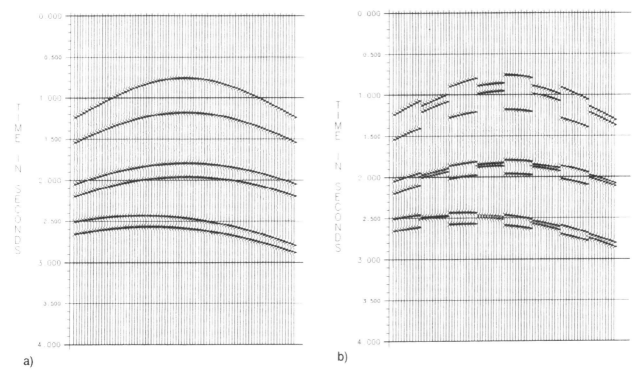

FIG. 7.9. Comparison of synthetic crosslines, (a) continuous shot lines, (b) staggered shot lines. For the purpose of this example the input data have been stacked (2-fold) without NMO correction. The model consists of three reflectors.

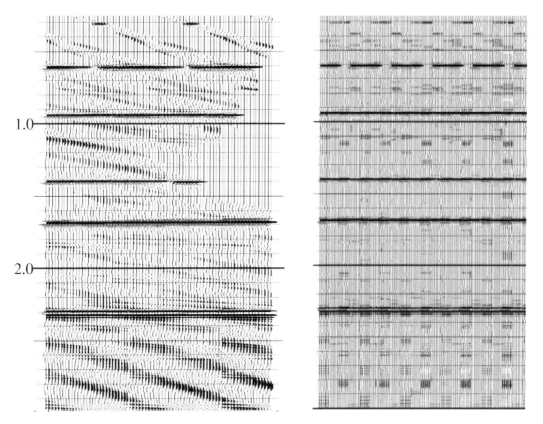

FIG. 7.10. Comparison of inline section with crossline section for brick geometry with coherent noise. Left: inline, right: crossline. (from Moldoveanu et al., 1999). Note choppy character of noise in the crossline section.

148   Chapter 7   Examples of 3-D symmetric sampling

Development Company (SPDC) in Nigeria. He felt personally addressed by this message, because a few years earlier SPDC had changed from narrow continuous geometry to narrow brick geometry for all 3-D surveys. He gave us the benefit of the doubt, and decided to acquire a small test geometry across part of the 3-D survey currently being acquired. At the next Geophysical Conference in 1993 he showed some preliminary results of the test. He framed his presentation between the "Bricklayer's prologue" and the "Bricklayer's epilogue," here reproduced as Figure 7.11. It was late 1994 before the data were fully processed and interpreted. Then the results appeared to be at least as good as the production data at about 60% of the cost. As a consequence, SPDC decided to shoot two production surveys with the "cross-spread technique," and once these proved to be successful, all 3-D surveys are now being acquired with the continuous shot-line technique.

Now I will describe the old and the new acquisition geometry, followed by some processing results and the key interpretation results. Some results were shown earlier in Vermeer (1998).

7.3.2 Acquisition geometry

The survey area is the Niger delta, which is characterized by mangrove swamp, jungle, and a multitude of narrower and wider creeks. The creeks provide for an easy means of transporting material, but would produce numerous gaps in the survey if these were considered as obstacles for shot and receiver placement. Therefore, an airgun array vessel provides the shots in water, and hydrophones replace geophones locally. The source on land is either deep single-hole dynamite or a shallow-hole linear dynamite array.

The conventional acquisition geometry used in

BRICKLAYER'S PROLOGUE

When that Decembre with his festal cheere
Had gi'en to Januar with snowes so vere
In year of '92, that of oure Loorde
Then sev'ral worthy knygtes with one accorde
Did then resolve to hold upon that chaunce
A geophysicale conferaunce

Now at this conferaunce was muche ado
On 3D sysymyke methodes new
Of how to lay the spredds on Goddes yrthe
And so to make the data bettre worthe
A Nombre-Smyth stood up, Gijsbert by name
And to the congregatioun exclaim
"Use not the Brikke spredd" he cried "To whit
The five dimens-ions cannot permit
The samplying one gets, it makes me pale
For Crosse-Spredds will thereby not entail"

At this the audience gave a roar
Gijsbert's dimens-ions increased from four
And "Lo" they cry "has come to pass agayn
Gijsbert invented yet a new domayn

A worthy Bricklayer, yclepte Paul
Stood up right then and so addressed the Halle
"The Brikke-Spredd" he sayd "Is common use
It startleth me to heare such abuse
For we are getting data wondrous faire
And soon I hope to showe it to you there
Next year" said he, "I will returne hale"
And so he did, and this then is his tale.

BRICKLAYER'S EPILOGUE

And so ye worthies, it has been youre fate
To heare of shotte lynes offset and shottes strait
But oure conclus-ion is at this tyme
To keep the Brikke spredd with offset lyne

But we admit full well the Crosse-Spredd
Should furthyr be investigatied
And so we didde and that wyth muche content
Conduct a certayn grate experiment
Full sixty Crosse-Spredds recorded there
With wondrous speed and most especial care
And now to Gijs the data I present
And urge him to make the tyme well spent

The Worke shoppe for processing is near
A year agayn we all will gather here
To learn if we must do and use therefore
The Crosse-Spredds in Nineteen Ninety Four.

FIG. 7.11. The bricklayer's tale by Paul Wood.

Nigeria in 1992 is described in Figure 7.12a. The template consists of four 6000-m active receiver lines spaced at 350 m. The distance between the shot-line segments is 400 m. Shot and receiver station spacings are 50 m. The maximum inline offset of this geometry is 3000 m, and the maximum crossline offset is 700 m. Aspect ratio = 0.23, LMOS = 403 m, inline fold = 7.5, and crossline fold = 2.

The main limitation for selecting an alternative geometry was the availability of only 480 channels. Fortunately, the target zone started below 1.7 s. Therefore the distance between the receiver lines could be doubled to 700 m without affecting the target levels too much. The selected test geometry is shown in Figure 7.12b. The maximum inline offset is reduced to 2000 m, and the maximum crossline offset is increased to 2100 m. Aspect ratio = 0.95, LMOS = 806 m, inline fold = 5, crossline fold = 3. The test, consisting of 10 shot lines and six receiver lines, was acquired such that 60 complete cross-spreads were gathered. It took only ten days to acquire the test.

Figure 2.16 (top) shows a plot of cumulative fold and trace density for the two geometries. The trace density plot can be compared with the histograms made using the actual recorded data shown in Figure 7.13. These curves illustrate that fold buildup is much faster in the narrow production geometry than in the wide test geometry. The azimuth distribution is very peaked for the narrow geometry and more evenly spread for the wide geometry. The peaks in the azimuth distribution of the wide geometry stem from the corners of each cross-spread. If the midpoint area of the cross-spread were circular, then the azimuth distribution would be completely flat.

7.3.3 Some processing results

After some initial processing by SPDC, including refraction statics, a copy of the test data was sent to Shell's research lab in The Netherlands. In the lab, we tried to apply some cross-spread oriented processing techniques, whereas SPDC applied their standard processing sequence to the test data. In the next section, the final processing results of SPDC are discussed. As part of the research work, much time was spent on surface-consistent residual statics, deconvolution, and on a new, algorithm for applying DMO. I can only show a few of the results here.

The quality of the seismic data acquired in this area is quite good. This is illustrated in Figure 2.14 by some time slices taken from one of the cross-spreads. These time slices also confirm that above the target zone (1.7 s) the geology is horizontal.

Figure 7.14 shows a diagnostic display used in the surface-consistent deconvolution process to check on quality of individual receivers and shots. Each trace in each one of the 60 cross-spreads produces a pixel of which the color represents some seismic attribute for that trace. In this case it is the absolute maximum sample value in a 1000-ms window starting just before the first break. The data have been arranged such that all pixels for the same shot position are along horizontal lines (a clever trick devised by Justus Rozemond), and all pixels for the same receiver position are along vertical lines. Note that surface consistency is shown clearly in this display. Weak receivers show as narrow vertical stripes crossing over between neighboring cross-spreads. Missing shots appear as white horizontal stripes.

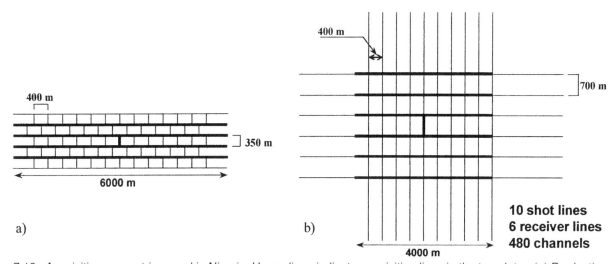

FIG. 7.12. Acquisition geometries used in Nigeria. Heavy lines indicate acquisition lines in the template. (a) Production survey geometry, (b) complete test geometry. Shot and receiver station spacings are 50 m in both geometries.

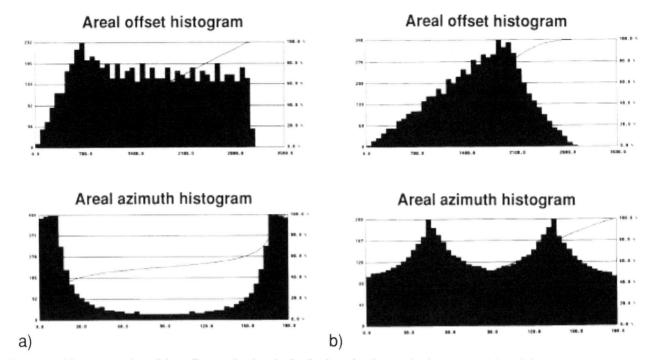

FIG. 7.13. Histograms describing offset and azimuth distributions for the production survey (a) and the test geometry (b).

The type of display shown in Figure 7.14 was also used for diagnosing the statics. Errors become immediately obvious upon inspection of a display showing all static corrections.

Figure 7.15 shows the benefit of dual-domain (f,k)-filtering. The production data did not allow (f,k)-filtering in the crossline direction, but the cross-spread data did. However, the difference between dual-domain filtering and single-domain filtering was no longer visible after application of DMO. Apparently, the DMO process also has a beneficial effect on the same noise which is tackled by (f,k)-filtering.

7.3.4 Interpretation results

SPDC carried out a careful comparison between the production survey and the test geometry, both processed with their standard processing sequence. Not surprisingly, the results at shallow levels were better for the production geometry: fold at shallow levels is higher for the production survey. On the other hand, the results at target level for the test geometry were at least as good as for the production survey, even though no special cross-spread oriented processing had been attempted. A comparison of some significant results is shown in Figures 7.16 and 7.17. In Figure 7.16, an illumination display of a target horizon is compared for the two geometries. The main features are the same, but the test geometry produced a cleaner looking result.

In Figure 7.17 the amplitudes are compared for the same horizon. Blue indicates the no-pick areas from the automatic tracker. In the test geometry the blue areas are smaller, producing better defined faults than in the production geometry. The irregularities around the edges of Figure 7.17b are caused by the edge effects of the small test survey. Figure 7.17a does not show these effects since it is just a small part of the larger survey.

Based upon the results of this interpretation, it was decided to shoot one of the next 3-D surveys with the continuous shot-line geometry. This was an easy decision since shooting with the new geometry parameters was considerably cheaper, because of the wider receiver line spacings and the straight shot lines.

7.3.5 Discussion

The question remains: Is the better quality at target level a result of the continuity of the shot lines, or is it due to the wider geometry, or both? It is impossible to give a definite answer to this question. One thing that may be said is that going from brick to continuous without changing the width of the geometry would likely have produced worse results than the original brick. This may be expected because of the worse stack response of the narrow geometry (see Figure 3.18), which is not compensated by the ability for filtering in the common-

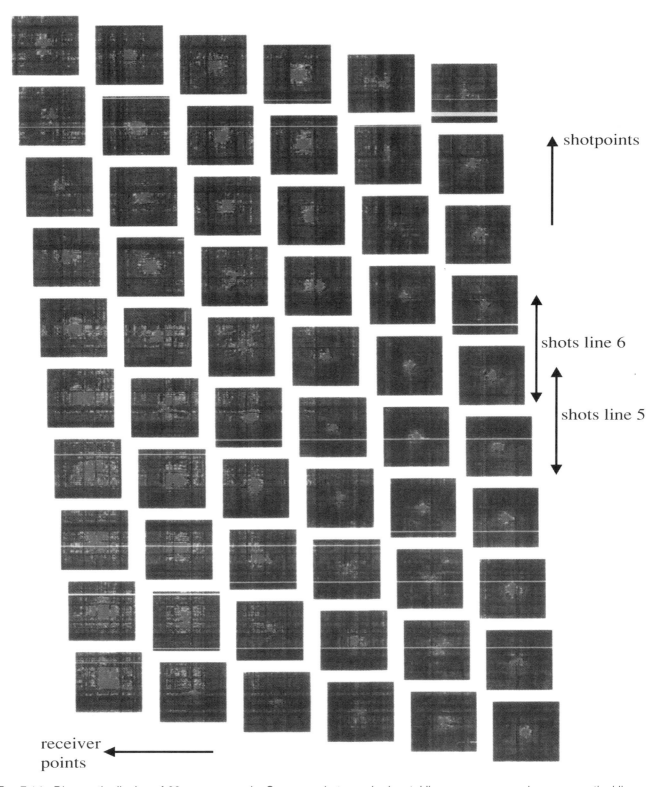

FIG. 7.14. Diagnostic display of 60 cross-spreads. Common shots are horizontal lines, common receivers are vertical lines. The shotpoints of adjacent shot lines partially overlap in this display. Likewise the receiver points of different receiver lines. Shown is maximum absolute sample value for a 1000 ms window starting just before the first break on each trace.

152   Chapter 7   Examples of 3-D symmetric sampling

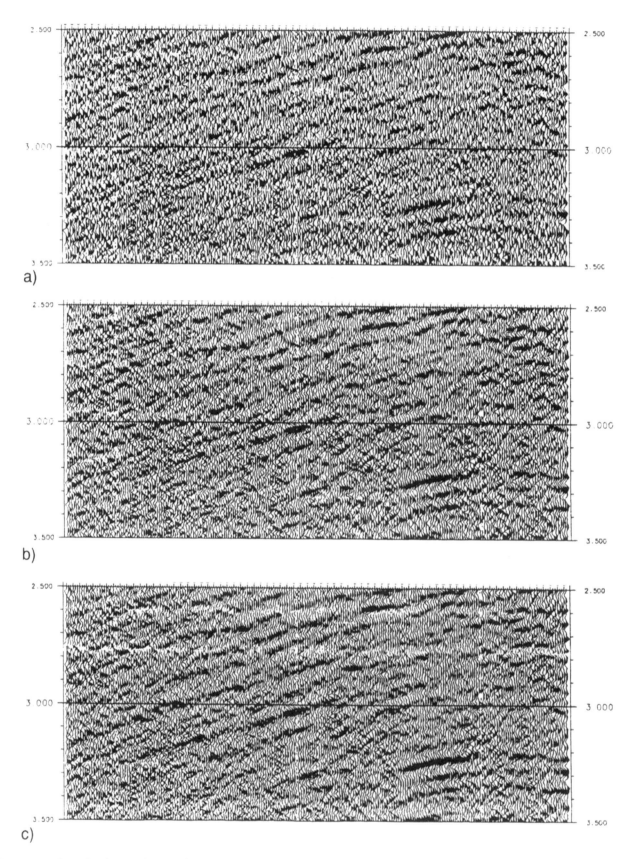

FIG. 7.15. Benefit of dual-domain filtering. (a) No filter applied, (b) after common-receiver *(f,k)*-filter, (c) after common-receiver and common-shot *(f,k)*-filter.

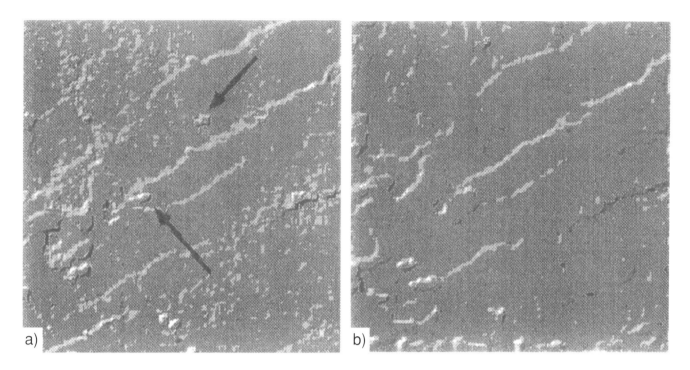

FIG. 7.16. Illumination displays for target horizon. (a) Brick geometry, (b) cross-spread geometry. The result for the cross-spread geometry is cleaner, hence more reliable than the result for the brick geometry.

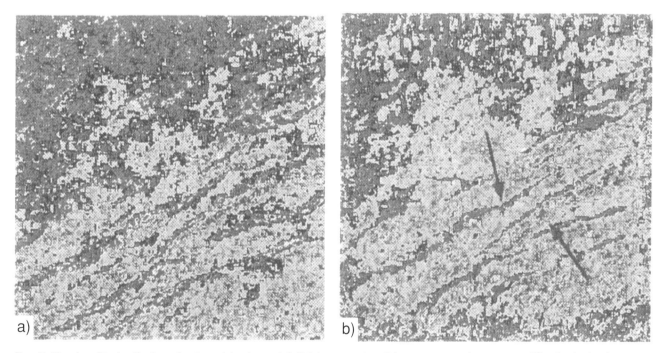

FIG. 7.17. Amplitude displays for target horizon. (a) Brick geometry, (b) cross-spread geometry. The faults in the cross-spread result are better defined.

receiver gathers. So, the width of the geometry is definitely a contributing factor.

I am convinced that the improved result is not only due to the wider geometry, but also to the greater spatial continuity provided by the continuous shot lines. Unfortunately, this data comparison cannot prove this conviction. Whether a continuous geometry is better than a bricked geometry might be tested using a wide geometry acquired with continuous shot lines. The test would be to compare two binated data sets: one data set consisting of every other shot line, the other with all shot lines, but now converted into a brick pattern.

An important lesson from this exercise is that deep targets allow wide line spacings; perhaps not always, but definitely in this case. Basically, the original geometry was oversampled as far as receiver-line interval and shot-line interval were concerned. After this exercise SPDC decided to increase fold from 15 to 30, which only marginally increased the acquisition cost as compared to the brick geometry.

7.4 Prestack migration of low-fold data

7.4.1 Introduction

Single-fold well-sampled 3-D data sets (minimal data sets) are suitable for migration. Although this property is always exploited when migrating stacked data, it is not generally appreciated that the property also applies to prestack data. Of course, the result of migrating single-fold prestack data might be quite noisy; multifold data are needed to suppress more noise. However, for imaging it is sufficient that the data set has been well-sampled, which means that, for each reflector, there is an illumination area corresponding to the midpoint area of the data set (see Section 2.5.3).

In this section, I will discuss migration of a single cross-spread followed by prestack migration of low-fold data.

7.4.2 Migration of a single cross-spread

A cross-section along the diagonal of a prestack time migrated cross-spread is shown in Figure 7.18. It is immediately clear that a single cross-spread does not illuminate much of the subsurface; the image extends across a small range only. Furthermore, this range becomes smaller and smaller for shallow levels. In the shallow center of the cross-spread the image suffers from the presence of ground roll. Apart from that, the image (where there is one) looks surprisingly good.

This single cross-spread already shows that the geology in this area is rather flat down to 1.7 s, with gradually increasing dips from there to deeper levels. Of course, the deeper steep events have migrated updip, away from the midpoint area of the cross-spread. Along the flanks of the image area, incomplete images are visible. These incomplete images signify the edge effects of the cross-spread. Part of these edge effects may still contribute to the image forming of the collection of cross-spreads, depending on the shot- and receiver line intervals. Another part of the edge effects is just noise, which has to be suppressed by the action of overlapping cross-spreads.

Each trace in the migration result shown in Figure 7.18 is composed of contributions from all traces in the original cross-spread. Each output sample is the summation of the amplitude values that can be found along the

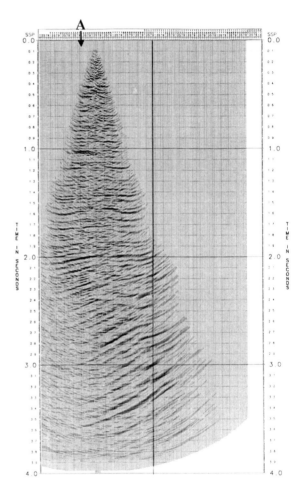

FIG. 7.18. Cross-section along the diagonal of a migrated cross-spread. The midpoint range of the cross-spread is to the left of the vertical black line. Note the updip shift of the deeper data and the edge effects on both sides. The arrow indicates the position of the output point for which time slices through the diffraction-flattened gather are shown in Figure 7.19.

diffraction traveltime surface for the position (t, x, y). It is instructive to look at those amplitude values before summation. This can be done by flattening the diffraction traveltime surface and then making time slices. (See also the discussion of migration as a two-step process in Sections 6.2.4 and 8.3.7, and the discussion of diffraction-flattened gathers in Section 10.4). Figure 7.19 shows a number of flattened diffraction traveltime surfaces for the strong reflection around 2 s. The zone of stationary phase of the strong reflector is located in the lower left corner of the cross-spread. This is the position where reflection and diffraction traveltime surfaces coincide and have about the same slope. The time slice at 2012 ms produces the maximum amplitude of the strong reflection. There the diffraction traveltime surface cuts through a large number of positive reflection amplitudes.

Figure 7.19 illustrates that the zone of stationary phase with its slowly varying amplitude is competing with many other amplitude values whose average amplitude value should be zero for the cleanest result. A good migration program should taper the deepest parts of the diffraction traveltime surfaces to suppress truncation effects.

Figure 7.19 also illustrates the need for equal shot and receiver sampling. If the shot interval was twice as large as the receiver interval, the near-circular zones of stationary phase would shrink into ellipses, whereas spatial aliasing would occur in the common-receiver gathers (in the direction of the short axis of the ellipses), leading to less clean results.

7.4.3 Low-fold prestack migration

Now, I will take five cross-spreads with partially overlapping midpoint areas for a test on low-fold migration. Figure 7.20 shows the arrangement of the input data. Along the diagonal through the five cross-spreads, the fold-of-coverage varies between 1 and 3. The migration result is computed along the heavy black line in Figure 7.20 for each of the five cross-spreads. The individual contributions of the five cross-spreads are shown side-by-side in Figure 7.21. Similar to the migration of a sin-

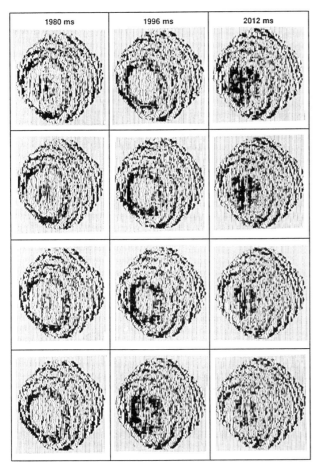

FIG. 7.19. Time slices through a cross-spread after flattening of diffraction traveltime surfaces for position A indicated with an arrow in Figure 7.18. Time is increasing downward in steps of 4 ms. The data of each slice are summed to form one output sample of the trace at A. The zone of stationary phase is in each time slice. The strong black loop around 2 s in Figure 7.18 is composed from the time slices at 2008-2016 ms with the maximum at time 2012 ms.

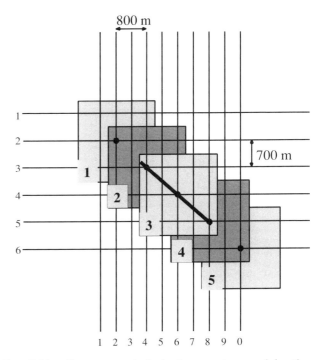

FIG. 7.20. Cross-spreads in test geometry used for the migration test. The heavy line in center indicates the position of migration results shown in Figures 7.21 and 7.22.

gle cross-spread in the previous section, all traces of each cross-spread can potentially contribute to the migration result, i.e., Figure 7.21 shows the result of 3-D migration.

Inspection of these results shows that cross-spread 5 hardly contributes, whereas cross-spreads 2 and 4 fill up the edges of the result for cross-spread 3. Cross-spread 1 contributes some images of steeply dipping events around 3 s. In other words, when these results are stacked, the image fold is never more than 2. Figure 7.22a shows the stack of the five migration results. It can be compared with the straight stack in Figure 7.22b. The stacking fold varies between 2 and 3. Below about 1.6 s, the migration result starts to look like real geology, which is quite remarkable for this very low-fold data. Above 1.6 s, edge effects and ground-roll effects disrupt the continuity of the result.

7.4.4 Discussion

In the previous two sections it is shown that 3-D prestack migration of low-fold data may produce quite reasonable results. Of course, it should be granted that the quality of the input data was very good. Nevertheless, this is not a unique situation, as shown by Figure 7.23, which is reproduced from a paper by Lee et al. (1994). Figure 7.23b shows the result of a test carried out by Mobil. The test consisted of a single shot line, intersected by a number of perpendicular receiver lines. According to the authors, the 3-D result is even cleaner than the high-fold 2-D result shown in Figure 7.23a. This is attributed to the absence of side-swipe energy in the 3-D result.

It may be concluded that in good data quality areas, low-fold 3-D data may be adequate for certain purposes, e.g., for reconnaissance 3-D. Low-fold 3-D may give an interpretable 3-D result at a cost comparable to a grid of 2-D lines, in particular for deep targets. It should be realized, that sparse acquisition should not be achieved by increasing shot station intervals, but by increasing shot- and receiver-line spacings while keeping the station spacings of shots and receivers the same (and adequate for the purpose).

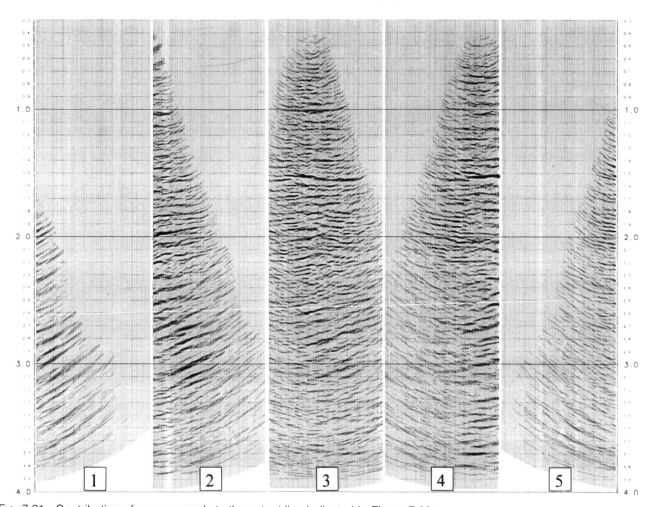

FIG. 7.21. Contribution of cross-spreads to the output line indicated in Figure 7.20.

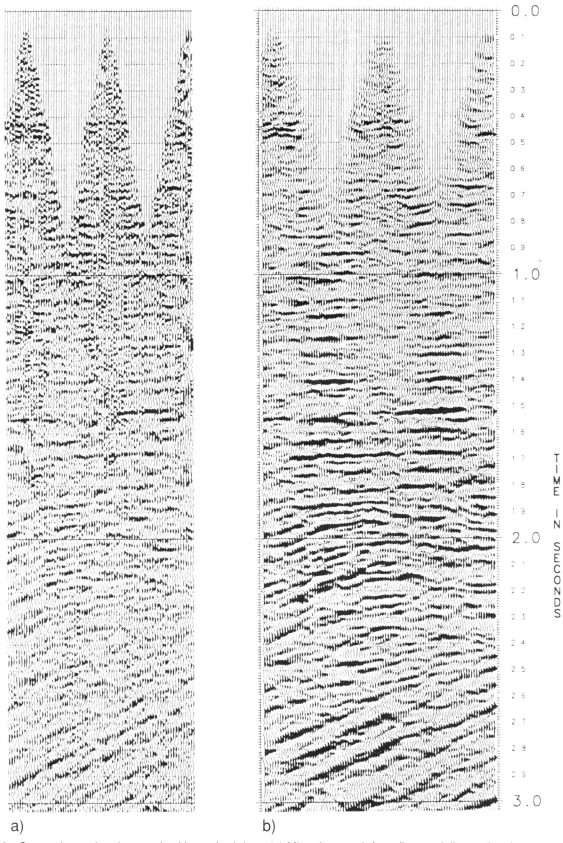

FIG. 7.22. Comparison migration result with stacked data. (a) Migration result from five partially overlapping cross-spreads indicated in Figure 7.20. (b) Corresponding stack (mostly 2-fold).

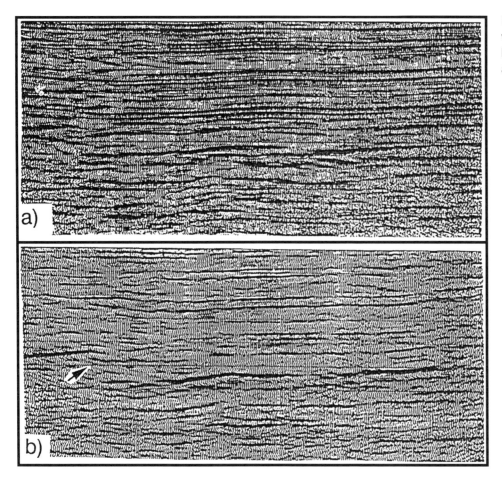

FIG. 7.23. Comparison conventional 2-D data (a) with low-fold 3-D (b) (from Lee et al., 1994).

Very often, the quality of the recorded traces at shallow levels is much better than at deeper levels. Again, this may mean that the required fold for shallow levels is much smaller than for deeper levels, and in some cases, 4-fold data may be adequate for mapping high-quality shallow data.

References

Berni, A. J., and Roever, W. L., 1989, Field array performance: Theoretical study of spatially correlated variations in amplitude coupling and static shift and case study in the Paris Basin: Geophysics, **54**, 451–459.

Krey, Th. C., 1987, Attenuation of random noise by 2-D and 3-D CDP stacking and Kirchhoff migration: Geophys. Prosp., **35**, 135–147.

Lee, S. Y., et al., 1994, Pseudo-wavefield study using low-fold 3-D geometry: 64th Ann. Internat. Mtg., Soc. Expl. Geophys., Expanded Abstracts, 926–929.

Moldoveanu, N., Ronen, S., and Michell, S., 1999, Footprint analysis of land and TZ acquisition geometries using synthetic data: 69th Ann. Internat. Mtg., Soc. Expl. Geophys., Expanded Abstracts, SACQ2.5, 641–644.

Regone, Carl J., 1997, Measurement and identification of 3-D coherent noise generated from irregular surface carbonates, *in* Palaz. I., Marfurt, Kurt J., Ed., Carbonate seismology: Soc. Expl. Geophys.

Smith, J. W., 1997, Simple linear inline field arrays may save the day for 3D direct-arrival noise rejection: Proc. Soc. Expl. Geophys., Summer Research Workshop.

Vermeer, G. J. O., 1990, Seismic wavefield sampling: Soc. Expl. Geophys.

——— 1998, 3-D symmetric sampling in theory and practice: The Leading Edge, **17**, 1514–1518.

Wright, S., and Young, J., 1996, Lodgepole 3-D seismic: Design, acquisition and processing: 66th Ann. Internat. Mtg., Soc. Expl. Geophys., Expanded Abstracts, 397–400.

Chapter 8
Factors affecting spatial resolution

8.1 Introduction

The theory of spatial resolution has been dealt with in great detail by various authors on prestack migration and inversion (e.g., Berkhout, 1984; Beylkin, 1985; Beylkin et al., 1985; Cohen et al., 1986; Bleistein, 1987), and on diffraction tomography (e.g., Wu and Toksöz, 1987). Despite all this work, the practical consequences of the theory are still open to much debate.

Von Seggern (1994) discusses resolution for various 3-D geometries, and concludes: "Uniform 3-D patterns, asymmetric patterns, and both narrow and wide swath 3-D patterns all produce nearly equivalent images of a point scatterer, without significantly better resolution in one or the other horizontal direction." These results were obtained using quite a coarse measurement technique; moreover, fold varied across the midpoint range. As a consequence, the considerable differences in resolution that do occur between different geometries were overlooked.

Neidell (1994) submitted that coarse sampling, if compensated by high fold (24-fold or higher), does not sacrifice resolution. His conjecture led to a flurry of reactions (Vermeer, 1995; Neidell, 1995; Ebrom et al., 1995b; Markley et al., 1996;, Shin et al., 1997).

Ebrom et al. (1995b) and Markley et al. (1996) investigate resolution using a tank model consisting of a number of vertical rods. The time slices at the level of the top of the rods are compared for various sampling intervals and folds of coverage. Whereas Ebrom et al. (1995b) showed that the resolution in the timeslice could be finer than the acquisition common midpoint (CMP) binning, Markley et al. (1996) conclude that finer CMP binning improves the image significantly compared to coarse binning with the same number of traces, thus contradicting Neidell's conjecture. Shin et al. (1997) illustrate that fold can partially compensate for coarse sampling.

This chapter modified from Vermeer (1999).

The issue of sampling is expanded further with the introduction of quasi-random sampling (Zhou and Schuster, 1995; Sun et al., 1997; Zhou et al., 1999). Zhou and Schuster (1995) and Zhou et al. (1999) demonstrate that quasi-random coarse sampling may lead to less migration artifacts than regular coarse sampling. Sun et al. (1997) conclude that migration of data sampled with the quasi-Monte Carlo technique can reduce the computational work load by a factor of four or more. These results might be interpreted as "random sampling is superior to regular acquisition for purposes of noise reduction" (Bednar, 1996), a statement that assumes that the (coherent) noise is coarsely sampled. Sun et al.'s (1997) conclusion is questioned in Vermeer (1998b).

Apart from the authors mentioned in the first paragraph, none of the above authors mentioned Beylkin's (1985) formula for spatial resolution, even though it had already been published. The present chapter uses Beylkin's formula to derive resolution formulas for simple cases, and to explain results obtained for various configurations. Lavely et al. (1997) and Gibson et al. (1998) also use Beylkin's formula as a starting point for resolution analysis.

Levin (1998) provides a lucid narrative of the resolution of dipping reflectors. This chapter, although not dealing explicitly with reflectors, confirms many insights offered in that paper, which is recommended for further reading.

In conventional seismic acquisition the measurements are carried out at or close to the surface, basically in one horizontal plane. This measurement configuration leads to quite a difference between the resolution in the vertical direction and the resolution in a plane parallel to the measurement plane. This chapter deals only with such configurations; hence, it does not discuss the resolution of measurements at various depth levels, such as made with vertical seismic profiling (VSP).

Resolution is about the resolvability of two close events. This resolvability is determined by the width of the main lobe of the wavelet, and by the strength of the side lobes relative to the main lobe. In this discussion, I will leave the effect of side lobes mostly aside and concentrate on measurements of the width of the wavelet after migration. [For a detailed discussion of the effect of side-lobes, see Berkhout (1984). Especially, if two events have different strengths, side lobes of the strong event may mask the main peak of the weak event.] The wider the wavelet, the larger the distance between two events needs to be for their resolvability. The smallest distance for which two events can still be distinguished is called the minimum resolvable distance.

The theory of resolution leads to a potential resolution (i.e., the best possible resolution for a given source wavelet, velocity model, shot/receiver configuration, and some position of the output point). The potential resolution can only be achieved if the wavefield is properly sampled. Next to potential resolution, this chapter also uses achievable resolution, which is defined as the best possible resolution that can be achieved in practice. Events which do not satisfy the velocity model, migration noise caused by coarse sampling, and other types of noise all affect resolvability, hence the achievable resolution is not as good as the potential resolution.

How to measure temporal resolution has been the subject of various papers. In a classic paper, Kallweit and Wood (1982) discuss how various criteria (Rayleigh, Ricker, Widess criteria) can be used to describe the width of a wavelet as a measure of temporal resolution. They conclude that (potential) resolution is proportional to maximum frequency (strictly speaking, to frequency bandwidth; Knapp, 1990). In this chapter their results are extended into the realm of spatial resolution, i.e., spatial resolution is proportional to maximum wavenumber, and the minimum resolvable distance is inversely proportional to maximum wavenumber.

This chapter starts with a summary of the main points on spatial resolution as made in Beylkin et al. (1985) and applies this theory to a constant-velocity medium. This leads naturally to similar resolution formulas (for 2-D data) as given in Ebrom et al. (1995a) with an extension to offset data. In the next part, I will illustrate various aspects of spatial resolution (aperture, offset, acquisition geometry) using a single diffractor in a constant-velocity medium (the same model as used in von Seggern, 1994). The width of the spatial wavelet after migration is used as a measure in the resolution comparisons. Finally, I will discuss why sampling is important, even though the sampling interval does not appear in the resolution formulas, and I will discuss the influence of fold. A poster version of this chapter was published in Vermeer (1998a).

8.2 Spatial resolution formulas

8.2.1 Spatial resolution—The link with migration/inversion

In the literature, true-amplitude prestack migration formulas have been derived for *single-fold* 3-D data sets with two spatial coordinates ξ_1 and ξ_2, and traveltime t or frequency f as the third coordinate. The coordinates ξ_1 and ξ_2 describe the shot/receiver configuration. That is, for fixed X and fixed Y, $\mathbf{x}_s = (X, Y, 0)$ and $\mathbf{x}_r = (\xi_1, \xi_2, 0)$ describe a 3-D common-shot gather, and $\mathbf{x}_s = (\xi_1, Y, 0)$ and $\mathbf{x}_r = (X, \xi_2, 0)$ describe a cross-spread. Note that these data sets are the same data sets encountered earlier as subsets of various 3-D geometries in Chapter 2, and which are also called minimal data sets (Padhi and Holley, 1997).

Beylkin (1985) and Beylkin et al. (1985) derive formulas to compute ("reconstruct") acoustic impedance contrast as a function of position $\mathbf{x} = (x, y, z)$ from seismic measurements with limited aperture. The limited aperture is defined by the range of $\xi = (\xi_1, \xi_2)$. They show that in this process, the observed data are transformed into reconstructed data using a mapping of (ξ_1, ξ_2, f) (the coordinates of the observed data) to (k_x, k_y, k_z) (the coordinates of the reconstructed data). The mapping is given by:

$$\mathbf{k} = f\,\nabla_x \phi(\mathbf{x}, \xi), \qquad (8.1)$$

in which $\mathbf{k} = (k_x, k_y, k_z)$ is the wavenumber vector in the reconstructed (migration) domain, and $\phi(\mathbf{x}, \xi)$ is the traveltime surface (also called migration operator) of a diffractor in $\mathbf{x}$ for shot/receiver pairs described by ξ. $\nabla_x \phi(\mathbf{x}, \xi)$ represents the derivative of $\phi(\mathbf{x}, \xi)$ with respect to the point of reconstruction (output point) $\mathbf{x}$; $\phi(\mathbf{x}, \xi)$ has to be computed from the background model (velocity model).

Equation (8.1) maps the 5-D traveltime surface $\phi(\mathbf{x}, \xi)$ to 3-D wavenumber. This mapping corresponds to the fact that, in prestack migration, each input trace described by ξ is used in the reconstruction of a volume of output points (x, y, z). Equation (8.1) determines the region of coverage D_x in the spatial wavenumber domain (the 3-D spatial bandwidth). Beylkin et al. (1985) state: "the description of D_x is, in fact, the estimate of spatial resolution." The larger the region of coverage in $\mathbf{k}$, the better the potential resolution.

To further explain the meaning of equation (8.1), it is worthwhile quoting Beylkin et al. (1985) (with minor modifications to reflect the notation used in this chapter):

The mapping equation (8.1) is of fundamental importance with respect to inversion algorithms. It shows how the total domain of integration (ξ_1, ξ_2, f) on which our data are defined is related to region of coverage in the domain of spatial frequencies.

To summarize, the spatial resolution at a given point **x** defined by the region D_x depends on
1) the total domain of integration, which is determined by the configuration of sources and receivers and the frequency band of the signal, and
2) the mapping equation (8.1) of this domain into the domain of spatial frequencies, which is determined by the background model and can be obtained numerically by raytracing. This mapping is different for each point of reconstruction.

Together (1) and (2) determine the limits on spatial resolution at each point of reconstruction given the configuration of experiment and the background model.

Beylkin's formula [equation (8.1)] makes analysis of potential resolution quite simple: It should be possible to explain many resolution tests by analyzing the spatial gradients of the diffraction traveltime surfaces $\phi(\mathbf{x}, \xi)$ in the given experiment configuration.

It is not (always) necessary to analyze the full coverage in **k**. As follows from Kallweit and Wood (1982), the maximum wavenumber [corresponding to maximum gradients of $\phi(\mathbf{x}, \xi)$] can give a fair indication of resolution, provided **k** = 0 is part of the wavenumber range.

The diffraction traveltime $\phi(\mathbf{x}, \xi)$ can be described as

$$\phi(\mathbf{x},\xi) = \tau(\mathbf{x},\mathbf{x}_s) + \tau(\mathbf{x},\mathbf{x}_r) = \tau_s + \tau_r, \quad (8.2)$$

where $\tau(\mathbf{x}, \mathbf{y})$ is the traveltime from surface position **y** to subsurface position **x**. Similarly, **k** can be written as the vectorial sum

$$\mathbf{k} = \mathbf{k}_s + \mathbf{k}_r, \quad (8.3)$$

where $\mathbf{k}_s, \mathbf{k}_r$ are the contributions of shot and receiver, respectively, to the wavenumber vector **k**. It can be shown that the directional derivatives of the traveltimes τ_s and τ_r with respect to **x** are, in fact, the directions of the corresponding raypaths in **x**. Hence $\mathbf{k}_s$ and $\mathbf{k}_r$ point in the direction of the raypaths at **x** (see Figure 8.1). Each shot/receiver pair in the geometry corresponds to a point **k** in wavenumber space. Taking all shot/receiver pairs of a configuration and all frequencies leads to a collection of points which determines the region of coverage D_x in wavenumber space.

This mapping of a geometry configuration to wavenumber space is also the subject of many papers dealing in particular with VSP- and crosswell resolution analysis (Devaney, 1984; Wu and Toksöz, 1987; Goulty, 1997; Lavely et al., 1997). Goulty (1997) provides a very

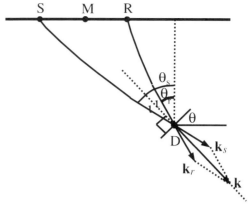

FIG. 8.1. Illumination of diffractor D by shot/receiver pair S/R. The directions of the raypaths at D determine the shot and receiver wavenumber components of total wavenumber **k**. SD and RD are also the reflection raypaths for a reflector through D with dip angle $\theta = (\theta_s + \theta_r)/2$. The raypaths make an angle $i = (\theta_s - \theta_r)/2$ with **k**.

readable description of this approach. Beylkin's formula describes this mapping in a concise way.

In zero-offset data, $\mathbf{k}_s$ and $\mathbf{k}_r$ coincide so that for this configuration |**k**| has the largest value. As a consequence, zero-offset data can produce potentially the highest resolution.

Before taking the next step, I want to mention that sampling considerations do not appear at all in the above discussion. Beylkin et al. (1985) assume, in fact, continuous variables ξ_1 and ξ_2. In other words, because, in practice, sampling is inevitable, sampling should be dense enough to allow accurate evaluation of the integrals involved in migration. The resolution that can be obtained in that case is the potential resolution as introduced earlier.

8.2.2 Spatial resolution formulas for constant velocity

It is illuminating to investigate D_x for a medium with constant velocity v and zero-offset geometry.

For a point $\mathbf{x}_s = \mathbf{x}_r = (\xi_1, \xi_2, 0)$, substitution of equation (8.2) into equation (8.1) leads to:

$$\mathbf{k} = 2(\frac{x-\xi_1}{d}, \frac{y-\xi_2}{d}, \frac{z}{d}) f/v, \quad (8.4)$$

where d is the distance from the coinciding shot and receiver to the subsurface point **x**. The vector in the parentheses is the unit vector pointing from $\mathbf{x}_s$ to **x**. The left side of Figure 8.9 depicts equation (8.4) graphically.

Now consider a 2-D zero-offset geometry laid out along the x-axis (see Figure 8.2). Then the maximum

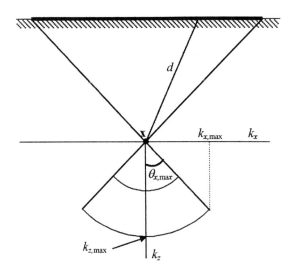

FIG. 8.2. Two-D zero-offset geometry (heavy line) laid out along x-axis with wavenumber space in subsurface point x. For each frequency, a circle arc with radius $|\mathbf{k}| = f/v$ forms the mapping of surface geometry to wavenumber space. Arcs are drawn for $|\mathbf{k}| = f_{max}/v$ and for $|\mathbf{k}| = 1/2 f_{max}/v$. Maximum wavenumber in vertical direction $k_{z,max} = f_{max}/v$, whereas maximum wavenumber in x-direction $k_{x,max}$ is limited by $\theta_{x,max}$.

values for k_x and k_z can be written as:

$$k_{x,max} = 2 f_{max} \sin\theta_{x,max} / v,$$
$$k_{z,max} = 2 f_{max} / v, \quad (8.5)$$

where $\theta_{x,max}$ is the angle between the vertical and the raypath from the output point to the farthest shot/receiver pair and f_{max} is maximum frequency.

Note the difference between horizontal and vertical resolution: k_x reaches its maximum for the maximum value of d in the x-direction, i.e., for a shot/receiver pair at maximum distance from $\mathbf{x}$, whereas k_z reaches its maximum for the minimum value of d, i.e., for a shot/receiver pair at zero lateral distance from $\mathbf{x}$, then $d = z$. A corollary of these observations is that horizontal resolution can be improved by using a larger migration aperture (migration radius), thus including a steeper part of the diffraction traveltime curves, whereas vertical resolution does not depend on aperture.

Kallweit and Wood (1982) show that a practical limit for temporal resolution, i.e., the minimum resolvable time interval R_t, is given by the tuning thickness of a zero-phase wavelet, which is the distance between peak and first trough (Rayleigh criterion). For a Ricker wavelet they show that

$$R_t = \frac{1}{2.6 f_p}, \quad (8.6a)$$

where f_p is the peak frequency of the Ricker wavelet. For a sinc wavelet Kallweit and Wood (1982) show that

$$R_t = \frac{1}{1.4 f_{max}} = \frac{c}{f_{max}}, \quad (8.6b)$$

where f_{max} is the maximum frequency, and the proportionality factor $c = 0.71$.

Analogously, for spatial frequencies, the minimum resolvable distance in a particular direction α follows from $R_\alpha = c/k_{\alpha,max}$. Using equation (8.5), this yields

$$R_x = \frac{cv}{2 f_{max} \sin\theta_{x,max}}, \quad (8.7a)$$

and

$$R_z = \frac{cv}{2 f_{max}}. \quad (8.7b)$$

These two equations may be rewritten to provide a relation between horizontal and vertical resolution as

$$R_x = \frac{R_z}{\sin\theta_{x,max}}. \quad (8.7c)$$

This relation is also given in Denham and Sheriff (1980). With $c = 1/2$, equations (8.7a) and (8.7b) lead to the same formulas for horizontal and vertical resolution as given in Ebrom et al. (1995a). For measurements based solely on peak-to-peak or peak-to-trough distances, $c = 1/2$ is too optimistic. However, "below the tuning thickness limit, amplitude information encodes thickness variations provided the entire amplitude variation is caused by tuning effects, and amplitude calibration then permits . . . thickness calculations for arbitrarily thin beds" (Kallweit and Wood, 1982).

[A different, but questionable formula for resolution, is presented in Safar (1985) and quoted in Neidell (1995). Using the same notation as above, equation (7) in Safar (1985) reads:

$$R_x = \frac{1.4 v}{4 f_{max} \tan\theta_{x,max}}, \quad (8.8)$$

which means that unlimited resolution would be achievable with unlimited aperture.]

Using similar reasoning as for the 2-D zero-offset gather above, it follows that for a 2-D common-offset gather (acquired along the *x*-axis) the minimum horizontally resolvable distance becomes:

$$R_x = \frac{cv}{f_{max}(\sin\theta_{s,max} + \sin\theta_{r,max})}, \quad (8.9)$$

where $\theta_{s,max}$ and $\theta_{r,max}$ are the angles of the vertical with the raypaths as indicated in Figure 8.1 for the shot/receiver pair with the largest distance of its midpoint to the output point. Note that equation (8.9) also applies to a 2-D common-offset gather acquired along a line parallel to the *x*-axis. In that case, the angles are measured in the plane through acquisition line and output point.

Equation (8.9) can also be written as (see Figure 8.1)

$$R_x = \frac{cv}{2f_{max}\sin\theta_{x,max}\cos i}, \quad (8.10)$$

where $\theta_{x,max} = (\theta_{s,max} + \theta_{r,max})/2$ (i.e., the maximum dip angle illuminated by the shot/receiver pairs), and $i = (\theta_{s,max} - \theta_{r,max})/2$ (the angle of incidence of the raypaths for the maximum dip angle).

Note the similarity between equations (8.7a) and (8.10): for $i = 0$, equation (8.10) reduces to equation (8.7a). Both equations show that the maximum horizontal resolution is closely coupled to the maximum dip angle that can be illuminated.

The vertical resolution that can be reached with a COV gather can be written as

$$R_z = \frac{cv}{2f_{max}\cos i}, \quad (8.11)$$

where *i* is now the angle for the shot/receiver pair with $\theta_s = -\theta_r$ (i.e., for constant velocity, this shot/receiver pair has its midpoint located vertically above the output point). Cos *i* in equations (8.10) and (8.11) describes the NMO stretch effect, which reduces f_{max} to $f_{max}\cos i$. As a consequence, for a given midpoint range the minimum resolvable distance achievable by offset data is larger than for zero-offset data (i.e., resolution is best for zero-offset data).

Equations (8.5)–(8.11) are also valid for media with $v = v(x, y, z)$, not just for constant velocity. The geometry of the raypaths at the subsurface point **x** fully determines the orientation of $\mathbf{k}_s$ and $\mathbf{k}_r$. In the formulas, *v* is the local velocity in **x**. Raytracing is necessary to link the geometry of the raypaths in **x** to the acquisition geometry at the surface.

Comparison of equation (8.10) with equation (8.11) shows that vertical resolution is better than horizontal resolution. Potential horizontal resolution also depends on the maximum illumination angle $\theta_{x,max}$, which in its turn depends on the choice of migration radius.

Before discussing spatial resolution measurements, I would like to make a link with discussions on migration stretch (Tygel et al., 1994; Levin, 1998). Figure 8.1 illustrates that each shot/receiver pair corresponds to a wavenumber vector **k**, which is normal to the plane illuminated by the shot/receiver pair. For a plane dipping in the *x*-direction with angle θ, $\mathbf{k} = (k_x, k_y, k_z) = 2f/v$ (sin θ cos *i*, 0, cos θ cos *i*), where *i* is the angle of incidence. The factor 1 / (cos θ cos *i*) is sometimes called the migration stretch factor, or vertical pulse distortion (Tygel et al., 1994). Similarly, the factor 1 / (sin θ cos *i*) might be called the horizontal pulse distortion. The larger θ, the larger k_x, hence the better the horizontal resolution; $\theta_{x,max}$ is determined by the range of input data, or, what is about the same, the migration radius. As argued in Levin (1998), the pulse distortion as a function of θ is only an apparent distortion, because the magnitude of **k** in the θ direction is not affected by it. Only the cos *i* factor (NMO stretch factor) affects all components of **k**, and means a reduction in resolution in all directions. An extensive discussion of these insights is given in Levin (1998).

A corollary of the discussion in the previous paragraph is that the vertical pulse distortion is not a good measure on which to base any migration stretch limitation. A distinction must be made between the sin θ effect and the sin *i* effect. The migration stretch limit should only depend on the NMO stretch factor and should not include the dip stretch effect.

8.3 Spatial resolution measurements
8.3.1 Procedure for resolution analysis

Next, I will illustrate various issues relating to resolution based on a model consisting of a single diffractor **d** = (0, 0, 500) in a constant-velocity medium with velocity = 2500 m/s. The source wavelet is a Ricker wavelet with peak frequency f_p = 50 Hz. The same model and isotropic source wavelet was used in von Seggern (1994). The starting point is a modified version of von Seggern's equation (1), which was derived from equation (21) of Cohen et al. (1986):

$$f(\mathbf{x}) = \iint d\xi_1 d\xi_2 h(\mathbf{x},\xi) p[\phi(\mathbf{x},\xi) - \phi(\mathbf{d},\xi)], \quad (8.12)$$

where *f*(**x**) is the image in **x**, *p*[*t*] is the source wavelet, and *h*(**x**, ξ) is the Jacobian of coordinate transformation corresponding to equation (8.1). The traveltime surface of the actual diffractor, the data, is φ (**d**, ξ) [equation (8.2)], whereas φ (**x**, ξ) is the traveltime surface of a diffractor in the output point, i.e., the integration path; $p[\phi(\mathbf{x}, \xi) - \phi(\mathbf{d}, \xi)]$ picks the value of the wavefield at the correct point in the source wavelet. Amplitude factors normally occurring in the migration formulas cancel in this case as the output point is close to the actual diffractor (von Seggern, 1994).

In von Seggern (1991), it was shown that, for a point scatterer, migration of surface data recorded with a Ricker wavelet as a source pulse produces a Gaussian spatial wavelet in the horizontal directions, but maintains the Ricker wavelet in the vertical direction. Figure 8.3 displays the source wavelet and the corresponding Gaussian along the same scale. The Gaussian represents the ideal horizontal wavelet.

In the following I will concentrate on measurements of the width of the spatial wavelet in the horizontal direction, this width being representative of the minimum resolvable distance in that direction.

8.3.2 2-D resolution in the zero-offset model

For a varying line length, a constant sampling interval of 25 m, and using coinciding shots and receivers along the *x*-axis, Figure 8.4 displays the amplitude of a horizontal trace at the depth of the diffractor (500 m). The maximum amplitude of all traces has been normalized to one. The ideal spatial wavelet is also displayed. It virtually coincides with the wavelet found for a line length of 6000 m. Figure 8.4 shows that limiting the line length (migration aperture width) leads to wider spatial wavelets. This wavelet stretch is an expression of the horizontal pulse distortion introduced earlier.

I will now introduce a measure of width of the various wavelets by defining the width of the ideal wavelet as 12.5 m (horizontal line in Figure 8.4). Figure 8.5 tests the hypothesis that this width is representative of the maximum wavenumber and of the spatial resolution. The squares indicate the measured widths of the wavelets shown in Figure 8.4, whereas the drawn line represents predicted widths according to

$$w = \frac{v}{4 f_p \sin\theta_{x,\max}}. \qquad (8.13)$$

The choice of proportionality factor 1/4 ensures $w = 12.5$ m for $\sin\theta_{x,\max} = 1$. According to equation (8.7a) the right-hand side of equation (8.13) is proportional to minimum resolvable distance ($f_{\max}$ is proportional to f_p). The near-perfect agreement between measured width and predicted width confirms the hypothesis.

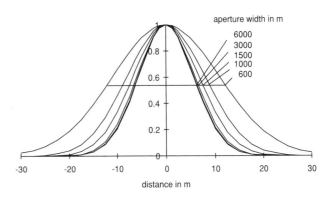

FIG. 8.4. Horizontal resolution in a 2-D zero-offset geometry for various apertures and a diffractor in (0, 0, 500). Starting with the widest, the wavelets correspond successively to aperture widths 600, 1000, 1500, 3000, and 6000 m. The horizontal line in the center of the figure indicates the level at which widths have been measured for Figures 8.5 and 8.6 (width of ideal wavelet is 12.5 m).

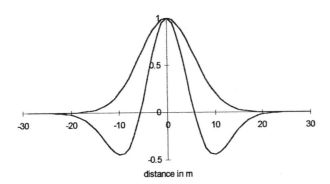

FIG. 8.3. The basic spatial wavelets used in this chapter. The Ricker wavelet and the Gaussian wavelet have been drawn for a peak frequency of 50 Hz and a velocity of 2500 m/s. The Gaussian wavelet is the narrowest achievable bell in prestack migration for the horizontal coordinates.

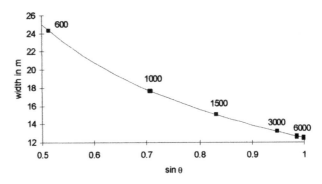

FIG. 8.5. Widths of spatial wavelets shown in Figure 8.4 plotted against sin θ, with θ being the maximum angle between diffractor and shot/receiver pairs. Each square is labeled with its corresponding aperture width. The drawn curve corresponds to equation (8.13).

8.3.3 2-D resolution in the offset model

In Figure 8.6, the results of different offset experiments have been brought together. As in Figure 8.5, the widths of the spatial wavelets are measured at the same normalized value (squares), and also computed on basis of a modification of equation (8.9) (solid curves):

$$w = \frac{v}{2f_p(\sin\theta_s + \sin\theta_r)}. \quad (8.14)$$

Each curve represents the results for a single midpoint range. In this case, the agreement between predicted value and measured value is not as good as for the zero-offset data in Figure 8.5. However, the main trends are caught reasonably well, with increasing discrepancies for increasing line lengths.

For line length 2500 m, the width of the spatial wavelet tends to decrease with increasing offset. For even wider apertures, the width becomes even smaller than the ideal width (12.5 m) corresponding to the input wavelet. I suspect that this is caused by nonlinear effects for large apertures. Line lengths of 2500 m and more are unrealistically long compared to the depth of the diffractor at 500 m. This causes distortion of the wavelet.

8.3.4 Asymmetric aperture

In the previous sections, the diffractor was placed at the center of the midpoint range. It is of interest to investigate what happens for an asymmetric configuration that may occur along the edge of a survey. Also, in single-fold 3-D data sets with limited extent (such as the cross-spread or a 3-D common-shot gather), the resolution may depend on the position of the output point with respect to the center of the data set.

Figure 8.7 describes a series of zero-offset experiments with constant midpoint range (500 m) and varying position of the diffractor. Figure 8.8 shows the resulting spatial wavelets for these experiments. The ideal spatial wavelet is also shown. The widest wavelet is obtained for the symmetric aperture (diffractor 1), whereas diffractors 2 and 3 lead to the "better resolution" represented by the next two wavelets. (Actually, resolution is better for reflectors dipping toward the left, but reflectors dipping toward the right are less well resolved.) The spatial wavelet for diffractor 3 is virtually the same as for a symmetric experiment with line length 1000 m (cf., Figure 8.4). With diffractor 3, we deal with a perfect one-sided operator producing a response, which (at least in the actual diffraction point) is identical in shape (but half its true amplitude) to the response that would have been

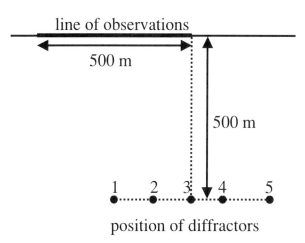

FIG. 8.7. Geometry for the asymmetry test.

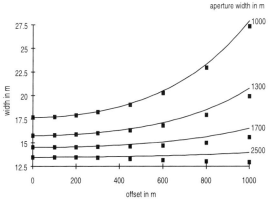

FIG. 8.6. Widths of spatial wavelets as a function of offset for line lengths 1000, 1300, 1700 and 2500 m. The drawn curves correspond to equation (8.14).

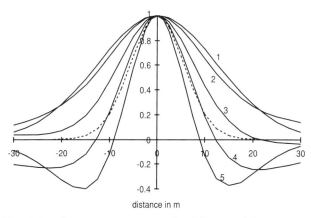

FIG. 8.8. Asymmetry test results. The spatial wavelets have been computed for the five diffractors shown in Figure 8.7, but have been plotted on top of one another for easier comparison. The width of the central loop becomes progressively smaller for diffractors 1–5. All curves are virtually symmetrical. The ideal spatial wavelet is drawn as a dashed line for reference.

obtained had the line extended also 500 m in the other direction.

For even larger aperture angles (diffractors 4 and 5) the central lobe continues to become smaller, at the expense of developing side lobes. For these diffractors, $k_x = 0$ does not occur in the wavenumber range anymore, leading to incomplete spatial wavelets.

These results reveal a limitation of the resolution analysis using the spatial wavelet of a diffractor only as measured along the horizontal through the diffractor. Analysis of the full image would show its asymmetry for asymmetric input (Margrave, 1997). Mapping the configuration in the wavenumber domain would also show the asymmetry.

8.3.5 3-D spatial resolution

Up to this point I have discussed spatial resolution results for 2-D input only. Next, I will compare the resolution of different minimal data sets. For a fair comparison, the midpoint areas of the different configurations are equal to 1000 × 1000 m in all experiments. The diffractor is chosen in the center of the configuration at a depth of 500 m. Figure 8.9 shows the wavenumber spectra (computed from Beylkin's formula) for four different minimal data sets for two different input frequencies. The four boxes all have the same scale and, for ease of comparison, the positions of two corresponding points are indicated. The zero-offset wavenumber spectrum lies on a sphere with radius $|\mathbf{k}| = 2f / v$ [cf., equation (8.4)]. For the wavenumber spectra of the other minimal data sets, $|\mathbf{k}| \leq 2f / v$, because of the NMO stretch effect. The 1000-m offset spectrum is strongly asymmetric; it is much wider in the crossline direction than in the inline direction.

It is interesting to note that a single input frequency gives rise to a wide range of horizontal wavenumbers, including $k_x = 0$ and $k_y = 0$. This should not be taken to mean that a single frequency is sufficient for optimal horizontal resolution (Vermeer, 1998a). It just means that the given midpoint range allows resolution in a wide range of directions (cf., Figure 8.1). For good resolution, it is still necessary to have a broad input spectrum, leading to a broad range of wavenumbers in all those directions which have been illuminated by the range of input data. In other words, it is necessary that a volume of wavenumbers is generated by the measurement configuration, rather than only a surface as is the case with a single frequency.

The maximum vertical wavenumber $k_{z,max}$ of the zero-offset data, the cross-spread data, and the common-shot data is reached in the center of the plot: $k_{z,max} = 2f / v$. For $f = f_{max}$, this value gives an upper limit to the potential vertical resolution of any data set. Note, however, that the cross-spread and the 3-D shot reach this high value only for an output point right below the center of the data set. For output points away from the center, the maximum vertical wavenumber will be smaller, with correspondingly smaller potential resolution. The value at the center for the 1000-m offset data can be derived from equation (8.11) (and $R_z = c / 2k_{z,max}$), and equals $2f / (v\sqrt{2})$. The maximum value of k_z is somewhat larger in this case and is reached some distance from the center (see Figure 8.9, second display from the left).

The projections on the horizontal wavenumber plane of the wavenumber spectra shown in Figure 8.9 are drawn in Figure 8.10. The spectrum for 600-m offset is included as well. Figure 8.10 allows the prediction of the outcome of resolution tests for the five minimal data sets. The zero-offset section shows the broadest wavenumber range, followed by the 600-m offset data. Note the strong asymmetry of the spectrum for the 1000-m offset data. The 1000-m offset, the cross-spread, and the 3-D shot all have the same maximum wavenumber along the k_x-axis. This does not mean that these three data sets all have the same resolution in x. The maximum wavenumber as a

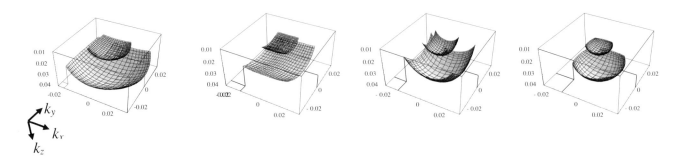

FIG. 8.9. Wavenumber spectra for four minimal data sets. All data sets have the same 1000 × 1000 m midpoint area with the diffractor in the center. The surfaces correspond to two constant input frequencies. From left to right: zero-offset gather, 1000-m COV gather, cross-spread, and 3-D shot.

function of k_y also plays a role. Maximum k_x does not vary as a function of k_y for the cross-spread, but it becomes smaller for the 1000-m offset gather and the 3-D shot; smallest for the 3-D shot.

Figure 8.11 shows the results of the numerical computation of the spatial wavelets for the five minimal data sets discussed in Figure 8.10. For ease of comparison, the wavelets are not shown in an areal sense; only the wavelets for the *x*-coordinate are shown. The wavelet for the 1000-m offset data acquired in the *y*-direction $[(h_x, h_y) = (0, 500)]$ nearly coincides with the wavelet for the inline 600-m offset. This confirms once more that the resolution of the COV gather is better in the crossline direction than in the inline direction. The sequence of wavelet widths shown in Figure 8.11 is predicted by the wavenumber ranges shown in Figure 8.10.

The worst potential resolution is obtained for the 3-D shot. At first sight, this might be surprising, because the diffraction traveltime surfaces as we know them are steeper for a common shot than for a zero-offset gather. However, this is the behavior of the diffraction traveltime curves on input, as a function of midpoint (x, y), whereas Beylkin's formula says that spatial resolution depends on the steepness of the traveltime curves as a function of the output coordinates.

The results of Figure 8.11 confirm that the maximum wavenumber is not sufficient to predict the resolving power of a 3-D data set. Rather than the maximum wavenumber, it is the average maximum wavenumber taken for all k_y that turns out to determine the resolution in *x*. This can be understood by realizing that the result of the 3-D experiment can be considered as the average of the results of many 2-D experiments, each 2-D experiment consisting of data with constant *y*. The 2-D data with the largest *y* have a maximum k_x which is (usually) smaller than the data with $y = 0$ and hence produce a wider spatial wavelet. Mathematically, the spatial wavelet of the whole 3-D data set is the normalized sum of the spatial wavelets of the contributing 2-D data sets.

The spatial wavelets shown thus far have all been normalized to the same maximum value to allow comparison of their relative widths. However, the discrimination against noise is also important. To get an idea about resolving power in the presence of noise, Figure 8.12 shows the "true amplitude" spatial wavelets for which no normalization has taken place. The small peak value and the relatively large tail value of the 3-D shot suggest that this configuration also scores worst as far as noise suppression is concerned. This aspect of geometry comparison is not pursued further in this chapter.

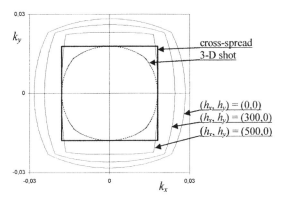

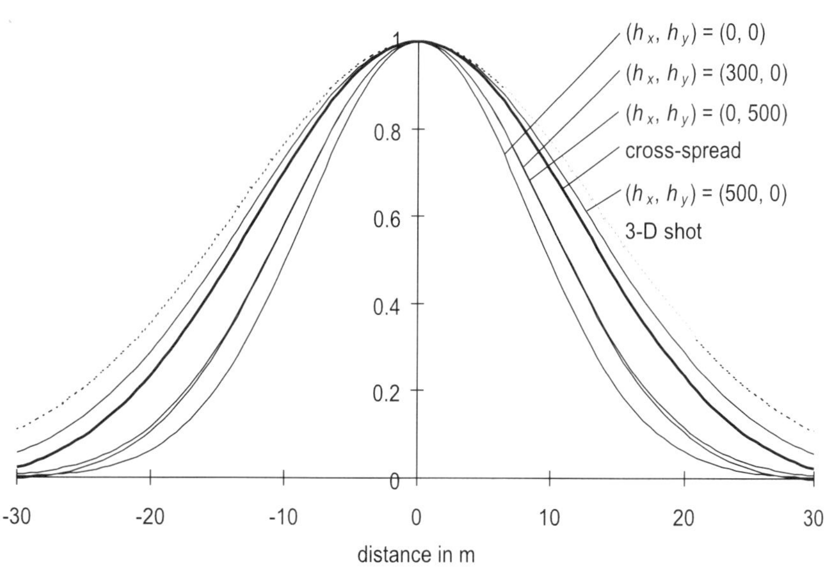

FIG. 8.10. Coverage in the horizontal wavenumber domain by five different minimal data sets with the same 1000 × 1000-m midpoint area. The largest wavenumbers are reached for the zero-offset section; hence, this section has the best spatial resolution.

FIG. 8.11. Spatial wavelets for various minimal data sets. The zero-offset gather produces the narrowest wavelet, the 3-D common-shot gather the widest. The curves for 600-m inline common offset and 1000-m crossline common offset nearly coincide. The relative widths of the wavelets confirm predictions based on Figure 8.10.

8.3.6 Sampling and spatial resolution

The formulas for spatial resolution do not contain the sampling interval, because these formulas have been derived for a continuous wavefield. If sampling takes place (which is inevitable, regardless whether we carry out modeling or field experiments), we will sample the integrands of the migration formulas such as equation (8.12). If sampling is not rapid enough to keep up with the variations of the integrand (i.e., the integrand is aliased), unreliable results are produced, and resolution will suffer (see also the next section).

Despite the obvious importance of adequate sampling, there has been much discussion on the relation between sampling and resolution (Neidell, 1994, 1995; von Seggern, 1994; Ebrom et al., 1995a, b; etc.). Some of the results even seem to indicate that resolution is not significantly impaired by coarse sampling.

Coarse sampling does not influence the resolution of some model experiments, because of the simplicity of the model. This can be illustrated with another simple experiment. In Figure 8.13, the spatial wavelets are shown for two 2-D geometries with the same line length of 1000 m, but different sampling intervals of 12.5 and 200 m. The wavelets are virtually identical except for the far end. The reason for this seemingly odd result is that the model only consists of the single diffractor. In output points close to the diffractor, the integrand in equation (8.12) varies only slowly as a function of ξ [the difference $\phi(\mathbf{x}, \xi) - \phi(\mathbf{d}, \xi)$ is a slowly varying function of ξ; the other elements in the integrand vary slowly as well]. Hence, in this case, the large sampling interval of 200 m is dense enough to follow the variations of the integrand.

A similar reasoning can be applied to the results in von Seggern, (1994, Figures 4 and 5). Those results seem to indicate even better resolution for the coarser sampling intervals, but that effect can be attributed to the fact that in that paper the effective spread length (the product of number of samples and sampling interval) of the experiments increases with increasing sampling interval.

8.3.7 Sampling and migration noise

In the previous section it was shown that coarse sampling does not have much effect on resolution as measured with a single scatterer. However, migration of coarsely sampled input data produces so-called migration noise. In this section, the relation between sampling and migration noise is investigated.

To understand the effect of sampling on the migration result (and hence on spatial resolution), it is useful to describe the migration process as a two-step procedure (see Figure 8.14). First, the data along the diffraction traveltime curves corresponding to the output point are flattened. This process converts all data contributing to that output point into a new data set, the diffraction-flattened gather, in which the diffraction produced by a diffractor in the output point is turned into a horizontal event (Figure 8.14b). A dipping event is turned into a bowl-shaped event with its apex at the position that has illuminated the output point, and with flanks that may be steeper than the dip in the input. The second step is to stack all these data into a single trace at the output point (Figure 8.14c).

The response of this second step can be described as

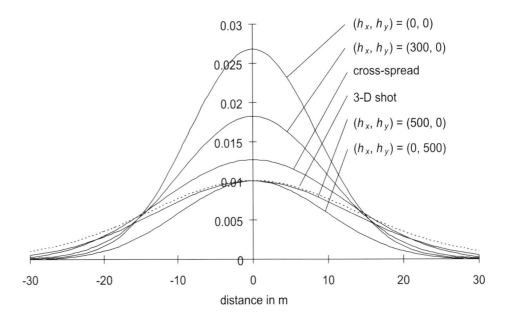

FIG. 8.12. "True amplitude" spatial wavelets for same configurations as in Figure 8.11. The two solid curves with the same maximum at 0.01 correspond to the inline and the crossline resolution of the 1000-m offset gather.

a stack operator that depends on sampling (see Figure 8.15). For regular sampling, this operator has a passband around $k = 1/d$, d being the spatial sampling interval. If the input data are coarsely sampled, it will contain energy above $k_N = 1/2d$. Then the migration operator moves some of this energy to higher wavenumbers and also to the passband at $k = 1/d$, allowing that energy to enter in the output. The stack operator of irregularly sampled data only shows a passband at $k = 0$ (d is not constant), and, hence, may better suppress energy above $k = 0$ than regularly sampled data. Therefore, random coarse sampling can be better than regular coarse sampling because it avoids the large peak in the response. On the other hand, if the input data are well-sampled, there will not be any energy moving all the way to the passband at $k = 1/d$. Instead, with regular sampling, suppression of energy in the flanks of the operator benefits from the very low response around $k = 1/2d$, whereas the reward for doubling the sampling density in random sampling is only a reduction of 3 dB in the overall response. Hence, regular dense sampling gives much better suppression above $k = 0$ than random dense sampling.

This reasoning is put to the test with the experiments described in Figure 8.16 for a horizontal event recorded by a 2-D zero-offset configuration. It shows vertical spatial wavelets with maximum amplitude normalized to one. [Equation (8.12) does not include a phase-shift correction, therefore the reflection at 500 m is no longer zero phase.] The three leftmost wavelets have been produced by migrating input data sampled at 12.5 m, 25 m, and 33.3 m. The sampling interval of the other two wavelets was 33.3 m on average with random shifts of

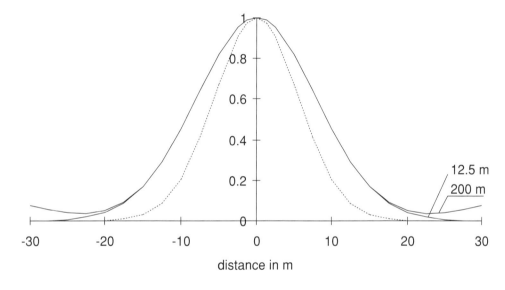

FIG. 8.13. Independence of spatial wavelet from spatial sampling. The two nearly coinciding outer wavelets correspond to five samples at 200 m and to 80 samples at 12.5 m. The narrow dotted curve is the ideal spatial wavelet. Six rather than five samples at 200 m (von Seggern, 1994) would give a narrower wavelet.

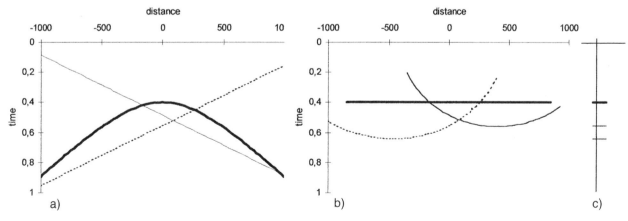

FIG. 8.14. Migration as a two-step process illustrated with 2-D zero-offset section. (a) Input showing diffraction (heavy curve) and two dipping events (thin curves). (b) In the first step, the input data are realigned according to the diffraction traveltimes in the output point. Shown is the realignment for the output point at $x = 0$, which is the position of the diffractor. (c) In the second step, the realigned data are summed (stacked) to form one output trace. The response of the second step depends on the sampling of the input data and is illustrated in Figure 8.15.

170 Chapter 8 Factors affecting spatial resolution

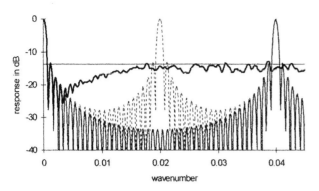

FIG. 8.15. Two-D stack responses of regular dense sampling (sampling interval 25 m, first alias band at $k = 0.04$, thin line), regular coarse sampling (sampling interval 50 m, first alias band at $k = 0.02$, dotted line), and random coarse sampling (sampling interval 50 m on average, average of 50 realizations, no passbands, heavy line). Horizontal line indicates level of random noise suppression. Note that random sampling removes strong peak(s), but cannot match rejection of regular dense sampling in the central part of the wavenumber axis.

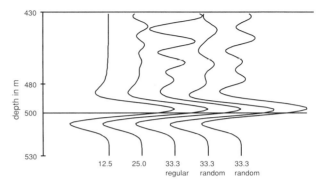

FIG. 8.16. Effect of sampling interval on migration noise for horizontal reflection. Input spatial sampling intervals are (from left to right): 12.5 m, 25 m, 33.3 m, and two random samplings with 33.3-m interval on average. The two rightmost curves (random sampling of input) show somewhat less migration noise than the central curve for which the input data were regularly sampled at 33.3 m. Note that regular sampling with a smaller sampling interval of 25 m (second curve from the left) produces less migration noise than the random input.

maximally 11.1 m on either side of the target sample points (the random shifts were generated using a uniform distribution). The figure illustrates that the event itself is (reasonably) well imaged in all cases, but that coarse sampling leads to migration noise above the event. The two rightmost wavelets illustrate the findings in Zhou and Schuster (1995) and in Zhou et al. (1999) that quasi-random sampling may reduce migration noise.

In practice (assuming that quasi-random sampling is a practical proposition, which I doubt), apparent velocities in the wavefield made up of reflections and diffractions may be larger than those of coherent ground roll events. In that case, the desired signal may be properly sampled by using a dense sampling, whereas the coherent noise is still undersampled. Under these conditions, the coherent noise would be better suppressed by quasi-random dense sampling, whereas the desired signal would be best served with regular dense sampling. This dilemma is not solved here.

The suppression of random noise, of course, is independent of the sampling regime; it would only depend on the number of samples contributing to each output sample.

8.3.8 Bin fractionation

Bin fractionation and flexi-bin are acquisition techniques for orthogonal geometries which achieve finer midpoint spacing than the natural bin size following from the shot and receiver station intervals. Figure 8.17 illustrates the bin-fractionation technique (GRI, 1994; Flentge, 1996). In the flexi-bin technique, a finer distribution of midpoints is achieved by choosing line intervals which are a noninteger multiple of the station intervals (Cordsen, 1993; Flentge, 1996). The question is, will the finer midpoint spacing lead to better resolution?

With the bin-fractionation technique, the same cross-spreads are acquired as with conventional acquisition with shot and receiver locations not staggered. The only difference are the sample positions. From the discussion in this chapter, it should be clear that potential resolution (being independent of sampling) cannot be improved with the bin-fractionation technique. If an improvement in resolution is to be achieved, it should be the result of less sensitivity to coarse sampling, i.e., bin fractionation should produce less migration noise for the same coarse sampling intervals.

The interleaving of cross-spreads using the bin-fractionation technique may be compared with the interleaving of zero-offset data sets. Two or more coarsely sampled but interleaved zero-offset data sets form a new zero-offset data set with finer sampling. The migration result of the combined data set will show less migration noise than each of the original zero-offset data sets, because their migration noises are largely in antiphase. However, overlapping and interleaved cross-spreads do not form a new and better sampled single cross-spread. Therefore, the migration noises of the cross-spreads generally will not be in antiphase with each other, and just

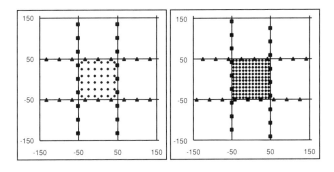

FIG. 8.17. Sampling schemes in orthogonal geometry. Left: conventional, right: bin fractionation. Squares and triangles represent shotpoint and receiver locations, respectively. Diamonds represent the midpoint positions. The distance between midpoints with bin fractionation is one quarter of the distance between the stations (in this example).

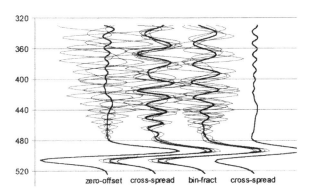

FIG. 8.18. Migration noise for different acquisition strategies, measured on a dipping event. The thin curves represent coarsely sampled configurations with a sampling interval of 33 m. From left to right: four zero-offset data sets, four regularly sampled cross-spreads, and four cross-spreads sampled as indicated in Figure 8.17 on the right. The heavy curves are the averaged results of each group of four coarsely sampled data sets. The rightmost curve is the result for a single cross-spread with 16.5-m shot and receiver station spacings. Note that bin fractionation does not lead to a significant reduction of migration noise.

reduce each other according to rules of fold. Even though the midpoint sampling has improved, the sampling of the subsurface (illumination) has not, in general, improved.

This reasoning is tested in Figure 8.18. It shows that coarsely sampled interleaved zero-offset sections lead to a significant reduction in migration noise when merged (leftmost curves). Also, a densely sampled cross-spread does not produce much migration noise (rightmost curve). On the other hand, regular coarse sampling of cross-spreads and staggered coarse sampling of cross-spreads produce similar amounts of migration noise, also after merging (central curves).

Claims (Cordsen, 1993; GRI, 1994; Flentge, 1996) that the finer midpoint sampling would lead to better resolution can be dismissed. Likewise, using a larger station spacing on the basis of bin fractionation would produce more migration noise, hence, reduce achievable resolution. These conclusions apply as much to flexi-bin acquisition as to bin fractionation.

8.3.9 Fold and spatial resolution

The analysis of spatial resolution as given in Beylkin et al. (1985) deals with single-fold 3-D data. As discussed above, it assumes implicitly that the temporal and two spatial coordinates have been sampled properly. If N-fold data are used, ideally the data can be split into N such well-sampled single-fold subsets (cf., Section 2.5). For each subset, the potential resolution can be analyzed. The resolution of the stack of the N migration results will be some average of the resolutions of the contributing subsets (in the absence of any noise that does not satisfy the velocity model; otherwise, such noises would influence the resolvability of close events). Since the best possible resolution for a given midpoint range can be obtained with a 3-D single-fold zero-offset gather, the resolution of the stack will not be as good as the resolution of that zero-offset gather. More on this subject can be found in Levin (1998), where minimal data sets are called "nonredundant data subsets."

If each contributing subset of an N-fold data set is undersampled, giving rise to migration noise for each subset, then the stack of the N single-fold migration results would reduce the noise. Now the achievable resolution (in any direction) of the stack of the N migration results should be better than the achievable resolutions of the contributing subsets. Yet, even with very large N, resolution cannot become better than the limit imposed by the maximum frequency in the input data. In an interesting physical modeling experiment, Markley et al., (1996) show that fold improves resolution of coarsely sampled data, but that the result cannot match the resolution of well-sampled single-fold data.

8.4 Discussion

All observations and conclusions in this chapter have been derived for a simple constant-velocity model. As such they provide valuable insight into various factors affecting spatial resolution, but what about more complex models? In my opinion, the results of this chapter can be used as a first-order approximation to more complex situations. In case of doubt about the applicability to more complex models, I recommend application of Beylkin's formula to those models. The main requirement is that the diffraction traveltimes can be computed for the given velocity model and measurement configuration (acquisition geometry and source wavelet). To avoid that fold will confuse the issue, it is important to investigate resolution for separate minimal data sets.

The treatment of resolution in this chapter did not include the effect of errors in the velocity. Of course, velocity errors will lead to mispositioning of the data, and velocity errors are likely to affect resolution as well (Lansley, 2000).

Theoretically, the best possible resolution (the potential resolution) cannot be improved by better sampling, because it already assumes perfect sampling. This truism applies to any measurement model, not just to the simple model investigated in this chapter. However, it tends to be overlooked in discussions on the relation between sampling and resolution. Neidell (1997) denies the truism: "According to the Huygens's approach, achievable resolution can be increased almost without limit if we increase the redundancy of the wavefield sampling." Indeed, redundancy may increase achievable resolution by reduction of noise and a more accurate evaluation of the migration integrals, but the limits set by Beylkin's formula (maximum frequency of the source wavelet and steepest time dips in the diffraction traveltime surfaces) cannot be tresspassed.

On the other hand, Beylkin's formula only sets limits on the range of wavenumbers. How this translates into minimum resolvable distance depends on the proportionality factor c. If amplitude information can be used [see remark following equation (8.7b)] or if additional information is available [e.g., well information (Levin, 1998), or smoothness of an interface], c may be considerably smaller than the value 0.71 following from the Rayleigh criterion. This elusiveness of c might be the reason for much confusion in resolution discussions.

The nature of the surface seismic acquisition technique causes a difference between vertical and horizontal resolution. It also causes a difference between the wavelets. In our case, the Ricker wavelet remains a Ricker wavelet in the vertical direction, but it turns into a Gaussian in the horizontal directions. Different wavelets lead to different resolution measurements (Kallweit and Wood, 1982). This difference leads to a complication when trying to compare horizontal and vertical resolution on the basis of measurements of the width of the main lobe of the wavelet. I have dodged this issue by comparing only wavelets in the horizontal direction for various situations; I only looked at the vertical direction to investigate migration noise. Beylkin's formula is available to compute the range of wavenumbers in (k_x, k_y, k_z)-space allowing a comparison of those ranges in x, y, and z.

The results for the bin-fractionation technique show that the sampling of the minimal data sets of the geometry (cross-spreads in this case) determines the achievable resolution, and not the sampling density of the midpoints. On the other hand, increasing the midpoint sampling density of the zero-offset gathers did help, because now the midpoint sampling also determines the sampling of the minimal data set. This raises an interesting question about some intermediate situations. In marine streamer acquisition, the fold-of-coverage is smaller than the number of different offsets (for single streamer, and source interval equal to or larger than the group interval). This means that each offset is undersampled, and full single-fold coverage can only be achieved by combining two or more neighboring offsets. Would the migration noise produced by the merged common-offset gathers be similarly reduced as for the zero-offset gather in Figure 8.17, or would it be more like the results for the two sets of cross-spreads shown in that figure? I suspect that the merged gather is close enough to a minimal data set to benefit from the denser midpoint sampling, but this needs confirmation by further research.

In the treatment of resolution for 3-D acquisition, I assumed perfect minimal data sets as input. In practice, perfect MDSs across the whole survey area do not exist. Instead, pseudominimal data sets may be found which constitute a more or less good approximation of ideal MDSs (see Chapter 10: common-offset gathers with discontinuous azimuths in multisource, multistreamer acquisition; offset-vector tiles in orthogonal geometry). These pMDSs suffer from spatial discontinuities, which produce irregularities in the diffraction traveltime curves used in migration. To what extent the spatial irregularities of these pMDSs influence achievable resolution is a matter of further research.

8.5 Conclusions

In this chapter, I have linked the description of spatial resolution given in Beylkin et al. (1985) to the more heuristic approach to spatial resolution as given in, for example, Ebrom et al. (1995a). The simple resolution formulas that apply to 2-D data provide a lower limit to the minimum resolvable distance that can be achieved with 3-D data.

Potential resolution (theoretically best possible resolution for a given geometry and a correct velocity model) is determined by the spatial gradients of the diffraction traveltime curves and the source wavelet. Beylkin's formula links these gradients to spatial wavenumbers.

Surface seismic data produce spatial resolutions which are different in the horizontal and vertical directions. In this chapter, only constant-velocity models have been investigated. For those models, horizontal resolution is determined mainly by aperture of the seismic experiment and by the maximum frequency in the source wavelet. The horizontal resolution also depends on the seismic experiment configuration: For the same range of midpoints, common-offset data have lower potential resolution than zero-offset data, and in the inline direction, resolution of common-offset data is lower than in the crossline direction. Cross-spreads have better potential resolution than 3-D common-shot gathers, but have, in general, worse resolution than common-offset gathers. This puts some ranking on the corresponding acquisition geometries. The vertical resolution does not depend on aperture, but does depend on maximum frequency and offset.

Potential resolution assumes perfect sampling. Sampling influences the correctness of the migration process to a large extent, because sampling is a way of approximating the migration integration formulas as derived for continuous shot and receiver variables. Invalid migration results are obtained as soon as the integrand in those formulas varies more rapidly than sampling can follow, i.e., as soon as the data are aliased along the integration paths.

Migration noise (caused by coarse sampling) can also be reduced by using quasi-random sampling instead of regular sampling. However, as dense regular sampling would minimize migration noise, quasi-random coarse sampling cannot match the quality obtainable with regular dense sampling.

Staggered sampling of the acquisition lines (the bin-fractionation technique) produces a denser sampling of midpoints, but it does not compensate for coarse sampling.

Noise in the data will reduce the achievable resolution. Therefore, increasing fold will virtually always improve achievable resolution, even though generally it would not improve potential resolution. This applies to noise in the form of ambient noise, ground roll, and multiples, as well as to migration noise caused by coarse sampling.

All results and conclusions are based on investigations using a simple constant-velocity model. As such it provides some valuable insights, which might also apply to more complex models.

References

Bednar, J. B., 1996, Coarse is coarse of course unless. . . : The Leading Edge, **15**, 763–764.

Berkhout, A. J., 1984, Seismic resolution—A quantitative analysis of resolving power of acoustical echo techniques: Geophysical Press.

Beylkin, G., 1985, Imaging of discontinuities in the inverse scattering problem by inversion of a causal generalized Radon transform: J. Math. Phys. **26**, 99–108.

Beylkin, G., Oristaglio, M., and Miller, D., 1985, Spatial resolution of migration algorithms: *in* Berkhout, A. J., Ridder, J., and van der Wal, L. F., Eds., Proc. 14th Internat. Symp. on Acoust. Imag., 155–167.

Bleistein, N., 1987, On the imaging of reflectors in the earth: Geophysics, **52**, 931–942.

Cohen, J. K., Hagin, F. G., and Bleistein, N., 1986, Three-dimensional Born inversion with an arbitrary reference: Geophysics, **51**, 1552–1558.

Cordsen, A., 1993, Flexi-bin 3-D seismic acquisition method: Can. Soc. Expl. Geophys. Ann. Mtg. Abstracts, 19.

Denham, L. R., and Sheriff, R. E., 1980, What is horizontal resolution?: 50th Ann. Internat. Mtg., Soc. Expl. Geophys., Reprints.

Devaney, A. J., 1984, Geophysical diffraction tomography: IEEE Trans. Geosci. Remote sensing, GE-22, 3–13.

Ebrom, D., Li, X., McDonald, J., and Lu, L., 1995a, Bin spacing in land 3-D seismic surveys and horizontal resolution in timeslices: The Leading Edge, **14**, No. 1, 37–40.

Ebrom, D. A., Sekharan, K. K., McDonald, J. A., and Markley, S. A., 1995b, Interpretability and resolution in post-migration time-slices: 65th Ann. Internat. Mtg., Soc. Expl. Geophys., Expanded Abstracts, 995–998.

Flentge, D. M., 1996, Method of performing high resolution crossed-array seismic surveys: US Patent 5 511 039.

Gibson, R. L., Tzimeas, C., and Lavely, E., 1998, Optimal seismic survey design for optimal imaging and inference of elastic properties: 60th Conf., Eur. Assoc. Geosc. and Eng., Extended Abstracts, ACQ1.6.

Goulty, N. R., 1997, Crosswell seismic reflection imaging of field data sets: The need for migration: First Break, **15**, 325–330.

GRI (Gas Research Institute), 1994, Staggered-line 3-D seismic recording: A technical summary of research conducted for Gas Research Institute, the U.S. Department of Energy, and the State of Texas by the Bureau of Economic Geology, The University of Texas at Austin.

Kallweit, R. S., and Wood, L. C., 1982, The limits of resolution of zero-phase wavelets: Geophysics, **47**, 1035–1046.

Knapp, R. W., 1990, Vertical resolution of thick beds, thin beds, and thin-bed cyclothems: Geophysics, **55**, 1183–1190.

Lansley, M., 2000, 3D seismic survey design: A solution: First Break, **18**, 162–166.

Lavely, E., Gibson, R. L., and Tzimeas, C., 1997, 3-D seismic survey design for optimal resolution: 67th Ann. Internat. Mtg., Soc. Expl. Geophys., Expanded Abstracts, ACQ2.2, 31–34.

Levin, S. A., 1998, Resolution in seismic imaging: Is it all a matter of perspective?: Geophysics, **63**, 743–749.

Margrave, G. F., 1997, Seismic acquisition parameter considerations for a linear velocity medium: 67th Ann. Internat. Mtg., Soc. Expl. Geophys., Expanded Abstracts, ACQ2.6, 47–50.

Markley, S. A., Ebrom, D. A., Sekharan, K. K., and McDonald, J. A., 1996, The effect of fold on horizontal resolution in a physical model experiment: 66th Ann. Internat. Mtg., Soc. Expl. Geophys., Expanded Abstracts, ACQ2.2, 36–38.

Neidell, N. S., 1994, Sampling 3-D seismic surveys: A conjecture favoring coarser but higher-fold sampling: The Leading Edge, **13**, 764–768.

——— 1995, Round Table, Is "coarse" the right course? Reply: The Leading Edge, **14**, 989–990.

——— 1997, Perceptions in seismic imaging, Part 4: Resolution considerations in imaging propagation media as distinct from wavefields: The Leading Edge, **16**, 1412–1415.

Padhi, T., and Holley, T. K., 1997, Wide azimuths—Why not?: The Leading Edge, **16**, 175–177.

Safar, M. H., 1985, On the lateral resolution achieved by Kirchhoff migration: Geophysics, **50**, 1091–1099.

Shin, Y., Higginbotham, J. H., and Sukup, D. V., 1997, A new viewpoint on aliasing seismic data for imaging: 59th Conf., Eur. Assoc. Geosc. and Eng., Extended Abstracts, A035.

Sun, Y., Schuster, G. T., and Sikorski, K., 1997, A quasi-Monte Carlo approach to 3-D migration: Theory: Geophysics, **62**, 918–928.

Tygel, M., Schleicher, J., and Hubral, P., 1994, Pulse distortion in depth migration: Geophysics, **59**, 1561–1569.

Vermeer, G. J. O., 1995, Round Table, Is "coarse" the right course?: The Leading Edge, **14**, 989–990.

——— 1998a, Factors affecting spatial resolution: The Leading Edge, **17**, 1025–1030, 1161.

——— 1998b, Discussion on: "A quasi-Monte Carlo approach to 3-D migration: Theory," Sun, Y., Schuster, G. T., and Sikorski, K., authors: Geophysics, **63**, 1475.

——— 1999, Factors affecting spatial resolution: Geophysics, **64**, 942–953.

von Seggern, D., 1991, Spatial resolution of acoustic imaging with the Born approximation: Geophysics, **56**, 1185–1202.

——— 1994, Depth-imaging resolution of 3-D seismic recording patterns: Geophysics, **59**, 564–576.

Wu, R-S, and Toksöz, M. N., 1987, Diffraction tomography and multisource holography applied to seismic imaging: Geophysics, **52**, 11–25.

Zhou, C., and Schuster, G. T., 1995, Quasi-random migration of 3-D field data: 65th Ann. Internat. Mtg., Soc. Expl. Geophys., Expanded Abstracts, 1145–1148.

Zhou, C., Chen, J., Schuster, G. T., and Smith, B. A., 1999, A quasi-Monte Carlo approach to efficient 3-D migration: Field data test: Geophysics, **64**, 1562–1572.

Chapter 9
DMO

9.1 Introduction

In this chapter the paper "DMO in arbitrary 3-D acquisition geometries" is reproduced (Vermeer et al., 1995). It describes the result of research carried out in 1992 within the context of Shell Research's project "Fundamentals of 3-D seismic data acquisition." The initial project results pointed to advantages of using wide geometries when an orthogonal geometry was chosen for the 3-D survey. However, there was a general feeling that DMO would produce best results for narrow-azimuth geometries. For instance, Beasley and Klotz (1992) wrote, "For DMO purposes, good offset distribution within each azimuth range should be a survey design goal," and: "... wide-azimuth surveys should be higher in fold than narrow-azimuth surveys to avoid artifacts from applying 3-D DMO." In 1988, den Rooijen had written a Shell report which advocated the use of narrow-azimuth geometry because of DMO. Because this prescription did not fit in with the budding theory of 3-D symmetric sampling, den Rooijen and I set out to investigate DMO in cross-spreads, and soon we found that DMO can indeed be applied successfully in those single-fold data sets.

In 1989, Padhi published the theory of DMO in cross-spreads and other minimal data sets in a Shell Oil research summary. Also in that summary the term minimal data set was introduced. Unfortunately, the significance of that work was not recognized on our side of the ocean, so we had to reinvent the wheel. Collins (1997a, 1997b, submitted to Geophysics in April 1994) elaborated on Padhi's work and Padhi and Holley (1997) provides a simplified version of Padhi's original paper.

Quite independently, Pleshkevitch (1996) also published a paper on DMO in cross-spread.

The paper by Vermeer et al. (1995) is reproduced in Section 9.2, since it fits in well with the general theme of this book: the link between acquisition geometry and imaging. The sampling problems discussed in Section 9.2.7 became the subject of a paper presented at the 1996 EAGE Conference (Vermeer, 1996). The results of a variety of programs of applying DMO to a synthetic data set consisting of some dipping events and also a horizontal event in a cross-spread were shown at that conference. The expanded abstract of that paper, supplemented with some figures shown in the oral presentation is reproduced as Section 9.3.

This chapter is rounded off with an epilogue. It discusses the reaction of the contractor's world to the problems discussed in the EAGE paper, and considers the use of pseudominimal data sets for an improved application of DMO and subsequent velocity determination.

The justification for putting this much emphasis on the application of DMO to single-fold data sets lies in the quality improvement that can be obtained. If good methods are available for applying DMO to single-fold data, one does not have to rely on multifold to iron out the artifacts. Eventually, it might even allow data acquisition with lower fold than would be necessary otherwise.

9.2 DMO in arbitrary 3-D acquisition geometries

9.2.1 Summary

Section 9.2 provides a theory for the application of 3-D dip moveout (DMO) to data with varying shot-to-receiver offsets and azimuths.

We will derive a general expression for the DMO-corrected time of a plane, dipping event in a constant-velocity medium. Inspection of that expression shows that DMO can be applied successfully to 3-D single-fold subsets of arbitrary 3-D acquisition geometries, provided those subsets are sampled alias-free.

We will illustrate this for the data of a cross-spread, the single-fold basic subset of the orthogonal geometry. In this data set the midpoints of the traces contributing to an output point fall along a hyperbola in the (x, y)-plane. The hyperbola contains exactly one shot/receiver pair that has illuminated the footpoint of the normal-incidence ray at the output point. This footpoint is found through DMO.

Correct sampling of the hyperbolas is difficult to achieve. Therefore, even in regularly sampled data sets, the result of 3-D DMO for data with varying shot-to-receiver azimuths is usually suboptimal.

9.2.2 Introduction

The dip moveout (DMO) operator is intended to correct for reflection point smear of traces in the same midpoint gather. It was originally devised for pure 2-D data, which are acquired with shots and receivers located on the same straight line. The extension to marine 3-D was straightforward insofar as common-offset gathers are also common-azimuth gathers. But, as far as we know, no satisfactory theory has been published that justifies the application of the DMO process to more general 3-D acquisition geometries, in which shot-to-receiver azimuths vary over all possible values. Despite this lack of a theoretical basis, 3-D DMO is often applied to land seismic data with surprisingly good results. For instance, Forel and Gardner (1988) demonstrated that 3-D DMO deals adequately with synthetic data having random azimuth variations. Similarly, Yao et al. (1993) used a physical model to show that the operation can handle wide azimuth data as well as narrow azimuth data. Those results and actual processing practice call for a theory of 3-D DMO in arbitrary geometries.

A noteworthy feature of 2-D DMO is that it "works" for single-fold common-offset gathers. For each output point, every gather always contains a single trace that has illuminated the same point on the reflector as the normal-incidence trace for the output point (assuming dense sampling). All other traces in the gather either contribute to the zone of stationary phase around that trace, or cancel each other along the flanks of the output operator. Here we show that a similar process operates in other single-fold data sets, particularly in the cross-spread.

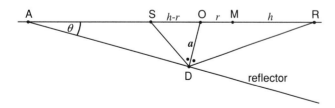

FIG. 9.1. Geometry of a plane dipping reflector. Putting d = OA and with OS : OR = SD : RD = AS : AR, it follows that $(h-r)(h+r)=dr$.

On our way to this result we will first derive a general expression for the time of a DMO-corrected event. This expression allows us to postulate a general criterion for successful 3-D DMO in single-fold 3-D data sets. Then we will derive, for the cross-spread, the locus of midpoints that contribute to an output point. We will prove that the single-fold cross-spread data are suitable for imaging with DMO. Finally, we will make some remarks about other geometries and discuss some sampling-related problems. In our work, we assume the reflectors have a constant dip in a constant velocity medium.

9.2.3 The time of a DMO-corrected event

Figure 9.1 illustrates the reflection point smear that DMO is supposed to correct. The shot/receiver pair (S, R) records a reflection from the depth point D which is posted at the midpoint M. The DMO-operation has to move the reflection to the normal-incidence point O and give it the normal-incidence time of O. As the subsurface dip is unknown, DMO is an imaging process in which all traces that can contribute to a particular output point O are moved to that point after application of the DMO-correction. We will call the collection of traces contributing to the output point a DMO panel.

If 3-D DMO is to be successful in imaging an event, then (in analogy to 2-D DMO) the DMO panel should meet three conditions:

1) It should have a point of stationary phase

2) The DMO-corrected time in that point should be equal to the normal-incidence time for the output point

3) The DMO panel should contain a well-sampled collection of traces around the point of stationary phase

We want to establish in this paper which single-fold data sets may produce DMO panels that satisfy these conditions. To achieve this, we first compute the DMO-corrected time for a trace moved from M to O (Figure 9.2). Suppose the subsurface contains a dipping reflecting plane:

$$n_x x + n_y y + n_z z - a = 0, \quad (9.1)$$

where, $(n_x, n_y, n_z) = (\sin \theta_0 \cos \varphi_0, \sin \theta_0 \sin \varphi_0, \cos \theta_0)$, and θ_0, φ_0 are reflector dip and azimuth. The normal-incidence reflection time in the origin O is

$$t_0 = 2a/c, \quad (9.2)$$

where c is the constant propagation velocity.

We now consider a shot at (x_s, y_s) and a receiver at (x_r, y_r) with corresponding midpoint (x_m, y_m) and half-offset $(h_x, h_y) = ((x_r - x_s)/2, (y_r - y_s)/2)$. The reflection traveltime is given by

$$t = \frac{2}{c}\left[(a - n_x x_m - n_y y_m)^2 - (n_x h_x + n_y h_y)^2 + h^2\right]^{1/2}, \quad (9.3)$$

where $h = \sqrt{h_x^2 + h_y^2}$. Application of the NMO-correction with dip-independent velocity c leads to

$$t_n = \frac{2}{c}\left[(a - n_x x_m - n_y y_m)^2 - (n_x h_x + n_y h_y)^2\right]^{1/2}. \quad (9.4)$$

As indicated in Figure 9.2, the shot/receiver pair can only contribute to the DMO output in the origin if the shot-to-receiver segment passes through O. We introduce r and φ: r is the distance from O to M; $r > 0$ if O lies between M and S, and $r < 0$ if O lies between M and R; $|r| < h$; φ is the angle measured from positive x-axis to $\overrightarrow{OR}$. Then, expressing x_m, y_m, h_x and h_y in terms of r, h, and φ and expressing n_x, n_y in terms of θ_0 and φ_0, we get

$$t_n = \frac{2}{c}\left[(a - r\sin\theta)^2 - h^2\sin^2\theta\right]^{1/2}. \quad (9.5)$$

Here θ is the apparent dip along azimuth φ; it is defined by $\sin\theta = \sin\theta_0 \cos(\varphi - \varphi_0)$.

The DMO-corrected time t_d is found by multiplying the NMO-corrected time t_n by the DMO-correction factor (Deregowski, 1982)

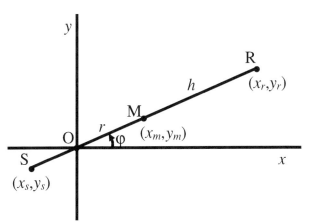

FIG. 9.2. Contributing source/receiver pair for output point in the origin.

$$t_d = t_n\sqrt{1 - \left(\frac{r}{h}\right)^2}. \quad (9.6)$$

Combining equations (9.5) and (9.6) yields the DMO-corrected time at the output point

$$t_d = \frac{2a}{c}\left[1 - \left(\frac{ar + (h^2 - r^2)\sin\theta}{ah}\right)^2\right]^{1/2}. \quad (9.7)$$

Note that this formula is valid for arbitrary shot-to-receiver offsets and azimuths, the only requirement being that the output point lie on the shot-to-receiver segment. Equation (9.7) shows that the DMO-corrected time at an output point for a plane, dipping event is always less than or equal to the true normal-incidence time at that point. The DMO-corrected time is *equal* to the normal-incidence time only if

$$ar + (h^2 - r^2)\sin\theta = 0. \quad (9.8)$$

In the case $\theta = 0$, the solution of equation (9.8) is $r = 0$, i.e., midpoint and normal-incidence point coincide for a horizontal reflector and also for shooting along strike. Otherwise, equation (9.8) can be written as $(h - r)(h + r) = d|r|$, where $d = a/|\sin\theta|$ (see Figure 9.1). This is exactly the same equation as that for a shot/receiver pair that has its reflection point at the footpoint of the normal-incidence ray at the output point. This equivalence expresses the property that DMO removes reflection-point dispersal (Deregowski, 1982).

Equation (9.8) can be solved only if r has a sign opposite to that of $\sin\theta$. Therefore, in order for DMO to image the normal-incidence event for positive and negative dips, the DMO panel must contain traces on both sides of the output point.

Now we postulate that the main criterion for a successful 3-D DMO in single-fold 3-D subsets of arbitrary acquisition geometries is that the subsets should be properly sampled. Proper sampling allows construction of DMO panels for each output point with the property that the midpoints of the traces in the DMO panel are distributed along a smooth curve passing through the output point. This curve is called the locus of contributing midpoints.

Basically, our criterion of proper sampling means that the spatial variables vary smoothly in the data set. This ensures that somewhere along the locus equation (9.8) is satisfied (the point of stationary phase), whereas elsewhere the DMO-corrected events follow a smoothly varying curve according to equation (9.7) (zone of stationary phase and flanks).

A formal proof of the suitability of certain 3-D single-fold data sets for 3-D DMO would include a description of allowable locations of the output point. Here we restrict ourselves to proving that equation (9.8) has a solution for the single-fold cross-spread.

9.2.4 Contributing traces in cross-spread

The cross-spread is a 3-D single-fold data set consisting of all traces that have a shot line and an orthogonal receiver line in common. For the time being, we consider the cross-spread as a continuous data set, i.e., shots and receivers occupy all positions along the acquisition lines.

The DMO panel at an output point O in a cross-spread consists of those traces whose shot-to-receiver segment passes through that point. Take any shot S along the shot line and connect it with O (Figure 9.3). The line SO intersects the receiver line at R. The corresponding midpoint M is a point on the locus of contributing midpoints, provided O lies between R and S, i.e., only midpoints in the same quadrant of the cross-spread as the output point can contribute to this point. Taking O as the origin of our coordinate system and (X, Y) as the center of the cross-spread, we can describe the locus by the equation

$$(x_m - \frac{X}{2})(y_m - \frac{Y}{2}) = \frac{XY}{4}. \tag{9.9}$$

Hence, the DMO panel is formed by traces whose midpoints lie on an orthogonal hyperbola passing through O, with asymptotes halfway between O and the shot and receiver lines.

Using equation (9.9), $X = \rho \cos \beta$, $Y = \rho \sin \beta$, and geometric relations (see Figure 9.3), one can parameterize r and h of a midpoint on the locus in terms of the shot-to-receiver azimuth φ:

$$r(\varphi) = \rho \, \frac{\sin(\beta + \varphi)}{\sin 2\varphi}, \tag{9.10}$$

and

$$h(\varphi) = \rho \, \frac{\sin(\beta - \varphi)}{\sin 2\varphi}. \tag{9.11}$$

9.2.5 The DMO-corrected time in the cross-spread

Equations (9.7), (9.10), and (9.11) together describe the time $t_d(\varphi)$ of the dipping event in the DMO panel. We rewrite equation (9.7) as

$$t_d(\varphi) = t_0 \sqrt{1 - \gamma^2(\varphi)}, \tag{9.12}$$

where

$$\gamma(\varphi) = \frac{ar + (h^2 - r^2)\sin\theta_0 \cos(\varphi - \varphi_0)}{ah}. \tag{9.13}$$

Substitution of the expressions for $r(\varphi)$ and $h(\varphi)$ into equation (9.13) gives

$$\gamma(\varphi) = \frac{-a\sin(\varphi + \beta) + \rho \sin 2\beta \sin\theta_0 \cos(\varphi - \varphi_0)}{a \sin(\varphi - \beta)} \tag{9.14}$$

The stationary point φ_s of the DMO panel, defined by

$$\left. \frac{\partial t_d(\varphi)}{\partial \varphi} \right|_{\varphi = \varphi_s} = 0, \tag{9.15}$$

is attained for $\gamma(\varphi_s)$ 0, i.e., $t_d(\varphi_s) = t_0$. We find

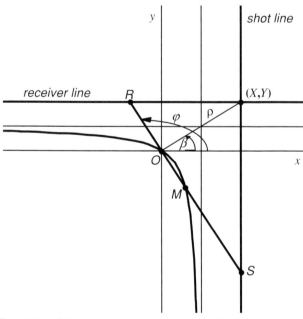

FIG. 9.3. Orthogonal cross-spread acquisition geometry. Locus of contributing midpoints of an output point O is a hyperbola through O.

$$\tan \varphi_s = -\frac{Y(a - 2n_x X)}{X(a - 2n_y Y)} = \qquad (9.16)$$

$$-\frac{\sin \beta (a - 2\rho \cos \beta \sin \theta_0 \cos \varphi_0)}{\cos \beta (a - 2\rho \sin \beta \sin \theta_0 \sin \varphi_0)}.$$

This shows that there is a stationary point at $t = t_0$, which proves that the proper normal-incidence time can be found by the application of DMO to cross-spread data, provided the extent of the cross-spread is large enough. Note that there is only one stationary point for each output point.

Figure 9.4 shows some graphs computed along the locus of midpoints of contributing traces as a function of DMO shift r. These graphs describe the DMO operation in orthogonal geometry.

9.2.6 Extension to other geometries

We have shown that the DMO operation can be applied successfully to single-fold cross-spread data in which the locus of contributing midpoints is an orthogonal hyperbola. Our derivations can easily be extended to geometries in which shot and receiver lines cross at arbitrary angles. Then the loci are oblique hyperbolas with asymptotes parallel to the acquisition lines.

In parallel geometry the hyperbolas reduce to straight lines, provided the distance between shot and receiver lines is kept constant. Then DMO can operate in common-offset gathers with constant shot-to-receiver azimuth. However, our analysis does not cover DMO in wide multisource, multistreamer configurations, because no alias-free single-fold 3-D subsets can be constructed for those geometries.

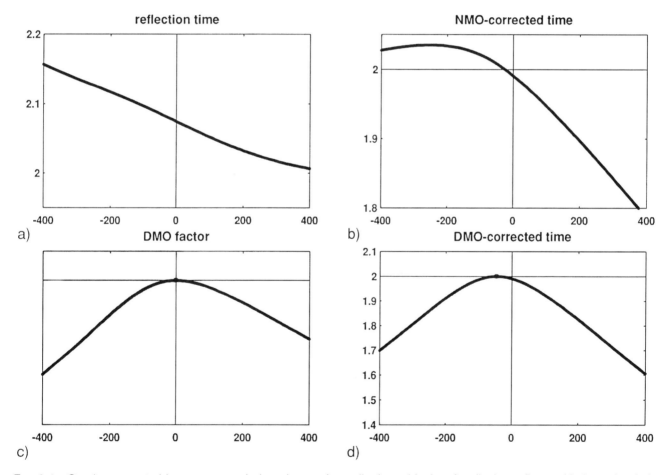

FIG. 9.4. Graphs computed in cross-spread along locus of contributing midpoints for dipping reflector. Horizontal axis is (signed) distance r from midpoint to output point. (a) Reflection time, equation (9.3); (b) NMO-corrected time, equation (9.5); (c) DMO factor, equation (9.6); (d) DMO-corrected time, equation (9.7). (Center of cross-spread $X = -500$ m, $Y = -300$ m; reflector $t_0 = 2$ s, $\theta_0 = 30°$, $\varphi_0 = 18°$).

It is not only oblique cross-spreads that are suitable for DMO; in fact, all 3-D single-fold data sets with an areal distribution of midpoints are suitable for 3-D DMO, provided they are alias free. These data sets include 3-D common-shot and common-receiver gathers as well as cross-spreads acquired with smooth, rather than straight, acquisition lines. Also the combination of a straight shot line with a feathered streamer produces a single-fold data set suitable for 3-D DMO.

9.2.7 Sampling problems

Conventional DMO programs use output bins rather than output points. All traces with a shot-to-receiver segment that cross the output bin may contribute a DMO-corrected trace to that bin (depending on the sampling along the shot-to-receiver segment). This is illustrated in Figure 9.5 for the cross-spread. In Figure 9.5a two hyperbolas are drawn, each of which is a locus of contributing midpoints for one corner of the bin. All midpoints that lie between the two hyperbolas have shot-to-receiver segments that pass through the bin; hence, they are potential contributors to the DMO panel for the bin. These midpoints are plotted in Figure 9.5b. Application of the DMO-correction to these traces leads to time jitter in the DMO panel.

An ideal DMO output would be obtained if the locus of midpoints could be properly sampled. However, resampling to obtain new samples along each and every locus of contributing midpoints is a very expensive exercise. A good compromise is to use finer sampling of the midpoints in the cross-spread. An alternative solution will be discussed in Section 9.4.1.

In marine data acquisition (with streamers) the DMO panels have to be equalized to correct for irregular sampling (Canning and Gardner, 1992; Beasley and Klotz, 1992). In land data acquisition (with the orthogonal geometry) it is not necessary to equalize DMO panels. Instead, regular alias-free cross-spreads should be acquired; alternatively, the cross-spreads should be regularized.

9.2.8 Conclusions

We derived an expression for constant-velocity DMO in arbitrary acquisition geometries. From this expression it follows that well-sampled single-fold 3-D data sets are suitable for 3-D DMO, irrespective of the shot-to-receiver azimuths in the data set. We have proved this for the cross-spread, the basic subset of the orthogonal geometry.

9.3 DMO in cross-spread: The failure of earlier software to correctly handle amplitudes

9.3.1 Introduction

The cross-spread is the basic subset of the orthogonal geometry. It is a single-fold 3-D data set, which is suit-

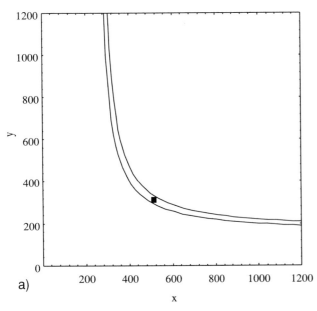
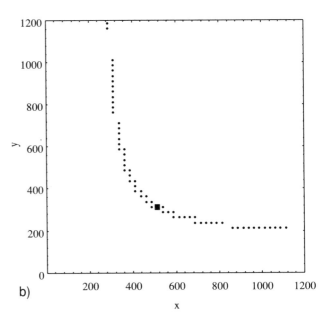

FIG. 9.5. Loci of contributing midpoints for one output bin, represented by the black square. (a) The area between hyperbolas contains all midpoints contributing to bin; (b) actual input midpoints.

able for DMO (Section 9.2). The midpoints of the traces contributing to the DMO result in an output point lie on a hyperbola (Figure 9.6). In a constant velocity medium, there is always a point on the hyperbola that has illuminated the footpoint of the zero-offset trace in the output point. This is the point of stationary phase for the output point. The other traces along the hyperbola either contribute to the zone of stationary phase or to the flanks of the DMO panel.

9.3.2 Sampling problem

The hyperbola in Figure 9.6 is computed with the assumption of continuous shot- and receiver coordinates. In actual fact these coordinates are sampled in a square grid (if shot and receiver intervals are the same). As a consequence, the hyperbola for an output point runs between the sample points. In Figure 9.7 the nearest midpoints to the hyperbola are drawn for four output points with the same x-coordinate but different y-coordinate. The corresponding shot-to-receiver segments of these points do run through the bin defined around the output point, but most segments do not run through the output point itself. Conventional 3-D DMO programs do not take these sampling problems into account. This bin smear leads to irregularly sampled DMO panels, loss of high frequencies, and erratic amplitude variations. The systematic deviation of the samples from the hyperbola in Figure 9.7a,b is typical for points close to the acquisition lines, and leads to systematic amplitude variations. Figure 9.8 shows the corresponding DMO panels (DMO-correction traces of one output point). The horizontal axis in these displays is r, the signed distance between the contributing midpoint and the output point. Note the irregular sampling of those panels.

9.3.3 Geometry effect

The DMO correction factor $\sqrt{1 - r^2/h^2}$, which squeezes the traveltimes of a trace [equation (9.6)], is not only dependent on half-offset h, but also on r. The behavior of this factor varies across the cross-spread. This leads to varying curvature of the events in the DMO panels of which Figure 9.8 is an example. Therefore, the width of the zone of data contributing to the amplitude in the output point (zone of influence, cf., Section 10.2) depends on the position of the image point with respect to the center of the cross-spread. So, even if it would be possible to perfectly sample the hyperbolas, there would still be a geometry effect to cope with. This effect is strongest close to the acquisition lines.

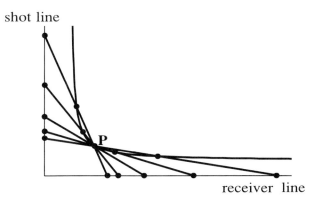

FIG. 9.6. The midpoints of shot/receiver pairs that can contribute to the DMO result in an output point P lie on a hyperbola through P.

9.3.4 Example

We sent a synthetic cross-spread to several seismic processing contractors. Their 3-D DMO results showed considerable variation, but also great similarities. Figure 9.9 illustrates the horizon slices after DMO for a horizontal reflector for six different DMO programs. If the DMO algorithm were perfect all amplitudes should be the same. However, all results have amplitudes, which show point-to-point variations caused by the sampling problem, and also variations with a longer wavelength. The latter are caused by a combination of sampling problems and the geometry effect. The result with the least jittery amplitude is due to a mixing effect of the DMO program causing loss of high frequencies.

9.3.5 The ideal 3-D DMO program

The ideal 3-D DMO program would compute traces along the hyperbola of each output point, and take the geometry effect into account. Unfortunately, proper resampling is a very expensive exercise. However, it turns out that improved results can be obtained using sampling theory (Section 9.4).

9.3.6 Conclusion

In theory cross-spreads are suitable for DMO. However, it is difficult to handle the seismic amplitudes correctly. There are two reasons for incorrect amplitudes produced by current 3-D DMO programs: shot-to-receiver segments do not pass through the center of the bins, and the cross-spread requires geometry-specific amplitude-correction factors. An alternative technique would be necessary to produce correct DMO amplitudes.

182 Chapter 9 DMO

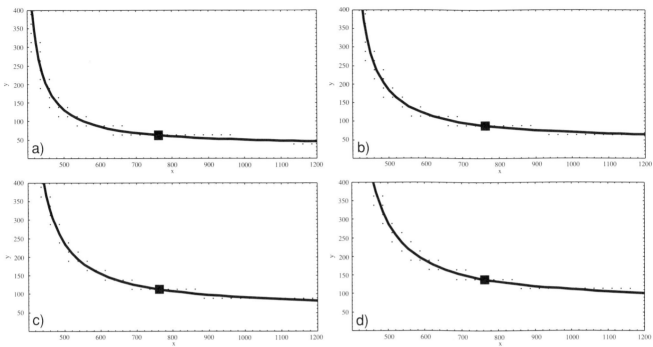

FIG. 9.7. Midpoints contributing to the bin around the output point.

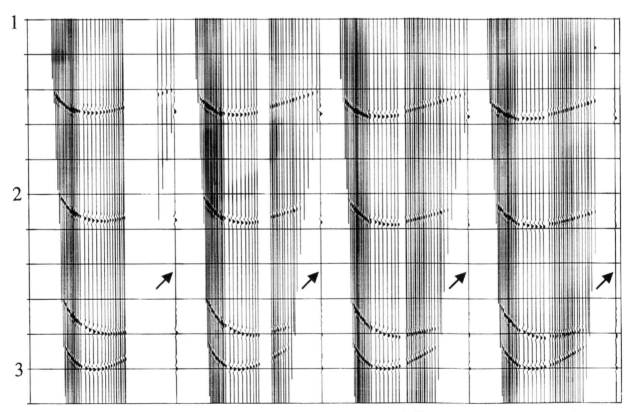

FIG. 9.8. DMO panels of synthetic data corresponding to midpoint positions in Figure 9.7. The leftmost panel corresponds to Figure 9.7a, etc. The arrows indicate the DMO output trace corresponding to each panel. Note the big gap in the leftmost panel: there are no midpoints in Figure 9.7a corresponding to this range. The irregularity of the DMO panels causes amplitude variations in the reflections and noise above the reflection events.

FIG. 9.9. Horizon amplitude map after DMO was applied to the same horizontal input reflection using different 3-D DMO programs. The bar in the top right corner of each map is caused by interference with a dipping event crossing the horizontal reflector. Amplitude differences for one program might be as large as a factor of two and standard deviation in amplitude, up to 20%. The smooth result of the center panel of the bottom row was achieved at the expense of the higher frequencies.

9.4 Epilogue

9.4.1 New DMO programs

Within a year after Figure 9.9 was shown at the 1996 EAGE Conference, various authors presented much improved results after adapting either the procedure (Cooper et al., 1996) or the DMO program (Beasley and Mobley, 1997a; Herrmann et al., 1997a).

Cooper et al. (1996) produced much improved results by first interpolating the cross-spread in the shot and receiver domains. After DMO, the data were high-cut filtered in x and y and resampled to the original sampling interval. This procedure highly reduces the loss of high frequencies and amplitude variations caused by the binsmear effect.

Beasley and Mobley (1997a,b) and Herrmann et al. (1997a,b) found an even better solution. They argued that a two-dimensional sinc function was required for every DMO-corrected trace to find its amplitude values at the surrounding bin centers. Of course, this theoretically correct solution would be horribly expensive if implemented to the full. As a compromise, they decided to use only the amplitudes of the sinc function at the four nearest neighbors. This gave a much-improved result over the earlier results. As a bonus, it turned out that this procedure produced better results for all 3-D DMO situations, because the line segment connecting shot and receiver never runs exactly through the bin centers, so that this bin-smear correction always applies.

All major contractors now offer this improved 3-D DMO algorithm as an option to their clients.

9.4.2 DMO in pseudominimal data set

Although the cross-spread is suitable for DMO, it has limited extent causing edge effects. The traces that need DMO most—the long offsets—are closest to the edges, and are most likely not completely imaged. Fortunately, the problem is less serious than it seems to be at first sight. The chances of reasonably good images are much improved by using OVT gathers as introduced earlier in Section 2.5.

In this case, OVT gathers are not really necessary for imaging the inside parts of the cross-spreads. The cross-spreads by themselves already give good images. However, the edges are more likely to produce good images when using tilings of matching OVTs. Although the locus of contributing midpoints will be discontinuous across the edges of the OVTs, continuous stretches of contributing midpoints are likely to be present in one or the other neighboring tile. The results might be further improved by applying equalized DMO (Beasley and Klotz, 1992) to the OVT gathers.

The use of OVT gathers would also come a long way toward the problem of velocity determination after DMO for data acquired with the orthogonal geometry. Each gather has a central offset, which could be taken as the representative offset of the gather. This procedure would be much better than splitting the data over fixed absolute-offset ranges, as is often done.

This section mentions some, as yet, untested but promising ideas. More work is necessary to test the ideas and improve on them.

References

Beasley, C. J., and Klotz, R., 1992, Equalization of DMO for irregular spatial sampling: 62nd Ann. Internat. Mtg., Soc. Expl. Geophys., Expanded Abstracts, 970–973.

Beasley, C. J., and Mobley, E., 1997a, Spatial dealiasing of the 3D Kirchhoff DMO operator: 59th Conf., Eur. Assoc. Geosc. and Eng., Extended Abstracts, A052.

——1997b, Spatial dealiasing of 3-D DMO: 67th Ann. Internat. Mtg., Soc. Expl. Geophys., Expanded Abstracts, SP3.7, 1119–1122.

Canning, A., and Gardner, G. H. F., 1992, Feathering correction for 3-D marine data: 62nd Ann. Internat. Mtg., Soc. Expl. Geophys., Expanded Abstracts, 955–975.

Collins, C. L., 1997a, Imaging in 3-D DMO, part I: Geometrical optics models: Geophysics, **62**, 211–224.

——1997b, Imaging in 3-D DMO, part II: Amplitude effects: Geophysics, **62**, 225–244.

Cooper, N. J., Williams, R. G., Wombell, R., and Notfors, C. D., 1996, 3D DMO for cross-spread geometry: A practical approach and application to multifold field data: 66th Ann. Internat. Mtg., Soc. Expl. Geophys., Expanded Abstracts, 1483–1486.

Deregowski, S. M., 1982, Dip-moveout and reflection point dispersal: Geophys. Prosp., **30**, 318–322.

Forel, D., and Gardner, G. H. F., 1988, A three-dimensional perspective on two-dimensional dip moveout: Geophysics, **53**, 604–610.

Herrmann, P., David, B., and Suaudeau, E., 1997a, DMO weighting and interpolation: 59th Conf., Eur. Assoc. Geosc. and Eng., Extended Abstracts, A050.

——1997b, Band limited spatial interpolation of the 3D Kirchhoff DMO operator: A key to amplitude preservation: 67th Ann. Internat. Mtg., Soc. Expl. Geophys., Expanded Abstracts, SP4.9, 1159–1162.

Padhi, T., and Holley, T. K., 1997, Wide azimuths—Why not?: The Leading Edge, **16**, 175–177.

Pleshkevitch, A. L., 1996, Cross gather data—A new subject for 3D prestack wave-equation processing: 58th Conf., Eur. Assoc. Geosc. and Eng., Extended Abstracts, P137.

Vermeer, G. J. O., 1996, DMO in cross-spread—The failure of existing software to handle amplitudes correctly: 58th Conf., Eur. Assoc. Geosc. and Eng., Extended Abstracts, X049.

Vermeer, G. J. O., den Rooijen, H. P. G. M., and Douma, J., 1995, DMO in arbitrary 3-D geometries: 65th Ann. Internat. Mtg., Soc. Expl. Geophys., Expanded Abstracts, 1445–1448.

Yao, P. C., Sekharan, K. K., and Ebrom, D., 1993, Data acquisition geometry and 3-D DMO: 63rd Ann. Internat. Mtg., Soc. Expl. Geophys., Expanded Abstracts, 552–554.

Chapter 10
Prestack migration

10.1 Introduction

The relation between acquisition geometry and imaging is of great interest as the leitmotiv of this book. First and foremost is the influence of sampling on a good imaging result. The relation is also apparent in the velocity-model updating procedure, when subsets of the data have to be selected for imaging. A good understanding of the properties of the acquisition geometry is necessary for successful true-amplitude migration. This chapter focuses on this relation between acquisition geometry and imaging. It is based on three earlier papers (Vermeer, 1998a, 1998b, and 2000).

The influence of sampling on the imaging result was already mentioned at various places in this book. In particular, Section 8.3.7 illustrated that coarse sampling generates migration noise. Section 4.5 mentioned that the size of the survey area depends on the migration distance and the zone of influence. Often, Fresnel zone is used in this context, but Fresnel zone has a very specific meaning and does not quite express the zone around the imaging point that is required for a complete image. The expression "zone of influence" is a better term for this requirement. It was introduced in Brühl et al. (1996) for modeling and can readily be extended to migration.

The process of velocity-model updating can be subdivided into two major steps: (1) the creation of images using subsets of the total data set, followed by (2) an analysis procedure to find an improved velocity model. The collection of all image traces for a given output point is called common-image gather (CIG). The analysis procedure first measures the imaged time or depth for a particular reflection. If this time or depth is the same for all images in a CIG, the velocity model is assumed to be correct (although this is not necessarily so). If the images are not horizontal in the CIG, the velocity model has to be updated.

For a successful velocity-model-updating procedure, it is essential that the images produced in step 1 are clean and do not suffer from artifacts. In parallel geometry, the obvious subset for creating CIGs is the common-offset gather. Firstly, it should produce clean images (usually a small range of offsets has to be taken as input to ensure complete coverage), and secondly, errors in velocity can be directly related to offset (Deregowski, 1990, Liu and Bleistein, 1995). Examples of horizon slices taken from migrated common-offset gathers are shown in Figure 5.12. As discussed before and shown in Figure 2.25, proper common-offset gathers cannot be extracted from an orthogonal geometry. This will pose considerable extra challenges for the velocity-model updating procedure to be used for this geometry.

It is tempting to use complete cross-spreads for imaging as each cross-spread is capable of producing clean images for a large part of the volume which it has illuminated. However, the area where clean images occur for a cross-spread is unpredictable without further analysis (it might be predicted using the current velocity model), and that area would be different for different overlapping cross-spreads. Using a tiling of adjacent cross-spreads as in Figure 2.20 would produce clean images in some places and strong artifacts in other places. A better alternative might be to use OVT gathers as described in Figure 2.21.

In the following, I will first discuss the zone of influence as an alternative to Fresnel zone. Next, I will discuss prestack migration using single MDSs, followed by a discussion of prestack migration using pMDSs. The discussion is illustrated with horizon slices of migrated dipping reflectors. For those situations where the pMDSs are not that good an approximation of COV gathers, the vector-weighted diffraction stack (Tygel et al., 1993) and the MITAS technique (Harris et al., 1998) are discussed as a means to estimate the offset of the image trace. This offset can be used in conventional migration-velocity analysis. Finally, true-amplitude migration will be discussed briefly.

10.2 Fresnel zone and zone of influence
10.2.1 Modeling

Fresnel zones were originally defined only for monochromatic waves. Brühl et al. (1996) show that the idea of *first* Fresnel zones can be readily extended to broadband data. It is defined as the area around a specular point, which leads to maximum (reflected) energy. Brühl et al. discuss Fresnel zones only from a modeling point of view, but their discussion can readily be expanded to migration (see next section). In modeling, the reflected energy is measured as a function of the radius of the reflecting circular disk. In migration, the energy of the migrated reflection can be measured as a function of the migration radius. In both situations, the energy starts from zero for zero radius, then increases to some maximum value, and next oscillates until some stable energy level is reached. In both cases the first Fresnel zone corresponds to the radius which produces the maximum energy. For modeling, this is illustrated in Figure 10.1.

The energy considered in Brühl et al. consists of a reflected wavelet from the circular disk itself, and a diffraction wavelet from the edge of the disk. Only if the radius of the disk is large enough will the two wavelets be separated, as illustrated in Figure 10.2. The length of the wavelet and the difference in traveltime between the specular point and the disk edge determine when the two wavelets are fully separated. Brühl et al. define the zone of influence as:

> ... the area on the reflector for which the difference between the reflection traveltimes and the diffraction traveltimes is less than the length Δt of the wavelet.

In the example of Figure 10.1c, the radius of the zone of influence I is a factor 1.8 larger than the radius of the Fresnel zone F.

10.2.2 Migration

The discussion in the previous section on modeling can readily be extended to migration. As a starting point, it is instructive to use the description of migration as a two-step process as discussed in Section 8.3.7 and Figure 8.14. In the first step, the whole seismic section is modified so as to flatten the diffraction traveltime curves in the output point. In this step the reflections are turned into bowl-shaped events (Figure 8.14b). In the second step, the whole modified section is stacked into one output trace (not mentioning phase shifts and weights). In this step, the apexes of the bowl-shaped events provide the image of the reflector at the output point, whereas the steeper parts of the bowls should cancel in the ideal situation (Figure 8.14c).

The question to be answered here is: When is the image of the reflector complete? Figure 10.3 provides the answer. In this figure a bowl-shaped event is enlarged to show various relationships. Let us assume for the time being that the event corresponds to a horizontal reflector. The insert in Figure 10.3 shows the behavior of the energy of the migrated result as a function of the migration radius. For zero migration radius the output energy must be zero. Then, by increasing the migration radius, constructive interference increases the total energy until a maximum value has been reached. At that point the migration radius equals the radius of the Fresnel zone in exactly the same way the maximum energy in modeling was reached for some radius of the reflecting disk. Increasing the radius further produces destructive interference, in the sense that the energy of the wavelet is reduced, until a point is reached after which an increase in migration radius does not change the energy of the image anymore, but only adds energy above the image. Hence, we can define the zone of influence for migration as:

> The zone of influence is the area in data space around the image point (point of stationary phase) for which the difference between the reflection traveltimes and the diffraction traveltimes is less than the length Δt of the wavelet.

Sun (1998) provides an insightful description of the effects of limited-aperture migration in 2-D migration. Figure 10.4 is copied from his paper. It shows the migrated image as it builds up as a function of increasing migration radius. The upward curving event in Figure 10.4 represents the truncation effect caused by the limited migration aperture. The amplitude of the image is largest in the fifth trace from the left. This position represents approximately the radius of the Fresnel zone. The radius of the zone of influence extends to the point where truncation effect and reflector image separate at about the ninth trace from the left. Sun (1999) extends Sun (1998) to 3-D.

The term Fresnel zone is universally used in the industry as a measure of how far to extend the migration radius, i.e., Fresnel zone is equated to zone of influence, whereas, in actual fact, the Fresnel zone is always smaller than the zone of influence. The zone of influence rather than Fresnel zone is to be included in 3-D survey design to define the migration fringe area.

The earlier discussion assumed a horizontal reflector. A dipping reflection is also turned into a bowl-shaped event, but now the apex is situated at the migration distance from the output point. Migration radius, as previously used, should be changed into distance from the

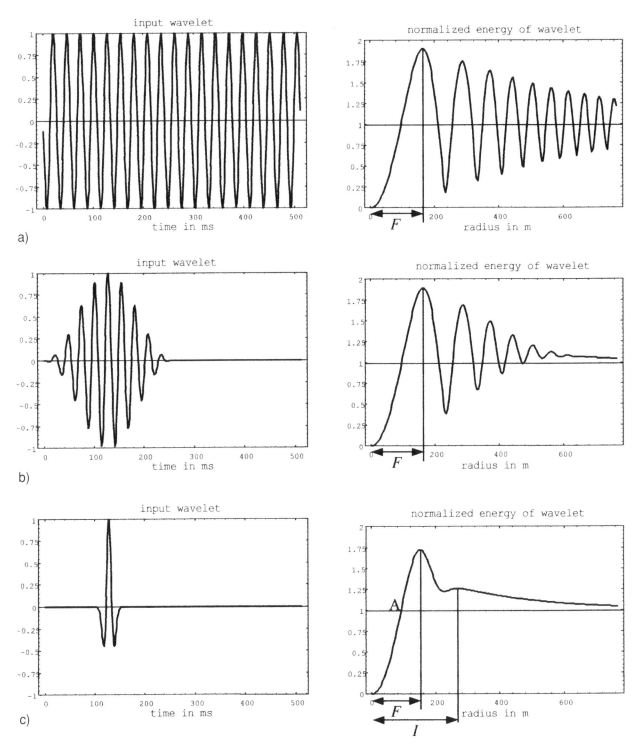

FIG. 10.1. Illustration of Fresnel zones for different wavelets. On the left the input wavelets are shown, all with a central frequency of 37.1 Hz; on the right the energy as a function of the radius of a circular reflector. The reflector depth is 1000 m, the velocity is 2000 m/s. The radius of the Fresnel zone F is in all cases defined by the maximum of the energy function. (a) Monochromatic wavelet, (b) narrowband wavelet, (c) broadband Ricker wavelet. In (c) an estimate I of the radius of zone of influence is indicated as well (modified from Brühl et al., 1996).

188   Chapter 10   Prestack migration

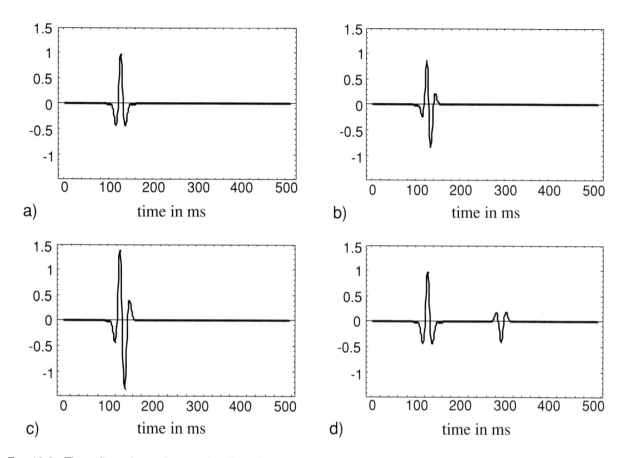

FIG. 10.2. The reflected wavelet as a function of the radius of a circular reflector. (a) Input wavelet; (b) reflected wavelet for smallest radius for which normalized energy equals 1 (point A in Figure 10.1c right); (c) reflected wavelet with maximum normalized energy, i.e., for a radius corresponding to the radius of generalized Fresnel zone; (d) reflected wavelet for a radius that is large enough to allow separation of the desired reflected wavelet and truncation effect, i.e., for a radius that is larger than the radius of the zone of influence. Note that only in (d) is the correct wavelet shape reproduced (after Brühl et al., 1996).

apex of the bowl-shaped event. Hence, to obtain a correct image of a dipping reflector, the migration radius and the migration fringe should at least be equal to the sum of migration distance and radius of zone of influence.

This discussion assumes zero-offset data. For prestack migration, it should be realized that migration distance and zone of influence also depend on offset. To determine their extent as a function of offset, raytracing may be carried out for COV gathers in parallel geometry or for various OVT gathers in orthogonal geometry. Next, the location of shots and receivers, which are required for complete imaging can be extracted from the midpoint areas corresponding to the zones of influence established for the various gathers. An even more elaborate scheme involves the use of the common focal point matrix as described in Bolte and Verschuur (2001) and Winthaegen et al. (2001).

The Fresnel zone and the zone of influence are closely coupled. They depend on the spectrum of the source wavelet, the depth of the reflector, plus the acquisition geometry. A 2-D line has a different Fresnel zone than a 3-D zero-offset section, because the interference effects are different (this is also evident from the different phase correction needed in 2-D and 3-D migration). The effect of an incomplete Fresnel zone also depends on the distribution of the missing energy, because the Fresnel zone depends on interference effects of the wavelet with itself (the zone of influence only depends on traveltime differences between reflection and diffraction). This is illustrated in Figure 10.5, which shows a horizon slice through the migration result for a horizontal reflector recorded with a zero-offset geometry. The area shown is equal to the midpoint area. On the edge of that area exactly half of the zone of influence is available for imaging. This leads to an amplitude of the event which is equal to half the amplitude as obtained with a complete zone of influence. Moving toward the inside of the area, the area of data contributing to the result increases, leading to an increased amount of energy until a maximum is reached. The distance from the edge to this point could be called the radius of the Fresnel zone for an edge. Further inside the illumination area, the amplitude drops again until a plateau is reached where the amplitude is constant. The edge of the plateau defines the radius of the zone of inluence. In the corners of the display of Figure 10.5 the interference patterns differ from the rest of the edges, because here data are missing in two directions.

10.3 Description of model experiments

In the following sections imaging results of some model experiments are shown. To avoid repetition, the parameters for the model experiments are described in this section. The model consists of a reflector with an easterly 15° dip in a 3000 m/s constant velocity medium. The depth of the reflector is approximately 3000 m in the center of the model (the horizon slice of Figure 10.8c was computed using a horizontal reflector at 3000 m in the same medium). In all models the source and receiver station intervals were 50 m, and an isotropic source emitted a 30 Hz Ricker wavelet. For orthogonal geometry, the additional parameters are 400 m source and receiver line spacings, and 2400 m maximum inline and crossline offset. Output traces were computed for a 25 × 25 m grid. The migration velocity was equal to the medium velocity.

10.4 Prestack migration with minimal data sets

By definition, all MDSs are suited for migration and capable of producing a single-fold image of the illuminated part of the subsurface. The migration result, i.e., vertical and horizontal resolution, is dependent on the

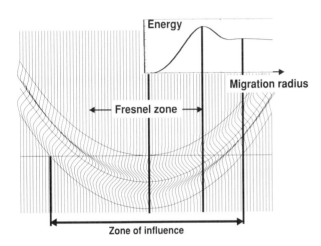

FIG. 10.3. Fresnel zone and zone of influence. A horizontal event is shown after flattening of the diffraction traveltime. Start and end of the wavelets is indicated by drawn curved lines. Data beyond the zone of influence cannot contribute to the migration result in the point of stationary phase. Data beyond the Fresnel zone is still needed for a complete image with correct phase and amplitude.

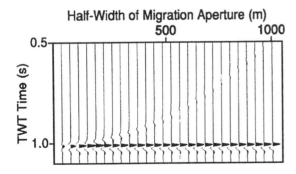

FIG. 10.4. The migrated image as a function of migration radius. Note the high amplitude of the image in the fifth trace from the left (from Sun, 1998).

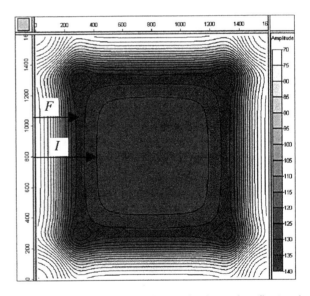

FIG. 10.5. Horizon slice for a horizontal reflector in migrated zero-offset volume of limited extent. The edge of the slice corresponds to the edge of the illumination area. Note that amplitude builds up from the edge until a maximum is reached, followed by a decrease in amplitude until a plateau value is reached. Distance F from the edge to maximum amplitude corresponds to the radius of the Fresnel zone, distance I from the edge to inner contour corresponds to radius of zone of influence.

source wavelet, the velocity model, and on the acquisition geometry. If, however, these data sets have been properly sampled, then the result will be independent of sampling (Section 8.3.6; Vermeer, 1999).

The dependence of the migration result on the acquisition geometry is illustrated with Figures 10.6 and 7, which allow comparison of illumination and imaging by a COV gather and by a cross-spread. Figures 10.6d and 7d represent the shape of the reflection traveltime surface after conversion to depth z according to the migration condition

$$(\sqrt{z^2 + s^2} + \sqrt{z^2 + r^2})/V = d/V, \qquad (10.1)$$

where s (r) is the distance from shot (receiver) position to surface position of output point (x, y, z), d is the length of the raypath from shot to receiver via the reflector, and V is the velocity of the medium. The left side of equation (10.1) represents the diffraction traveltime surface for the output point (x, y, z) as shown in Figures 10.6c and 7c; the right side of the equation represents the reflection traveltime surface across the MDS as shown in Figures 10.6b and 7b. Figures 10.6d and 7d may also be said to describe the reflection traveltime surface after flattening of the diffraction traveltime surface (i.e., the first step of migration as a two-step process described in Section 8.3.7), i.e., these data sets are diffraction-flattened gathers. Time slices through a diffraction-flattened gather of a cross-spread acquired in Nigeria are shown in Figure 7.19.

The migration result in the output point (x, y, z) is just the (weighted) horizontal summation of all data described by the contoured surfaces of Figures 10.6d and 7d. The apex of this surface in the diffraction-flattened gather corresponds to the depth of the reflector z in the output point. It is the point of stationary phase in the migration integral. The heavy curve in Figures 10.6d and 7d is a depth contour 60 m above the apex. If the length of the seismic wavelet is 60 ms (= 60 m for V = 2000 m/s), then all reflections inside the heavy curves contribute to the migration result at depth z. As discussed in Section 10.2.2, the area inside the heavy curve may be called "zone of influence." All energy outside of the zone of influence contributes only to the flanks of the migration operator. This energy should cancel in the migration summation, which it does to a large extent, provided that the data are properly sampled.

Figure 10.8 shows migrated horizon slices for different MDSs. Figure 10.8a shows the result of migrating a zero-offset section. Because a small input data set was chosen, the image shows edge effects. Figure 10.8b is the same as Figure 10.8a but now for a COV gather as input

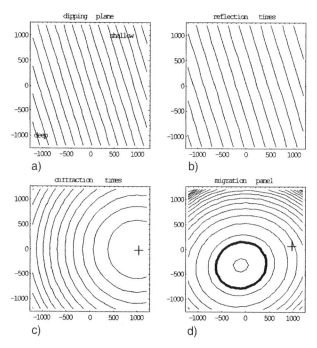

FIG. 10.6. Illumination and imaging with minimal data sets. The MDS is a COV gather with 1500-m offset and shot-receiver azimuth parallel to the x-axis. The model consists of a single 30° dipping reflector at around 2000 m in a medium with constant velocity 2000 m/s. Contour plots are shown for a 2400 × 2400 m midpoint area. (a) Depth contours, (b) reflection traveltimes, (c) diffraction traveltimes for a point R on the reflector with surface coordinates (0, 1000), (d) reflection times after diffraction-flattening in output point R. In (c) and (d) the (x, y)-coordinates of R are indicated by a +. Contour interval is 100 m in (a) and (d) (thin lines), and 100 ms in (b) and (c). The heavy line in (d) represents an extra contour at 60 m above the apex of the depth surface. It might be taken as the boundary of the zone of influence.

with offset-vector (2375, 0) m. The amplitudes are larger than in Figure 10.8a, because the zone of influence is larger for larger offset. Figure 10.8c and 8d show the migration result for a cross-spread, Figure 10.8c for a horizontal reflector, and Figure 10.8d for the same model as in Figure 10.8a. Note again the edge effects. In the interior of the horizon slice of Figure 10.8c the amplitude varies slightly because of changing offset; the interior amplitude in Figure 10.8d varies because of depth and offset variation.

True-amplitude migration (Gesbert, 2002) of the MDSs used for Figure 10.8 would have produced constant amplitudes inside the full-image areas with equal amplitude for all four examples.

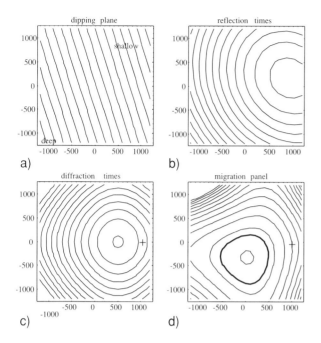

FIG. 10.7. Same as Figure 10.6, but now the minimal data set is a square cross-spread. Note the different shape of the zone of influence.

It is important to realize that the edge effects seen in Figures 10.8a and 8b can be pushed as far out as desirable by choosing a larger survey area, whereas the location of the edge effects seen in Figures 10.8c and 8d is fully determined by the maximum useful offset.

10.5 Prestack migration with pseudominimal data sets

10.5.1 Parallel geometry

Marine 3-D acquisition is most frequently carried out using multisource, multistreamer configurations. Also OBC data are often acquired with a configuration in which the shot lines are parallel to the receiver lines. In these parallel geometries the crossline component of the offset vector (= distance between source track and streamer track) is different for the various midpoint lines, which are acquired in one vessel pass. This means that common-offset gathers are not common-offset-vector gathers.

The discontinuities in the crossline offset lead to irregular illumination as illustrated in Figure 2.11 for various multisource, multistreamer configurations. Between vessel passes, large gaps in illumination may exist when shooting downdip and overlaps when shooting updip. Feathering may compound the problem, whereas antiparallel acquisition (sailing adjacent vessel passes in opposite directions) and sailing strike to the steepest dips reduce the impact of the discontinuities in crossline offset (Vermeer, 1997; Brink et al., 1997; Section 5.3.2). Figure 10.9 illustrates the behavior of the diffraction-flattened gather of a dipping event for equal inline offsets. Figure 10.9a shows the diffraction-flattened gather for ideal input, Figures 10.9b and 9c show the effect of the discontinuities in the geometry on the diffraction-flattened gather. The differences between Figures 10.9b and 9c lead to amplitude and phase variations of the migrated event.

Results of migrating single-fold data extracted from various multisource, multistreamer configurations are shown in Figure 5.12.

10.5.2 Orthogonal geometry

Orthogonal geometry poses a much larger problem to migration-velocity analysis than parallel geometry. In the first place, COV gathers cannot be assembled from that geometry, and in the second place, the MDSs of this geometry, the cross-spreads, have limited extent.

Figure 2.25 illustrates that it is impossible to generate single-fold COV gathers from orthogonal geometry. Not all offsets are present everywhere, moreover they have a wide variety of azimuths. Figure 10.10 shows a horizon slice for absolute offsets ranging from 700 to 850 m. The corresponding fold-of-coverage varies between 0 and 4. Very strong amplitude variations are the result of migrating this collection of data, which is not really suitable for image analysis.

The simplest way to generate single-fold coverage across the entire survey, i.e., a pMDS, is to make a tiling of cross-spreads (MDSs) with adjacent midpoint areas. In a regular geometry, it is possible to construct as many single-fold tilings as the fold count. However, even though the midpoint coverage of each tiling can be complete and regular in this way, the illumination of the subsurface will not be regular, because of the discontinuities in shot-receiver azimuths across the edges of the cross-spreads. This irregular illumination is illustrated in Figure 2.15. Figure 10.11a illustrates the discontinuities in the traveltime surface of the same dipping event across four adjacent cross-spreads, and Figure 10.11b shows that the diffraction-flattened gathers are discontinuous across the edges as well.

A characteristic of tiling with cross-spreads is that reflection times behave smoothly in the inside areas of each cross-spread, but may show large discontinuities from cross-spread to cross-spread. An alternative to tiling with cross-spreads is tiling with OVTs, as described in Sections 2.5.2 and 2.5.4. The advantage of

192   Chapter 10   Prestack migration

this pMDS is that there are no big jumps in shot-receiver azimuth in this data set as in the cross-spread tiling, particularly if the unit cell is small. A disadvantage is that there are a lot more edges, all of which produce discontinuities in the diffraction-flattened gathers.

Figure 10.12 illustrates imaging with OVT gathers. Figure 10.12a shows the traveltime surface, and Figure 10.12b the contours after diffraction flattening for one output point. Comparison of this figure with Figure 10.11b suggests that OVT tiling is more robust than cross-spread tiling. An additional advantage of OVT tiling is that it is easier to handle the shallow data (just apply the muting scheme as described in Section 2.6.4).

Figure 10.13 shows illumination by various OVT gathers. Except for Figure 10.13f, the reflector always dips in an easterly direction. Figure 10.13a used the OVT from the upper right corner of each cross-spread with average **h** = (1000, 1000). In this case the spatial discontinuity between the OVTs translates in vertical illumination gaps and horizontal overlaps. The reverse is the case with the OVT from the opposite side of the cross-spread shown in Figure 10.13b. The illumination by these two

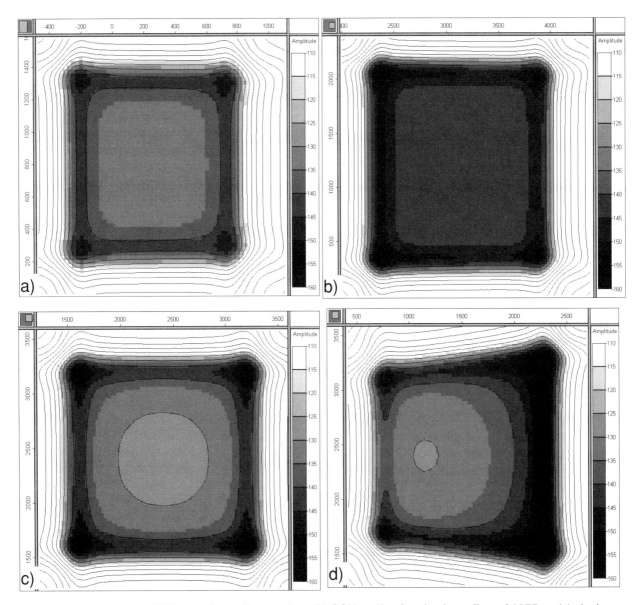

FIG. 10.8. Horizon slices for MDSs. (a) Zero-offset section, (b) COV section for absolute offset of 2375 m, (c) single cross-spread, (d) single cross-spread. All displays for reflector with 15° dip, except (c), which shows the result for a horizontal reflector.

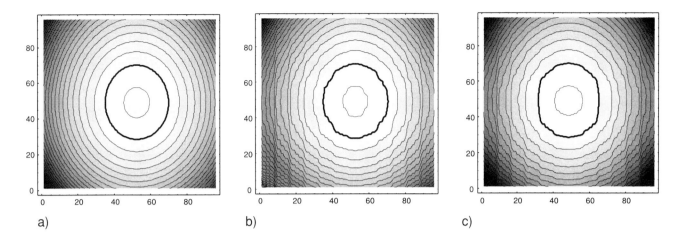

FIG. 10.9. Comparison of ideal geometry with two sources, four streamers geometry for constant-velocity medium and north-south 30° dipping reflector at around 2000 m. Shown are contours for the reflection times converted to depth according to equation (10.1) for an inline offset of 2000 m. Contour interval is 50 m, heavy contour at 60 m above the apex. (a) COV gather result, (b) two sources, four streamers geometry for output point in the middle of a vessel pass, (c) same as (b) for output point on the edge between two vessel passes. Vertical alignments in the contours are visible where the crossline discontinuities are largest.

OVT gathers is the most discontinuous of all possible gathers. It is interesting to see that their combination leads to an almost regular 2-fold illumination as shown in Figure 10.13c.

Figure 10.13d shows that illumination by complete cross-spreads is more continuous overall. However, the overlaps and the gaps are larger than in the case of the OVT gathers. Figure 10.13e and f show illumination by pairs of rectangles at the far end of the receiver line [average $|\mathbf{h}| = (1100, 0)$]. In Figure 10.13f the reflector makes an angle of 45° with the receiver line. In Figure 10.13e two-fold and zero-fold illumination alternate in thin horizontal strips, whereas everywhere else illumination is single-fold. In Figure 10.13f the irregularities are spread even more thinly.

Figure 10.14 shows migration results corresponding to Figure 10.13. Each figure shows a horizon slice through a migrated reflection. Not unexpectedly, the images show a clear correspondence to the illumination areas. Although the relative amplitude variation in Figures 10.14a and b is quite large, it is still about 50% less than in Figure 10.10. Yet, the OVTs in the far corners of the cross-spread have the largest discontinuities of all OVTs; note also that the offsets used in Figure 10.10 are much smaller than in Figure 10.14a and b. Combining the two opposite far corners gives a much improved image as shown in Figure 10.14c. It is interesting to note that the amplitude variation in Figures 10.14e and f is smaller than in Figures 10.14a and b, which means that OVT gathers composed from OVTs along the acquisition lines produce better images, hence are most suitable for application in velocity-model updating (also because the range in absolute offset in these tiles is smaller than in tiles away from the acquisition lines). The very weak amplitudes in the center of Figure 10.14d reflect the illumination gap shown in Figure 10.13d.

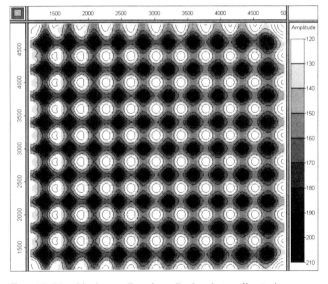

FIG. 10.10. Horizon slice for all absolute offsets in range 700–850 m of orthogonal geometry. Irregular illumination leads to strong amplitude variations in this display. Over a short distance, five contours cover a range of 100 amplitude units, as compared to an average amplitude of about 170.

194 Chapter 10 Prestack migration

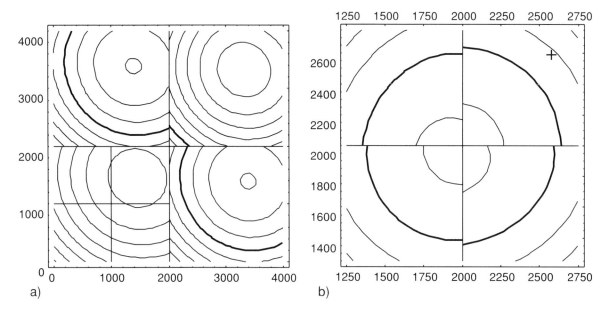

FIG. 10.11. Single-fold imaging with tiling of cross-spreads. (a) Traveltime surface of 15° dipping event across four adjacent cross-spreads, contour interval 100 ms. The heavy line represents the same contour value across the cross-spreads. Small cross indicates the axes of the lower left cross-spread. (b) Diffraction-flattened depth contours for output point at +. Contour interval is 50 m. The image points for the four cross-spreads do not coincide but lie close to the "four-corners point." The heavy contour indicates a level of 60 m above the depth in the output point. Due to the large discontinuities across the edges of the cross-spreads the amplitude in + cannot be correct.

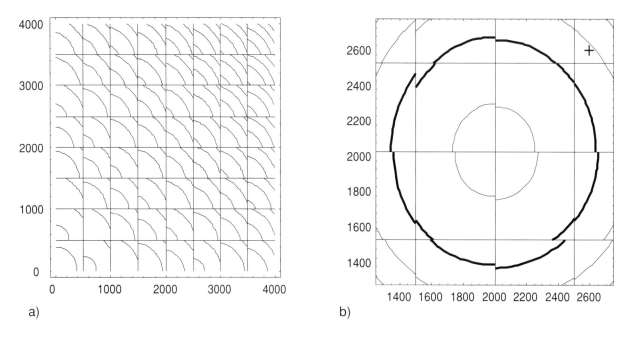

FIG. 10.12. Single-fold imaging with OVT gathers. Shown are top-right tiles 1/16th the size of a cross-spread. Same subsurface as in Figure 10.11. (a) Traveltime surface across 4×16 tiles, contour interval 50 ms; (b) diffraction-flattened contours for output point in +. Contour interval is 50 m. The image points for the offset-vector tiles do not coincide but lie close to the center of the picture. The heavy contour indicates a level of 60 m above the depth in the output point. Note the larger number of discontinuities than in Figure 10.11b, though the discontinuities in the center are smaller than in Figure 10.11b.

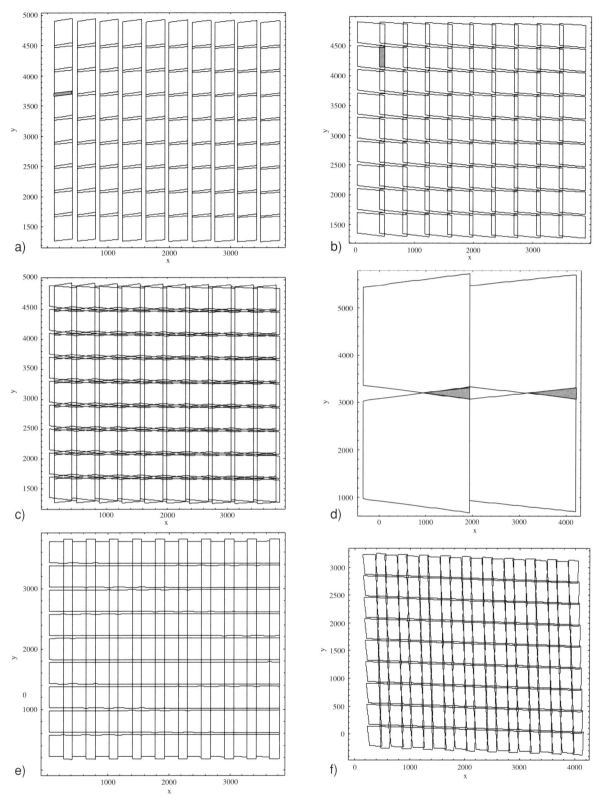

FIG. 10.13. Illumination of an east-dipping reflector by OVT gathers. (a) OVT of the upper right corner in cross-spread, (b) OVT of lower left corner, (c) superposition of (a) and (b), (d) adjacent cross-spreads, (e) rectangles from the far end of the receiver line (cf., Figure 2.22), (f) as (e) with dip azimuth of 135°. Grey areas in (a), (b), and (d) indicate overlapping illumination areas.

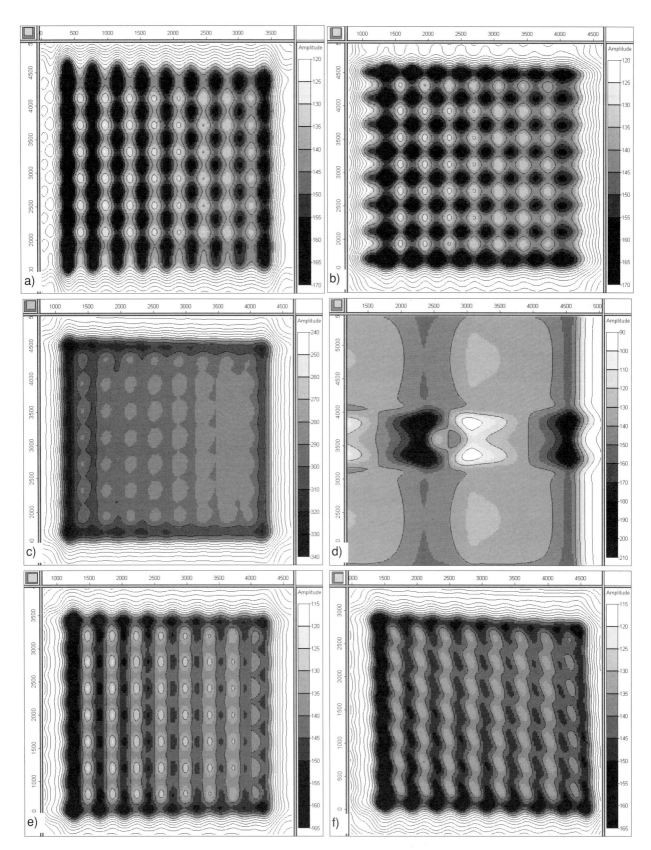

FIG. 10.14. Imaging of a reflector with 15° dip by OVT gathers. Each figure shows a horizon slice for the corresponding situation in Figure 10.13. Contour interval in (a), (b), (e), and (f) is 10 amplitude units; in the 2-fold coverage of (c), and in the adjacent cross-spreads of (d), contour interval is 20 amplitude units.

Gesbert (2002) shows that the acquisition footprints in Figure 10.14 can be reduced by true-amplitude migration taking geometry effects into account. This paper (which is recommended for further reading) also illustrates that migration smiles are reduced.

10.5.3 Irregular geometries

Often, acquisition geometry, even if nominally regular, is very irregular in practice. In other cases it may be regular, but coarsely sampled. Then it is impossible to collect properly sampled MDSs from the geometry, and the construction of pMDSs may be quite impractical. As a consequence, the conditions for good single-fold images are not met. Firstly, the zone of influence around each point of stationary phase is not well sampled, so that amplitude and phase of the image are not correct. Secondly the flanks of the migrated-depth surface are not well sampled either, leading to incomplete cancellation, i.e., migration noise, further reducing the possibility of picking reliable images. This reasoning underlines the importance of proper sampling techniques in acquisition.

However, even if it is impossible to generate reliable single-fold images, it may still be possible to obtain reasonable images from the total data set. Statistical averaging of noise and amplitude variations has to compensate for the irregular sampling. Velocity-model updating of such data has to resort to geological knowledge and velocity scanning as discussed in, e.g., Schmid and Bouska (1997). Amplitude compensation for irregularities can be achieved by true-amplitude migration (Albertin et al., 1999; Gesbert, 2002), discussed in Section 10.7.

10.6 Velocity-model updating

Whether cross-spreads or OVT gathers are used for imaging, the problem remains that the offset of the imaging trace (i.e., the trace at the point of stationary phase) is not known without further action. This is caused by the variation in offset that occurs across a cross-spread and still occurs across the OVT gather. Earlier I proposed to use the vector-weighted diffraction stack (Vermeer, 1998b) to determine the offset corresponding to each image. Tura et al. (1998) applied that technique for AVO analysis. They did not use it to determine the offset in the image (they were using common-offset gathers as input, hence knew the offset already), but to determine reflection coefficient and reflection angle. The recipe of the vector-weighted diffraction stack is given in Tygel et al. (1993) who expanded an earlier idea from Bleistein (1987). Unfortunately, the vector-weighted diffraction stack is quite sensitive to noise, because it depends on measurements made on prestack data.

A better way to find the offset in the image point might be a modification of an idea proposed in Harris et al. (1998). In their MITAS procedure they consider the volume of data being used to build a single image trace. The procedure consists of the following steps:

1) Flatten the diffraction traveltime curves in the input volume, i.e., create a diffraction-flattened volume; this will lead to bowl-shaped events for the reflections (cf. Figures 10.6d, 7d, and 7.19)
2) Stack the new volume in two orthogonal directions; this will improve the signal-to-noise ratio of the data to be analyzed
3) Determine the points of stationary phase of the major reflections in both stacks; the two points for each reflection will determine the position in the input volume of the image point

Harris et al. (1998) use this procedure to determine an area around the image point that will be included in the imaging process, whereas the data outside this area will be discarded. In this way migration aliasing noise is avoided and a cleaner image can be produced, in particular for coarsely sampled data. However, knowing the position of the image trace also means that its offset can be retrieved and be used for further analysis in the velocity-model updating procedure.

Using OVT gathers for this analysis provides the best chances for clean images and also allows the determination of the offset in the image point. Yet, the irregularities associated with the spatial discontinuities in the OVT gathers may still hamper an accurate analysis, especially if the image point is close to the edge of an OVT. To compensate for that situation, carry out the analysis as well for OVT gathers based on OVTs shifted over (*SLI*/2, *RLI*/2). Again, to minimize the amount of work to be carried out for this analysis, restrict the analysis to discrete locations and to specific target horizons.

10.7 True-amplitude, prestack migration of regular and irregular data

In this section, a synthesis is made of ideas described by Harris et al. (1998, see previous section), Albertin et al. (1999), Bloor et al. (1999), Rousseau et al. (2000), and Gesbert (2002), supplemented with some further ideas.

Albertin et al. (1999) write that for most acquisition geometries, even if acquired in a rather regular way, it is difficult to give an analytic expression of the Beylkin determinant (Bleistein, 1987), needed in true-amplitude migration. Instead, they introduce the idea of measuring the dip angles being illuminated in the output point by all the shot/receiver pairs in a data set. The dip angle illuminated by a single shot/receiver pair and its corresponding wavenumber vector is illustrated in Figure 8.1. All shot/receiver pairs together (Figure 8.2) determine the range of dips that can be illuminated by the data set. Albertin et al. (1999) propose to equalize the distribution of angles across the unit sphere in the output point by weighting according to the local density of wavenumber vectors on the unit sphere. Based on earlier work by Jaramillo (1998) and Jaramillo and Bleistein (1999), they show that this is equivalent to applying Beylkin's determinant. This true-amplitude migration technique not only corrects for irregular geometry but also for refraction effects in the overburden.

Bloor et al. (1999) apply Albertin's method to data acquired with a spider-web geometry (see Section 4.3.6). They show that this technique will lead to considerable improvement of data quality. In this application no distinction is made between data with different offsets or coming from different subsets: each shot/receiver pair in the total data set contributes its own angle and its own point on the unit sphere.

Rousseau et al. (2000) apply true-amplitude migration to the MDSs of the acquisition geometry. They illustrate this with common-offset data retrieved from a parallel geometry. Gesbert (2002) applies true-amplitude migration to a suite of single-fold data sets and shows that this reduces the acquisition footprint. The residual footprint depends on the deviation of the input data from true minimal data sets. Applying true-amplitude migration to subsets of the data makes the imaged data suitable for better AVO analysis, and does not mix up effects from widely different shot/receiver pairs. On the other hand, the low fold of an MDS may easily lead to gaps in the range of dips being illuminated. Weighting of the traces around such gaps has two effects: (1) if the gap occurs in the flat part of the bowl-shaped reflection events (after application of diffraction traveltime surface flattening, see previous section), then weighting will ensure a better amplitude of the image, but (2) if the gap occurs in the steep part of the bowl-shaped reflections, weighting of the traces will increase aliasing artifacts. This is clear from Figure 3 in Rousseau et al. (2000), where not only reflection amplitude is improved by weighting but an artifact caused by some missing inlines is enhanced as well.

The ultimate synthesis of all ideas is to use OVT gathers (pairwise, as discussed for AVO analysis) to establish the point of stationary phase using the Harris et al. (1998) method, and to apply aperture limitation around that point, followed by true-amplitude migration in the remaining area (where aliasing does not occur).

10.8 Discussion

Next to the introduction of the zone of influence as a better alternative to the Fresnel zone, this chapter has highlighted the use of pMDSs as single-fold data sets to be used in velocity-model updating and in amplitude regularization. For orthogonal geometry it seems best to choose two-fold subsets consisting of two OVT gathers with opposite offset vectors as input data sets for these applications (cf. Figure 10.14c). The acquisition footprints can be minimized by true-amplitude migration taking geometry effects into account (Gesbert, 2002). Also, gathers based on OVTs centered around the acquisition lines as illustrated in Figure 10.14e show smaller artifacts. Of course, the artifacts can be further reduced by reducing the acquisition line intervals, but this has significant impact on the cost of a survey. It should be realized that the artifacts of the OVT gathers, which are not situated along the edges of a cross-spread, are fully compensated when adding the result of migration of all input data. In the final output, only artifacts associated with the cross-spread edges remain, and these are reduced by the averaging effect of fold. The artifacts in parallel geometry are far less severe than in orthogonal geometry; this may be a reason (together with the easier processing overall) to select parallel geometry if one can afford it, and whenever possible (e.g., desert areas).

This chapter has only scratched the surface of geometry-related imaging problems. Additional work is needed to investigate the proposed velocity-model updating technique (or anything else that might work). In particular tests with erroneous migration, velocities are still called for. Yet, I hope that the interested reader is stimulated to try out some ideas and will expand on them.

References

Albertin, U., et al., 1999, Aspects of true amplitude migration: 69th Ann. Internat. Mtg., Soc. Expl. Geophys., Expanded Abstracts, SPRO11.2, 1358–1361.

Bleistein, N., 1987, On the imaging of reflectors in the earth: Geophysics, **52**, 931–942.

Bloor, R., Albertin, A., Jaramillo, H., and Yingst, D., 1999, Equalised prestack depth migration: 69th Ann. Internat. Mtg., Soc. Expl. Geophys., Expanded Abstracts, SPRO11-3, 1362–1365.

Bolte, J. F. B., and Verschuur, D. J., 2001, Applications of the 3D common focal point matrix in pre-stack data analysis for migration aperture and acquisition design: 63rd Conf., Eur. Assoc. Geosc. and Eng., Extended Abstracts, P164.

Brink, M., Roberts, G., and Ronen, S., 1997, Wide-tow marine-seismic surveys: Parallel or opposite sail lines: Presented at the 29th Annual Offshore Technology Conf., OTC8317.

Brühl, M., Vermeer, G. J. O., and Kiehn, M., 1996, Fresnel zones for broadband data: Geophysics, **61**, 600–604.

Deregowski, S. M., 1990, Common-offset migrations and velocity analysis: First Break, **8**, 225–234.

Gesbert, S., 2002, From acquisition footprints to true amplitude: Geophysics, **67**, 830–839.

Harris, C. C., Marcoux, M. O., and Bickel, S. H., 1998, MITAS, migration input trace aperture selection: 68th Ann. Internat. Mtg., Soc. Expl. Geophys., Expanded Abstracts, SP 10.7, 1373–1376.

Jaramillo, H., 1998, Seismic data mapping: Ph.D. thesis, Colo. School of Mines.

Jaramillo, H. H., and Bleistein, N., 1999, The link of Kirchhoff migration and demigration to Kirchhoff and Born modeling: Geophysics, **64**, 1793–1805.

Liu, Z., and Bleistein, N., 1995, Migration velocity analysis: Theory and an iterative algorithm: Geophysics, **60**, 142–153.

Rousseau, V., Nicoletis, L., Svay-Lucas, J., and Rakotoarisoa, H., 2000, 3D true amplitude migration by regularization in angle domain: 62nd Conf., Eur. Assoc. Geosc. and Eng., Extended Abstracts, B-13.

Schmid, R., and Bouska, J., 1997, 3D prestack depth migration and velocity analysis for sparse land data: 67th Ann. Internat. Mtg., Soc. Expl. Geophys., Expanded Abstracts, ST13.3, 1809–1811.

Sun, J., 1998, On the limited aperture migration in two dimensions: Geophysics, **63**, 984–994.

——1999, On the aperture effect in 3D Kirchhoff–type migration: Geophys. Prosp., **47**, 1045–1076.

Tura, A., Hanitzsch, C., and Calandra, H., 1998, 3-D AVO migration/inversion of field data: The Leading Edge, **17**, 1578–1583.

Tygel, M., Schleicher, J., Hubral, P. and Hanitzsch, C., 1993, Multiple weights in diffraction stack migration: Geophysics, **58**, 1820–1830.

Vermeer, G. J. O., 1997, Streamers versus stationary receivers: Presented at the 29th Annual Offshore Technology Conference, OTC8314, 331–346.

——1998a, Fold, Fresnel zones and imaging: Presented at the I/O Workshop.

——1998b, Creating image gathers in the absence of proper common-offset gathers: Exploration Geophysics, **29**, 636–642.

——1999, Factors affecting spatial resolution: Geophysics, **64**, 942–953.

——2000, A strategy for prestack processing of data acquired with crossed-array geometries: Proceedings, 20th Mintrop seminar.

Winthaegen, P. L. A., Bolte, J. F. B., and Verschuur, D. J., 2001, Time lapse seismic data selection for acquisition and processing by Fresnel zone estimation: 63rd Conf., Eur. Assoc. Geosc. and Eng., Extended Abstracts, P667.

Index

(f,k)-domain, 51, 79
(f,k)-filtering, 14, 32, 42, 71, 79, 92, 97, 144-145, 150
(f,k)-processing, 67
(f,k)-spectrum, 8, 62, 64
3-D microspread, 49, 98, 141-146
3-D subset, 1, 2, 4, 17, 18, 21-24, 28, 171, 175, 178, 180
3-D velocity filtering, 40, 49, 67, 71, 92
5-D wavefield, 1, 17-18, 20-23, 25, 43-46
acquisition footprint, 42, 84, 100, 104
acquisition imprint, 41, 110
airgun array, 108, 120, 148
Alba, 123, 137-139
alias peak, 62-63
alias protection, 27
aliasing, 2, 8, 11, 13, 16, 28-29, 32-33, 60, 63, 65, 72, 78-79, 107, 124, 142, 155, 174, 197-198
Amoco, 137
amplitude variation with direction, 44-45, 81, 116
amplitude variation with offset (AVO), 13, 40, 42, 44-46, 70-71, 81-82, 92, 94, 104-105, 107, 197-199
antiparallel acquisition, 3, 108-114, 191
apparent velocity, 32, 49-52, 62, 65, 78-79, 92, 125-126, 136, 142, 144
areal geometry, 2-4, 18-21, 27-28, 36, 70, 74, 86-87, 100, 104, 116-117, 120-121, 134-137
array
    airgun, 108, 120, 148
    areal, 27, 32, 49, 53-61, 85, 89, 146
    circular, 2, 56-58, 60
    field, 2, 9-17, 27, 28, 32, 33, 49, 52-61, 78, 79, 85, 86, 89, 98, 100, 141, 146, 149
    geophone, See array, field
    hydrophone, 116-120
    linear, 2, 9, 13, 29, 32, 49, 53-55, 60, 85, 89, 146, 148
    overlapping, 62
    receiver, See array, field
    shot, See array, field
array element, 9-14, 53-55, 60
array length, 9, 11, 13, 54, 60
array response, 11, 14, 53-56, 60-61
aspect ratio, 22, 32, 67, 100, 149
asymmetric sampling, 3, 11-15, 69, 72, 90, 92, 117, 130, 136
attribute analysis, 1, 69, 87, 90-95, 98-99, 149
AVO, See amplitude variation with offset
azimuth dependency, 2-3, 45, 46, 54, 60, 67, 70, 72, 81, 107, 116, 134, 136, 137
azimuth distribution, 12, 86, 149-150

background model, 160-161
back-scattered noise, 32, 50
basic sampling interval, 9, 27-28, 52-53
basic signal sampling interval, 27-28, 52-53, 71, 78
basic subset, 2-3, 21-22, 24, 27, 28, 36, 46, 70, 72, 75, 76, 92, 98, 99, 141, 175, 180
Beylkin determinant, 197-198
bin-fractionation technique, 4, 170, 172-173
bin size, 37-38, 84-86, 107, 127, 170
bin smear, 181, 183-184
box test, 146
BP, 120
brick-wall geometry, 3, 20-21, 25, 63, 65-67, 72, 74, 90, 98, 100, 146-148, 153, 154
Bullwinkle, 107
buried cable, 120

case history, 63, 141-158
center-spread
    geometry, 12, 88, 103, 138
    shooting, 10-14, 22, 25, 28, 29, 32, 37, 71, 88, 116, 135, 137
checkerboard effect, 14, 19, 29
Cheops pyramid, 21
Chevron, 117, 137
circle shoot geometry, 20, 74, 106
coarse sampling, 4, 25, 53, 79-80, 82, 100, 107, 142, 159-160, 168-173, 185, 199
coherent noise, 12-13, 28, 50, 61-63, 97, 107, 142, 147, 158-159, 170
common-image gather (CIG), 185
common-offset-vector (COV) gather, 21-24, 27, 29, 36, 39, 77-78, 112, 114, 126-129, 131-133, 136, 163, 166-167, 185, 188, 190-193
continuous wavefield, 1, 5, 8, 9, 14, 17-27, 37, 75, 168
converted waves, 125-138
coordinate system
    midpoint/offset, 5-7, 14-15, 21
    shot/receiver, 5-7, 11-12, 17, 21
crossed-array geometry, 18, 36, 72, 73, 136
crossline fold, 36-38, 65, 70, 84-85, 88-89, 110, 134, 149
cumulative fold, 35, 149

de-aliasing, 29
deepest horizon to be mapped, 3, 69, 75, 81
Definitions
    3-D symmetric sampling, 27
        areal geometry, 27-28
        line geometry, 28-36
    antiparallel acquisition, 108
    areal geometry, 18
    aspect ratio, 22
    basic subset, 21
    bin-fractionation technique, 170
    brick-wall geometry, 20
    CIG, 187
    circle shoot geometry, 20

common-image gather, 185
common-offset-vector gather, 21
continuous wavefield, 20
COV gather, 21
crossed-array geometry, 18
crossline roll, 88
cross-spread, 22
cross-spread geometry, 20
diffraction-flattened gather, 129
DMO panel, 176
double-zigzag geometry, 20
effective spread length, 168
flexi-bin technique, 170
full-swath roll, 89
illumination
    area, 37
    fold, 37
image
    area, 37
    fold, 37
inverted zigzag, 20
largest minimum offset, 41
level of full fold, 84
line geometry, 18
LMOS, 41
MDS, 22
midpoint/offset coordinate system, 5
migration apron, 86
minimal data set, 22
minimum maximum offset, 41
mixing, 32
multiline roll, 88
nontranslational geometry, 131
offset distribution, 104
offset sampling, 104
offset vector, 21
offset-vector tile, 37
orthogonal geometry, 18
OVT, 37
OVT gather, 39
parallel geometry, 18
pMDS, 36
prism wave, 74
proper sampling, 21
pseudominimal data set, 36
random geometry, 18
resolution
    achievable, 160
    potential, 160
seisloop geometry, 20
shot repeat factor, 90
shot salvo, 22
shot/receiver coordinate system, 5
single-line roll, 88
slanted geometry, 18
spatial discontinuity, 40

spider-web geometry, 20
swath (in land acquisition), 88
swath (in OBC acquisition), 133
symmetric sampling, 9
template, 22
total fold-of-coverage, 36
translational geometry, 131
triple zigzag geometry, 20
unit cell, 36
vector fidelity, 119
zag-spread, 21
zig-spread, 21
zigzag (single) geometry, 20
zone of influence, 129
depth of shot hole, 98
differential
    feathering, 3, 29, 32, 36, 106-107, 109-110, 112
    moveout, 33, 62, 66-67, 71, 82-84, 108
diffraction tomography, 159
diffraction-flattened gather, 129-131, 154-155, 168, 192-196, 199
Digiseis, 19, 117
dip moveout (DMO), 4, 21, 41, 43-48, 71, 82, 106, 112-113, 122, 138, 149-150, 175-184
dip or strike, 70, 103-107
dip shooting, 106-107
direct wave, 50, 53, 55, 58
DMO panel, 176-181
DMO shift, 44, 179
DMO-corrected trace, 71, 180-181, 183
double-zigzag geometry, 20, 65-67, 71-72, 84
downdip shooting, 11-12, 29-30, 113-114, 129, 135, 137, 191
dragged bottom cable, 119, 125
dual-sensor technique, 3, 103, 117-118, 121
dual-domain filtering, processing, 3, 67, 71-72, 75-76, 92-93, 150, 152
dynamite, 11-12, 20, 89, 148

edge effect, 4, 22, 25, 27, 33-34, 38-39, 42-44, 84, 97, 112, 132, 150, 154-156, 165, 183, 188, 191-196, 199-200
end-on
    geometry, 103
    shooting, 11-13, 28
energy distribution, 2, 14, 67, 79
equalized DMO, 184
Exxon, 18

feathering angle, 31, 109-110
filtering, dual-domain, 3, 67, 71-72, 74-75, 92-93, 150, 152
first-break picking, 40-42, 75
flexi-bin technique, 170
focal-beam analysis, 98
fold buildup, 33, 86, 149
fold-of-coverage, 2, 21, 36-38, 41, 62, 82, 84-85, 113, 127, 132-133, 136, 155, 172, 191
fold-taper zone, 3, 87
four-component (4-C), 3-4, 74, 103-104, 118-123, 125, 132, 134, 137-139
fracture density/orientation, 69, 118, 136
Fresnel zone, 4, 185-189, 198-199
full sampling, 18
full-resolution recording, 11
full-swath roll, 3, 33, 88-90, 97, 100, 118

gas chimney, 103, 107, 118-119, 125
Gaussian wavelet, 164, 172
geometry
    areal, 2-4, 18-21, 27-28, 36, 70, 74, 86-87, 100, 104, 116-117, 120-121, 134-137
    brick-wall, 3, 20-21, 25, 63, 65-67, 72, 74, 90, 98, 100, 146-148, 153, 154
    center-spread, 12, 88, 103, 138
    circle shoot, 20, 74, 106
    crossed-array, 18, 33, 72, 73, 136
    double zigzag, 20, 65-67, 71-72, 83
    end-on, 103
    inverted zigzag, 20
    line, 2-3, 19-24, 27-28, 32, 72, 146, 162, 188
    multiline roll, 3, 39, 90, 99-100
    narrow-azimuth, 83, 88, 175
    nontranslational, 131-132, 134
    orthogonal, 2-4, 18-202
    parallel, 2-3, 18-202
    parallel versus orthogonal, 70-71, 135-136
    random, 20-21
    seisloop, 20
    single zigzag, 20
    single-source single-streamer, 29
    slanted, 3, 18, 22-23, 70, 72, 88, 90-95, 100-101, 131
    slanted versus orthogonal, 72-73, 90-95
    slash, 72
    spider-web, 20, 74-75, 198
    target-oriented, 20, 70, 74

translational, 131
triple zigzag, 20
wide versus narrow, 2, 22, 32, 35-36, 66-67, 71, 83, 86, 89, 100, 113, 149-150, 154, 176
wide-azimuth, 21, 39, 70, 88, 176, 193
zigzag, 3, 18-24, 27, 33, 35, 42, 63, 65-67, 70-72, 83, 88, 100, 103
zigzag versus orthogonal, 63, 65-67, 70-72, 100
geometry imprint, 72, 89-90, 92, 104-105, 117
geophone array, 9-11, 13, 29, 32-33, 55-56, 58, 60-61, 85-86, 89
geophone coupling, 5, 13, 119-122
geophone string, 56
ghost, 103, 116-118
ground roll, 2-3, 5, 9, 11, 16-17, 28, 32, 40-41, 49-50, 52-53, 63, 67, 70-71, 74, 78, 80, 84-85, 93, 98, 141-142, 154, 156, 170, 173

hands-off seismic data acquisition, 17
hexagonal sampling, 27, 29, 86, 116, 136
horizon slice, 32, 45, 104, 112, 181, 188-196, 198-199

illumination
    analysis, 99, 138
    area, 3, 33, 37-38, 97, 100, 113, 127, 132-135, 154, 189, 197, 199
    fold, 34, 37-38, 84, 113, 127, 132-136, 138
    gap, 106, 112-113, 134, 193, 195, 199
image area, 37, 154, 197
image fold, 37-38, 84, 127, 156
infill shooting, 113, 115
inline fold, 37, 44, 65, 84-86, 88, 89, 149
interpolation, 28, 40-41, 79-80, 97-98, 107-108, 125, 135-136
intra-array statics, 9, 79, 85, 136, 141
inverted zigzag geometry, 20
irregular illumination, 3, 28-33, 70, 84, 103, 105, 109-113, 125, 128, 132-137, 193-195, 197

Jacobian, 163

land-type geometry, 2, 17, 69-100, 115-116
largest minimum offset (LMOS), 28, 40-41, 72, 80-81, 83, 88, 146, 149

line geometry, 2-3, 18-24, 27-28, 32, 72, 146, 162, 188
line interval (spacing), 3, 20, 25, 29, 32-33, 35-36, 69-71, 75, 81-91, 99, 112, 134, 137, 146, 150, 154, 156, 158, 170, 192, 200
line turn, 115
linear array, 2, 9, 13, 29, 32, 49, 53-55, 58, 85, 89, 146, 148
linear event, 54
low-fold
    data, 2-3, 19, 62-63, 72, 108, 141, 154-158
    migration, 112-114, 141, 154-157, 192-200
low-velocity noise, 2-3, 41, 49-51, 79

magnetotellurics, 81
make-up shots, 97-98
maximum crossline offset, 2-3, 22, 25, 27-28, 32-33, 35, 44, 69, 72, 81, 83-86, 88-90, 92, 94, 137, 146, 149, 189
maximum frequency, 7-9, 52, 70-71, 75-78, 129, 136, 142, 160, 162, 171-173
maximum inline offset, 2, 22, 27-28, 30, 32-33, 35, 44, 72, 81, 83-86, 88, 92, 137, 141, 149, 189
maximum offset, 75, 78, 81-86, 88-90, 146
maximum useful offset, 22, 27, 33, 75, 92, 191
maximum wavenumber, 7-9, 53, 76-77, 160-162, 164, 166-167
mega-bin technique, 27
Mesa, 99
microspread, 98, 141-146
midpoint coverage, 20, 36, 38, 97, 106, 116, 127, 133-134, 191
midpoint/offset coordinate system, 5-7, 21
midpoint/offset wavenumber domain, 14-15
migration
    aperture, 78-79, 83, 129, 160, 162, 164-166, 173, 186, 198-199
    apron, 4, 86
    artifact, 1-4, 39, 75, 83, 92, 97, 100, 132, 135, 137, 159, 187, 198
    distance, 45, 87, 185-186, 188
    fringe (apron), 4, 86, 190
    low-fold, 141, 112-114, 154-157, 192-200
    noise, 79, 82, 107, 112, 160, 168, 170-173, 185, 197

operator, 3, 71-72, 74, 78-79, 112, 130, 135, 160, 166, 169, 190
prestack, 1, 3-4, 17, 21, 25, 29, 37, 44-45, 71-72, 80-81, 86, 132, 154-157, 159-160, 164, 188-199
radius, 3-4, 74, 86-87, 135, 162-163, 186, 188-189
stretch, 163-164
minimal data set (MDS), 1-3, 22, 29, 36-38, 46, 72-73, 78, 97-98, 100, 125, 127-132, 154, 160, 166-167, 171-172, 175, 187, 190-194, 198-200
minimum maximum offset, 40-41, 81
minimum resolvable distance, 76, 160, 162-164, 172-173
MITAS, 185, 199
mixing, 11-12, 32, 181
Mobil, 80, 156
model-based survey design, 99
Monte Carlo technique, 159
multicomponent acquisition, 122-125, 123-124, 131-132, 138
multiline roll geometry, 3, 39, 88, 90, 99-100
multiples, 33, 49, 62-67, 70-72, 82-84, 92, 108, 118, 173
multisource, multistreamer configuration (MS/MS), 19, 30-32, 36, 63, 80, 103, 107-109, 111, 115, 172, 180, 191
muting, 40-42, 70, 81-82, 85-86, 90, 195

narrow-azimuth geometry, 83, 88, 175
nearest neighbor, 42-43, 184
Neidell's conjecture, 159
Niger delta, 148
Nigeria, 3, 80, 89, 141, 146-151, 190
NMO stretch, 3, 33, 71-72, 76-77, 81-82, 92, 163, 166
noise
    coherent, 12-13, 28, 50, 61-63, 97, 107, 142, 147, 158-159, 170
    low-velocity, 2-3, 41, 49-51, 79
    water-borne, 116
noise suppression, 1-3, 5, 13-14, 17, 28, 32-33, 49-67, 70, 75, 79, 82-85, 98, 117, 130, 146, 168-170
Noordoostpolder, 141
Nyquist, 9, 11, 79

obstacle, 3, 18-19, 87, 96-99, 103, 105-106, 115-117, 148
ocean-bottom cable (OBC), 2-3, 20,

69, 74, 103-104, 117-123, 125, 130, 133, 137-139, 191
ocean-bottom seismometer (OBS), 74, 104, 119-121
odd/even effect, 11, 14, 104
offset
 distribution, 2, 14, 17, 20, 25, 28, 33, 41-43, 61-63, 67, 69, 71-72, 75, 83, 86, 90, 92, 99, 104-105, 150, 175
 sampling, 1-2, 62, 71, 104-105, 110, 117, 136, 146
 vector, 2, 5, 21, 39-41, 43-45, 191
offset-vector tile (OVT), 2, 4, 36-46, 72, 84-85, 89, 92, 132-133, 136, 172, 183-185, 187, 190, 194-200
Omni, 99
orthogonal geometry, 2-4, 18-200
oversampling, 8, 35, 87, 111
OVT gather, 38-44, 75, 79, 81, 90, 132, 184, 187, 190, 197-200

parallel geometry, 2-3, 18-200
patch, 18-19, 117-118
patent, 18, 22, 74, 116
physical modeling, 171
polar plot, 53, 55-57, 60
pore fill, 69, 118
porosity, 69
positioning error, 106-107, 117, 144, 172
preplanning, 96
prism wave, 74, 106
proper sampling, 1-2, 7, 20-21, 45, 75, 78, 136, 177, 197
pseudominimal data set (pMDS), 2, 4, 35-39, 77-78, 84, 99, 112-114, 132, 172, 175, 184, 187, 193-195, 197, 199-200

quad/quad geometry, 19
quasi-random sampling, 159, 170, 173

random geometry, 20-21
Rayleigh
 criterion, 76, 160, 162, 172
 wave, 49, 142
receiver array, *See* array, field
receiver-line interval, *See* line interval
reciprocity, 2, 8-9, 14, 18, 40, 97, 99, 108
rectangular sampling, 28
reflection point smear, 176
reflection-time picking, 41, 43, 45
Reflex, 99
regular fold, 36, 44, 75, 83-85, 90, 96-97, 132

regularization, 40-42, 135, 138, 181, 200
remotely operated vehicle (ROV), 74, 103, 119-121
repeatability, 3, 110, 118, 120-121
resampling operator, 9, 28, 32
resistivity imaging, 81
resolution, 1, 3, 13, 17, 33, 38, 45-47, 69-72, 74-80, 82, 92, 99, 112, 125, 127-130, 132-138, 159-174, 192
 achievable, 76, 78, 160, 171-173
 horizontal, 76-78, 135, 137, 162-164, 166, 172-174, 192
 potential, 70, 78, 82, 92, 160-161, 163, 166-167, 170-173
 spatial, 3, 45-46, 159-174
 temporal, 160, 162
 vertical, 77-78, 135, 159, 162-163, 166, 172-174, 192
Ricker wavelet, 162-164, 172, 187, 189
rolling of geometry, 88
 cross-line roll, 88
 full-swath roll, 3, 36, 88-90, 97, 100, 118
 multiline roll, 3, 40, 88, 90, 92, 94-95, 99-100
 single (one)-line roll, 3, 40, 88, 90-93, 97, 100
 zipper, 88
running mix, 11-12

scatterer, 2-3, 49-52, 54-55, 57, 59-60, 70, 119, 126, 144, 146, 159, 164, 168
seisloop geometry, 20
shallowest horizon to be mapped, 3, 69-70, 75, 80-81, 83, 86, 99
shear wave, 49, 74, 78, 121, 125-138, 142, 144
Shell, 1, 120, 141, 148-149, 175
Shell Petroleum Development Company (SPDC), 148-150, 154
shooting direction, 11-12, 21, 29-30, 103-107, 113-114, 116, 122, 129, 135, 137, 193
short-offset 3-D, 116
shot array, *See* array, field
shot density, 71, 88
shot repeat factor, 89-90, 118
shot salvo, 22, 25, 90
shot-generated noise, 83
shot-line interval, *See* line interval
side-scattered noise, 32, 60
side-swipe energy, 156
signal-to-noise ratio, 7, 18, 45, 69, 80, 82-83, 86-87, 92, 100, 116, 197
single-sensor acquisition, 53, 60, 100

single-source, single-streamer geometry, 29
slanted geometry, 3, 18, 22-23, 70, 72, 88, 90-95, 99-100, 131
slanted spread, 22-23, 72-73, 92, 131-132
slash geometry, 72
smearing, 144
smoothness criterion, 96-97
source wavelet, 18, 20, 76, 118, 160, 163-164, 172-173, 190, 192
sparse acquisition, 158
spatial continuity, 1, 3, 17, 25, 27, 32, 39-40, 42, 46, 69, 72, 74-76, 83, 92, 96-97, 99-100, 143, 154
spatial discontinuity, 1-3, 25, 37, 39-43, 75, 100, 132, 135, 172, 195, 199
spacial interpolation, 29, 41-42, 80, 97, 107-108, 125, 135-136
spatial wavelet, 160, 164-167, 169-170
spider-web geometry, 20, 74-75, 200
spiral survey, 106
spread length, 25, 33, 36, 81-82, 84-85, 89, 168
stack-array, 2, 5, 13, 16-17, 84
Statfjord, 137-139
statics
 computation, 42, 75, 89
 corrections, 5, 122, 150
 decoupling, 70, 89
 intra-array, 9, 79, 85, 136, 141
 refraction, 149
 residual, 41, 43, 149
station interval (spacing), 9-11, 16, 20, 25, 60, 71, 73, 75, 78-79, 83-88, 96-97, 100, 104, 118, 141-142, 146, 149, 158, 170-171, 191
stationary
 phase, 129-130, 155, 176-179, 181, 188, 191-192, 199-200
 recording system, 3, 19, 28, 32, 103-105, 113, 115-121
Statoil, 103, 119-120, 125, 137
steerable streamer, 32, 110, 113, 115
Strathspey, 116
strike shooting, 103, 105-107, 110, 112-114, 177, 193
striping, 32, 74, 92, 104, 110, 134
subsurface diagram, 6, 8
SUMIC, 74, 103-104, 118-122, 125
surface diagram, 5-6, 8
swath
 land acquisition, 25, 27, 70-71, 88-90, 92, 101, 159
 OBC acquisition, 117, 133-134, 137
 VHC acquisition, 116

sweep, 76, 98, 141

Teal South, 74, 120, 134
template, 3, 22-23, 25, 87-91, 93-95, 97, 99, 149
testing, 98
Texaco, 116
tiling, 33, 38-41, 84, 92, 132-133, 136, 184, 187, 193-196
time-lapse, 74, 110, 120, 134
topography, 87, 96, 98
total fold-of-coverage, 36-37, 84-85, 89, 134
trace density, 35, 44, 71, 83, 113, 149
traveltime
    contours, 21, 24, 32, 50, 71, 110, 126-127, 130-131, 192, 195-196
    surface, 2, 7, 21, 23, 49-52, 55, 60, 72, 125, 127, 129, 155, 160-163, 172, 192, 194-196, 200
triple zigzag geometry, 20

true-amplitude migration, 21, 72, 130, 160, 187-188, 200
tuning thickness, 162

undersampling, 79
undershooting, 19, 107, 115
unit cell, 36, 40, 42-43, 45, 80, 84, 86-87, 90, 92, 194
updip shooting, 11-12, 29-30, 113, 135, 137, 193

Valhall, 123, 137-139
vector fidelity, 119
vector-weighted diffraction stack, 187, 199
velocity
    analysis, 18, 41, 44, 107, 193
    filter, 42, 49, 52, 60, 67, 71, 94
velocity-model updating, 41, 187, 199-200
vertical hydrophone cable (VHC), 3, 74, 104, 116-117, 121
vertical pulse distortion, 163

vibrator, 27, 35-36, 55, 57, 71, 76, 85, 89, 141
VSP, 19, 74-75, 116, 119, 159, 161

WAS technique, 89
water-borne noise, 116
water-bottom reverberation, 117
wavenumber spectrum, 8, 28, 79, 127-129, 166, 167
wide versus narrow, 2, 22, 33, 35, 37, 66-67, 71, 83, 86, 89, 100, 113, 149-150, 154, 175-176
wide-azimuth geometry, 21, 39, 70, 88, 176, 193

zag-spread, 24-25, 33, 35, 71
zig-spread, 21-23, 27-28, 33, 35, 71-73, 131-132
zigzag geometry, 3, 18-25, 27, 33, 35, 42, 63, 65-67, 70-73, 84, 88, 100, 103
zipper, 88
zone of influence, 4, 83, 87, 113, 129-132, 134, 181, 187-193, 199-200